A Critical Companion to Zoosemiotics

BIOSEMIOTICS

VOLUME 5

Aims and Scope of the Series

Combining research approaches from biology, philosophy and linguistics, the emerging field of biosemiotics proposes that animals, plants and single cells all engage insemiosis – the conversion of physical signals into conventional signs. This has important implications and applications for issues ranging from natural selection to animal behaviour and human psychology, leaving biosemiotics at the cutting edge of the research on the fundamentals of life.

The Springer book series *Biosemiotics* draws together contributions from leading players in international biosemiotics, producing an unparalleled series that will appeal to all those interested in the origins and evolution of life, including molecular and evolutionary biologists, ecologists, anthropologists, psychologists, philosophers and historians of science, linguists, semioticians and researchers in artificial life, information theory and communication technology.

For further volumes:
http://www.springer.com/series/7710

Dario Martinelli

A Critical Companion to Zoosemiotics

People, Paths, Ideas

Springer

Dario Martinelli
University of Helsinki
Institute of Art Research
Faculty of Arts
PL 35 (Vironkatu 1)
00014 Helsinki
Finland
dario.martinelli@helsinki.fi

ISBN 978-90-481-9248-9 e-ISBN 978-90-481-9249-6
DOI 10.1007/978-90-481-9249-6
Springer Dordrecht Heidelberg London New York

Library of Congress Control Number: 2010929209

Cover illustration: Cover illustrations by Rūta Saulytė (www.saulyte.com)

In-text drawings: Dario Martinelli

Printed on acid-free paper

Springer is part of Springer Science+Business Media (www.springer.com)

To Smile and his/her mum.

Preface

There are other forms on intelligence on Earth, Doctor. Only human arrogance would assume the message must be meant for man.
(Dr. Spock, Star Trek IV)

There are many reasons why this is a good time for writing a Critical Companion to Zoosemiotics; reasons that pertain to each of the words carefully chosen for the title of this book.

First of all, Zoosemiotics. Why write a book about this subject? Pretty soon after these lines have been written (in 2013), zoosemiotics will turn 50. As is well known, people at the end of their first half-century often feel the need to sit down for a while and make a sort of self-evaluation of their life, look back and look forward, look around, look inside. Writing a "companion", that is, something which aims to cover the majority of the topics related to a given subject, can be seen as just this sort of activity: discussing the status of the art of zoosemiotics as it turns 50. Discussing its values, impact, innovations (if any), and limitations.

Despite its popularity within the semiotic environment, zoosemiotics is still surrounded by a certain curiosity, the kind of curiosity that manifests itself in attitudes like scepticism and "exoticism". One goes from questions like "Do they *really* communicate?" to comments like "Oh! I know what you mean, you should see how smart my dog is". It is all understandable, of course. Zoosemioticians are the first not to have awfully clear ideas on what zoosemiotics is exactly, how far can they go, and so on. Less than fifty is very young, scientifically speaking, for a discipline to answer its most important questions, and moreover, it should be admitted that Sebeok's work, outside semiotics, was not as influential as it probably would have deserved to be. In this sense, a widespread curiosity about the zoosemiotic discipline is more than comprehensible.

But this is not the whole story. There is a growing interest not only about zoosemiotics, but also about the whole area of non-human animals studies. Finally, after decades of prejudices, those studies caught the interest of *cognitive sciences* (see the most recent trends in ethology and, partly, zoosemiotics itself), *human sciences* (is it still fair to call them just *human*?), and more generally are now also approached according to a wider perspective. Most of the competencies so far collected on non-human animals have been specialistic, punctual, *microscopic*, and thus not so open and interdisciplinary. Zoology focuses largely on the anatomy, the structure, and the particulars of animals. Classical ethology organizes animal behaviour in patterns and ethograms. TV constantly broadcasts documentaries showing, say, a group of lions in the African savannah, dealing with the usual two or three situations (hunt, reproduction, territory defence), or – as it is becoming trendier and trendier in channels such as Animal Planet – exceeds in the opposite sense, i.e., making a reality candid-camera-like spectacle out of animal abilities and actions. What has clearly been missing for a long time, in the discussion on non-human animals, are the good old philosophical questions (the *macroscopic* issues). As far as non-human animals are concerned, it seems that we either (believe we) know things for sure, or we do not. Very little seems to be in the middle. It is not so often that we have had "doubts", in the philosophical sense of the term.

But we now feel an urge to *doubt*, when discussing other animals. We feel the urge to define and refine them, as concepts. It is true that human sciences are generally hardly practical and empirical, but it is also true that biological sciences are a bit too uninterested in theoretical reflection. History has proven on various occasions (and keeps on doing so) that these forms of opposite yet sadly complementary superficiality have the primary effect of slowing down, rather than supporting, the evolution of human knowledge. We also start to realize that to study other animals under a humanistic perspective helps us, as humans, to learn more about ourselves. For a start, it reminds us that we ourselves are animals, thus, at least on a basic level, certain principles that are applicable to non-humans are of scientific interest for humans too. And that also means addressing new questions, or reformulating old ones. So, communication, signification, representation are all zoological phenomena, rather than simply anthropological ones. Therefore: what is really communication? What is signification? Where do they come from? What are the behavioural processes implied in their production? All these questions (and attemps to answer them) are good reasons for wanting to write about zoosemiotics.

Not only. This can also be an opportunity to open a new chapter for the field. A self-evaluation, after all, has mostly this purpose: we find what did not work out in our life, we try to explain why, and we seek for change. It will be seen later that there are quite a few things, within zoosemiotics, that should probably undergo a positive revision.

Most of all, finally, a field like zoosemiotics *deserves* attention, a work of this type is simply worthwhile. As a field of inquiry, one has to assume, it has made a strong contribution to animal studies, and it has still a lot to say and give. In fact, this companion was written largely with the idea of *adding* something to the subject, rather than restating the already-known.

Which introduces the second word, "Companion". Why a companion, then, if the intention is to say something new? The most obvious reply is that a companion to zoosemiotics was simply never written before, and this is already enough a reason for wanting to write one. But that would not be the end of the story. It is very possible that scholars approaching zoosemiotics need a working tool of this type.

And so do scholars who are already acquainted with the field, but feel the need to have a practical point of reference to quickly find out about a certain topic, or to compare their own hypotheses with.

In this sense, the aim is still to perform the proper job of offering sources and information about most ("all" would be impossible, of course) of the topics of interest for this field. In a way it is the foundations of this book that make it a Companion in the truest sense of the word. At a general level, the reader should be able to find nearly all s/he needs to form a fair picture of what zoosemiotics is, does, and aims to.

However, on top of this layer, a few more were developed, in directions that it is probably appropriate to call "Critical". Zoosemiotics, like all respectable fields of inquiry, has its own problems, controversies, on-going debates. Those, obviously, deserve a particular treatement. Not only: some other aspects, even thought commonly accepted by the community of zoosemioticians, are probably still rather "critical", and cannot be taken so easily for granted. Finally, other aspects may be overlooked, or not given the amount of attention they deserve. All these became *layers* built over the basic "Companion" structure, and came to form the word "Critical" in the title of this book.

It is therefore proposed here a consciously "unbalanced" account of zoosemiotics. An account which priviledges the problematic over the granted, the controversial over the institutionalized, the adventurous over the safe. Most of all, however, the idea is to offer as many sources as possible for a satisfactory introduction to zoosemiotics. This is why, this book is "packaged" with different forms of reference, schematic and discoursive. Foreword and references apart, the work is divided in five main parts. In Chapter 1, "Introduction to Zoosemiotics", a hopefully extensive presentation of the field is provided, with a particular attention to its definition and its main problematics. A little coda to this part will gather a few definitions of zoosemiotics, as collected from very different sources, from Sebeok to Wikipedia. Chapter 2, "Ethological Zoosemiotics", describes the most traditional and important area of inquiry of zoosemiotics, i.e., the actual semiosis among non-human animals. Elements of systematics of the field shall be presented, the main theoretical issues, and a specific example of this type of research. Chapter 3, "Anthropological Zoosemiotics", on the other hand, focuses on the increasingly popular area of investigation of the human-other animal relationship, an area that includes applied zoosemiotics and the diverse cultural representations of the non-human animal. Chapter 4, "A Glossary of People, Paths and Ideas", offers a few hundreds entries for topics and scholars related to zoosemiotics. The intention is to provide the reader with a very practical tool for consultation, introducing (or deepening, if they are already mentioned in any of previous parts) the most important, and/or critical, key-terms and people of the field. The companion will also have an Chapter 5, entitled "Does Zoosemiotic Have An Ethical Agenda?", which explores the possibility of including ethical reflections on the zoosemiotic program, following the example of other branches of semiotics, which are currently addressing questions of this type in their own field of inquiry. It is not a "traditional" topic in zoosemiotics, but it is a specific author's intention that of making it an increasingly regular presence within the semiotic discussion.

It has been a labour of passion, which hopefully will be of some help for the readers.

Helsinki, Finland

Dario Martinelli

Acknowledgments

In the process of writing this book, I was supported and helped by a number of people I wish to sincerely thank. These are: Catherine Cotton, Ria Kanters, Ineke Ravesloot and the other friends at Springer, Marcello Barbieri, the students of my courses, Lina Navickaitė-Martinelli, Tom Pankhurst and Helena Telkänranta.

Contents

Chapter 1
Introduction to Zoosemiotics

1.1 What is Zoosemiotics?

Zoosemiotics is a field of inquiry introduced in 1963 by Thomas Albert Sebeok. That is the year when the term and a first definition make their first appearance, initially as a compromise between ethological and semiotic research (in the beginning, Sebeok was convinced that "zoosemiotics" had to be meant mostly as an umbrella term, gathering different scholarly approaches to animal communication). A synthetic definition of zoosemiotics, in the light of its most recent developments, can be today that of *the study of semiosis within and across animal species*. A spectrum of different possible definitions of the term has been attempted (in the next chapter of this book), but at the end of the day it is probably safe to trace a common ground in the way just mentioned.

The implications of this definition are crucial. First of all, the focus of zoosemiotics is not simply communication (which is what people normally expect to be the actual goal of semiotics), but rather the broader *Semiosis*, i.e., following Charles Morris, the process in which something is a sign to some organism. Communication, the process in which a sign is coded and transmitted from a sender to a receiver, is thus to be considered a special, therefore smaller, case of semiosis.

By consequence, zoosemiotics is interested in at least three important semiotic phenomena:

(a) *Signification*, occurring when the receiver is the only subject taking part in the semiosis, and a true sender is missing. In other words, zoosemiotics studies here the way animals make sense out of each other, or out of their environment;
(b) *Representation*, occurring when the sender is the only semiotic subject. In this case, zoosemiotics studies here the way animals construct sense and, often but not always, *offer* it to somebody else; and
(c) *Communication*, occurring when sender and receiver take both part in the semiotic phenomenon, and therefore the above-mentioned "sense" (or text) is exchanged, understood or misunderstood.

This specification, besides its paradigmatic necessity, is also needed because the first and foremost objection that can be (and has been in several occasions) put

D. Martinelli, *A Critical Companion to Zoosemiotics*, Biosemiotics 5,
DOI 10.1007/978-90-481-9249-6_1, © Springer Science+Business Media B.V. 2010

forward against studies like zoosemiotics is: can the message exchange that occurs among non-human living forms be really called communication? If there is no big problem in accepting that what happens among such animal species like dogs or dolphins is actually a communication process, the sign exchanges occurring among insects or reptiles, are more often than not surrounded by much scepticism, among semioticians.

The research target of zoosemiotics, and in fact of the whole semiotics, is thus a different broader one. Semiosis can be defined as the action of signs, or, as mentioned already, the process in which something is a sign to some organism. What normally happens is that semiosis is identified with communication, just because the latter is the most evident and predictable manifestation of the former.

This also means that another concept that is strictly related to the notion of communication, i.e., *intentionality*, is not necessarily part of a semiotic phenomenon. A common tendency, which comes straight from the linguistic/semiological tradition, is that of considering only intentional messages as worthwhile of semiotic analysis. For semiotics to be *interested* in a given message, that message must be somehow *meant*. This is still the main trend among those who see zoosemiotics as a threat for (what they consider to be) the real identity of semiotics. Such an identity – it is maintained – remains that of text analysis: it is acceptable that a painting, a building, a sonata are considered "texts", along with literary works, but the alarm signals of vervet monkeys, the song of a blackbird, or the dance of the honeybee cannot.

This stand, however, presents two main difficulties. On the one hand, to limit semiotics to the sole communicative/intentional dimension means to classify not only non-human semiosis as semiotically-uninteresting, but also most of the human semiosis as well, starting from body-language, proxemics, plus several types of anthropological and social interactions. And this might be rather peculiar if one thinks that such a position is advanced by those semioticians who have a strong linguistic background, i.e., they relate to a field that has an enormous interest for non-verbal human semiosis.

On the other hand, more importantly, it must be said that a concept like "intentionality" is not clearly-defined at all. What does it mean to *have an intention*? Is it the same as *to want* something, or is it something else? And how do one detect the existence of an intention in our thoughts/actions? In a famous article on primate deception, Richard Byrne and Andrew Whiten (1991), traced a 0–3 scale of intentional behaviour. The 0 level (observed when the animal reaches a goal in a completely random manner) is the only one that cannot be considered intentional. Level 1 (the simple aiming to a goal) is already an *intention.*

What seems to create the misunderstanding, in a common-sensical definition of "intention", is that, in human interaction, we witness the *awareness* and then the *verbalisation* of the intention: we are aware that we want something and we can say it. This is a process that is often confused with intentionality itself, which – in turn – could be nothing else (or nothing more) than the main characteristic of the sign: referring to something else than themselves (Dennett 1996: 48). Whether this (reductive?) definition is acceptable or not, it is clear that most of the time, when talking about intentionality, we tend to refer to the wrong phenomenon, i.e., the awareness of the intentionality, which is evidently a different matter. Intentionality itself seems to be a much more obscure concept. Peirce himself would extremely often use the term semiosis in his writings, but very seldom would he employ terms like communication and intentionality.

Putting the argument less seriously, one could say that the object of semiotics is also a sentence like "Go from A to B", and not only a sentence like "*I* am telling *you* to go from A to B, and *I mean it*".

Let us go back to the initial definition of zoosemiotics. It was said that this discipline studies semiosis *within* and *across* animal species. This means that there is a range of semiotic phenomena that may be called "intraspecific", and another category that should be called "interspecific". By intraspecific, it is meant the kind of semiosis occurring within one single animal species (or community, being the concept of species still a bit problematic, to a certain extent), i.e., within a group of animals that supposedly share a fairly similar perception of the world and similar ways to codify it. By interspecific, on the other hand, it is meant the kind of semiosis occurring between different species (or communities), i.e., between groups that do not share the above-mentioned perception and codification of the world, if not to a very basic extent (this latter normally being the very ground for establishing a – temporary or not – common code). It is a rather important distinction, because it implies a (sometimes radical) change of methodologies, and an address to a (sometimes radically) different order of problems.

Thirdly, the use of the term "animal species" in the definition here provided is intended to cover the entire Animal Kingdom, i.e., the human species as well. This means not only that a part of human semiotic behaviour (more or less, what transcends the linguistic domain, although the notion of language itself is "critical", as it will be shown later) easily falls under the zoosemiotic domain, as ethology had already shown, but also that zoosemiotics investigates a field of knowledge that include both natural and cultural elements, and that – ultimately – the critical notion of Culture is to be considered a part of the critical notion of Nature.

This is probably one of the most important, and courageous, statements of zoosemiotics, as – among other things – it represents the attempt (which by now can be deemed fully successful) to extend the attention of semiotic research to the realm of the non-human, starting exactly from the assumption that a great deal of characteristics that we thought were typical of human semiosis, are in fact to be widely reconsidered.

Throughout almost its entire history, indeed, semiotics has always been an anthropocentric and logocentric discipline, with an exclusive emphasis on human- and language-related issues (this despite the fact that the earliest conscious examples of semiotics consisted in the medical observation of the body – symptomatology, diagnostics, etc. –carried out by the likes of Hippocrates or Galen of Pergamon). John Locke, in the seventeenth century, used the word "semiotics" for describing the "doctrine of signs". It was once again a human-centered enterprise. Yet,

> While his prime concern was with those signs of our ideas "which men have found most convenient, and therefore generally make use of," that is, "articulate sounds" or verbal signs, Locke was fully aware that other creatures, such as birds, also have perception, "retain ideas

> in their memories, and use them for patterns", in brief, that they are comparably served by signs. (Sebeok 1990: 37)

The real turning point, in terms of the scope of this companion, appeared in the nineteenth century, as Charles S. Peirce gave a first clear acknowledgement of the semiotic nature of the non-human world (to him, the sign was a connective element not only in all experience and thought, but in the whole universe), but it is not until the biologist Jakob von Uexküll that the first, important, specific argumentation in support of what is nowadays known as biosemiotics, appears, that is, the study of semiosis in living forms. Uexküll's *Institut für Umweltforschung*, founded in 1926 at Hamburg University, investigated the perceptive environment of animals (i.e., their Umwelt, as shall be soon discussed at length). Though not a semiotician, and probably never intending to be one, Uexküll brought to attention a number of topics of fundamental (bio) semiotic interest, and later his son Thure, and Thomas A. Sebeok, introduced his work to the semiotic community, labelling the German biologist with the infamous term "cryptosemiotician" (a fate shared with nearly all great contributors to human knowledge, in one moment or another of the ever-in-progress construction of semiotic history).

All this was going on while the then-dominant school of semiotics, the so-called semiology, of Saussurean tradition, made it very clear that the discipline was a natural continuation of Linguistics, or even – as Barthes had put it – just a part of it. In Sebeok's account:

> In an independent but parallel tradition, amplified by F. de Saussure's heritage, semiotics, alias semiologie, has remained steadfastly anthropocentric, intertwined with language, le patron general of Saussure's programmatic science. Many linguists later tended to more or less agree: thus L. Bloomfield asserted that "Linguistics is the chief contributor to semiotic," and U. Weinreich called natural languages "the semiotic phenomenon par excellence." But it was the prominent French critic Roland Barthes who – like W. H. Auden's "linguist who is never at home in Nature's grammar" – carried this glottocentricity to its preposterous (but perhaps playfully conceived) conclusion by turning Saussure's formulation topsy-turvy with his declaration that "linguistics is not a part of the general science of signs, even a privileged part, it is semiology which is a part of linguistics…" The validity of this paradoxical inversion of the customary order of things can be contemplated only, if at all, at the price of throwing all of comparative semiotics overboard by dividing the animate world into two unequal classes – speechless vs. language-endowed – and then consigning the sign behavior of well over two million extant species of animals beyond the semiotic pale. (Sebeok 1990: 38)

Back on the "opposite front", other signals of an upcoming new field of inquiry came from Charles Morris, the truest follower of Peirce, and from the oncologist Giorgio Prodi, who termed the study of biological codes "Nature Semiotics", and from Friedrich S. Rothschild (1962: 777), who first actually used the term "biosemiotics" in a scientific context:

> This approach presupposes acceptance of our position that the history of subjectivity does not start with man, but that the human spirit was preceded by many preliminary stages in the evolution of animals. The symbol theory of psychophysical relation bridges the gulf between these disparate avenues of research and unites their methods under the name of

> *biosemiotic*. We speak of biophysics and biochemistry whenever methods used in the chemistry and physics of lifeless matter are applied to material structures and processes created by life. In analogy we use the term biosemiotic. It means a theory and its methods which follows the model of the semiotic of language. It investigates the communication processes of life that convey meaning in analogy to language. (Rothschild 1962: 777)

One year later, as already mentioned, Sebeok coined the term and developed the theoretical paradigm of a specific biosemiotic field named "zoosemiotics", somehow inaugurating a new phase for semiotic history, a phase in which non-human semiotics is no longer ignored or underrated. This event certainly represents *the* milestone in zoosemiotic history, and shall be deepened later on.

Sebeok mantained that

> The process of message exchanges, or semiosis, is an indispensable characteristic of all terrestrial life forms. It is this capacity for containing, replicating, and expressing messages, of extracting their signification, that, in fact, distinguishes them more from the nonliving – except for human agents, such as computers or robots, that can be programmed to simulate communication – than any other traits often cited. The study of the twin processes of communication and signification can be regarded as ultimately a branch of the life science, or as belonging in large part to nature, in some part to culture, which is, of course, also a part of nature. (Sebeok 1991: 22)

Later, he added that "the life science and the sign science thus mutually imply one another". (Sebeok 1994: 114)

These reflections introduce a number of important key-terms and concepts, that should be considered the pillars of the bio- and zoosemiotic disciplines:

(1) The concept of semiosis, i.e., the action of signs, is the real target of semiotics;
(2) All life forms are semiotic. Thus, semiosis is primarily what distinguishes life from non-life;
(3) Culture and Nature are not concepts in opposition, but in fact the former is part of the latter;
(4) If life science intersects with sign science, then semiosphere and biosephere are probably synonyms.

In addition, Jesper Hoffmeyer pointed out the centrality of semiosis in biological studies. To Hoffmeyer, the biggest contribution that biosemiotics can make to the life sciences is the emancipation of sign and semiosis as the crucial elements in life: semiosis is the "most pronounced feature of organic evolution", and signs are the "basic units for studying life".

> The most pronounced feature of organic evolution is not the creation of a multiplicity of amazing morphological structures, but the general expansion of "semiotic freedom", that is to say the increase in richness or "depth" of meaning that can be communicated (Hoffmeyer 1996: 61).

And:

> The sign rather than the molecule is the basic unit for studying life (Hoffmeyer 1995: 369)

This position is somehow antagonistic to that held by Marcello Barbieri, who is on the contrary a supporter of a code-based biosemiotics, an approach that focuses on three important aspects:

> One is the idea that the cell is a duality of genotype and phenotype, i.e., a biological computer made of genetic software and protein hardware. The crucial point is that a computer contains codes but is not a semiotic system because its codes come from a codemaker, which is outside the system.
> The second basic concept is the idea that all biological novelties are generated by natural selection, i.e., by an agent, which is outside the cell just as the human mind is outside the computer. But if the cell is a biological computer assembled by natural selection, it is perfectly legitimate to say that it is not a semiotic system, and this justifies Florkin's statement that there is no real meaning in it. Ultimately, that leads to the physicalist thesis that there is no real code either at the molecular level, and that molecular semiosis is merely an illusion. The computer model of the cell, in short, keeps semiosis out of the cell, and this is why the first true model of molecular semiosis was the idea that every cell is a trinity of genotype, phenotype, and ribotype, i.e., the idea that the cell contains an internal codemaker [...] This was complemented by the idea that coding is not reducible to copying, and, therefore, that natural selection (based on copying) and natural conventions (based on coding) are two distinct mechanisms of evolution [...]
> Another important contribution to code-based biosemiotics came from the discovery of an increasing number of organic codes. That development started with the unveiling of the sequence codes by Trifonov [...] and has grown slowly but steadily ever since [...]
> The „code based" approach to biosemiotics, in short, is a road that started with the recognition of semiosis at the molecular level and worked its way up by extending the concepts of code and meaning to the higher levels of biological organization. At about the same time, however, there was also another road to biosemiotics that was being developed. A road that went exactly the other way round, i.e., that started at the higher levels and worked its way down towards the lower ones. (Barbieri 2008: 594)

This discussion is at present the hottest one within biosemiotics, and it is not within the scope of this book to declare a preference. It must be however pointed out that the notion of sign-based biosemiotics is more sympathetic to an idea of the field as a humanistic-oriented one, while code-based biosemiotics heads clearly in the direction of natural empirical sciences. The goal of this book is certainly to find a compromise between human and natural sciences in a paradigm that hopefully takes the best from both, yet one perceives that this compromise, on the part of a discipline like zoosemiotics (i.e., a branch of "semiotics"), requires more effort in the direction of biology, than the other way round. In that sense, there is a natural sympathy towards any contribution that will help in keeping the discipline within the necessary empirical premises, and will prevent any risk of metaphysical drift (a sympathy that will be more explicit in the paragraph "The ever present Cartesian dualism").

In any case, both Barbieri's and Hoffmeyer's schools make a strong case for the centrality of semiosis in biological processes, and for the intimately interdisciplinary nature of biosemiotic research. As Barbieri himself states:

> There have been historical disputes between the two versions but [...] they are not incompatible, and both share the idea that every living creature is a semiotic system, i.e., that semiosis (the production of signs) is fundamental to life. (Barbieri 2008: 577)

1.1.1 Characteristics of Zoosemiotics

Life semiosis, in general, can take place either *within* a living being and/or *between* two or more of them. The former case is named **endosemiotics**, or – according to the specific cases – protosemiotics, microsemiotics, cytosemiotics, etc. Endosemiotics involves the message exchange among cellular organelles, cells, tissues, organs and organ systems. When Hippocrates, for instance, analyzed the symptoms of a given disease as signs of the disease itself, he was basically a forerunner of endosemiotics.

Endo- is a Greek preposition that stands for *Inside*: the internal semiosis that takes place in an organism's body is extremely complex, it involves at least four channels (the chemical, the thermic, the mechanical and the electrical: further on it shall be discussed in detail what precisely channels are, but for now a preliminary definition would be: "a method or system used to send and receive information"), and is probably the most intense semiotic activity a scholar may find. A single human body consists of about 25 trillions cells, a number which – alone – is 2,000 times more than the entire human population on this planet. Plus, all these cells have direct or indirect connections with each other through more than one modality. The number of messages exchanged inside one single living organism is unimaginably high:

> Millions of so-called receptors capable of recognising specific signal molecules in the cell environment are located in the membranes of each of our cells. These receptors function as communication channels through which our cells, tissues and organs are persistently communicating with each other all around the body. Especially interesting is the recent discovery that receptors on the surface of immune cells are capable of decoding the messages exchanged among nerve cells and vice versa. The psycho-somatic integration of the nervous system, the immune system and the endochrinological system in a healthy organism is the result of this gigantic semiotic interaction among many thousand billions of cells, each of which is capable of interpreting a limited range of molecular signs. Disease my be seen then as the result of erroneous communication among our body parts. We fall ill because our cells cannot quite succeed in uniting to create us. (Hoffmeyer, in Bouissac 1998: 84)

Probably, the earliest forms of semiosis in our bio/semiosphere are those of the *prokaryotes*, one-celled nucleus-less micro-organisms, better known as bacteria.

Along with endosemiotics, there is *exosemiotics*, which regards the entire spectrum of message exchange between two or more complex organisms. Fields of exosemiotics are mainly *phytosemiotics* (whose object is the semiosis among plants), *micosemiotics* (semiosis among fungi) and, finally, *zoosemiotics* (semiosis among animals). A relevant branch of zoosemiotics is obviously *anthroposemiotics*, i.e., the semiosis among the human animals. Some people still consider anthroposemiotics as a separate field, to be lined alongside the other three, claiming that the distinctive feature is the presence of culture in anthroposemiotics only, but the classification is incorrect, both scientifically (the human being *is* an animal, not a distinct entity) and conceptually (as it will be shown further, the notion of culture, unless meant very narrowly,[1] is not alien to other animals as well).

---

[1] So narrowly that several human communities would be excluded too.

Finally, the globalistic conception of life as a whole as sign processes, that is, the interpretation of the fully and intimately semiotic nature of life, mostly promoted by the American semiotic tradition (Peirce and Sebeok first of all), led a few biosemioticians to the slightly pretentious belief that the entire Universe can be analysed and interpreted semiotically:

> In its most radical version biosemiotics sees itself as "general semiotics", while traditional semiotics studying human sign systems is seen as just a special part hereof, anthroposemiotics. This understanding may eventually be coupled to a cosmological vision of evolution as a general tendency of our universe to strengthen the autonomy of the semiotic sphere relative to the physical sphere on which it depends. In system Earth this might further be seen as a trend in organic evolution towards the formation of species with increasingly sophisticated umwelts, or in other words towards a general growth of semiotic freedom, a trend which has reached its temporarily richest expression in the art, religion and science of human cultures. (Hoffmeyer, in Bouissac 1998: 85)

It should be said quite frankly that it is very hard to consider this (increasingly popular) trend worth of too much attention, as the idea of biosemiotics expressed here is that of an approach that exclusively relies upon empirical bases. Biosemiotics, as well as zoosemiotics, is trying to build up a reputation as a serious field of research, both within semiotics (wherein it is often accused of being a bit too pretentious) and in relation with other fields of knowledge (wherein it is often labelled as too metaphysical and abstract). In both cases, the presence of this-is-how-the-universe-works and I-give-you-the-Answer types of approach does not help at all. This certainly sounds like quite a strong stand, and some may even consider it a bias. Be that as it may: at least this book will not be accused of being hypocritical.

Sebeok's very introduction of zoosemiotics into the scientific world ("The term zoosemiotics – constructed in an exchange between Rulon Wells and me – is proposed for the discipline, within which the science of signs intersects with ethology, devoted to the scientific study of signalling behaviour in and across animal species", 1963: 465) was obviously far from being the first attempt to study non-human signalling behaviour: leaving aside a series of philosophical reflections, as those provided by Porphyry, Locke or Hume, it was the impact of Darwin on animal studies, and particularly two of his late works, *The Descent of Man* (1871), and *The Expression of Emotions in Man and Animals* (1872), that radically changed the scientific perception and conceptualization of animal semiosis.

Still, Sebeok opened a door that scholars were rather hesitant to open. When one compares pre- or non-semiotic definitions of animal communication, such as those of Cullen ("Animal communication evokes a change of behaviour in another individual", 1972: 101), or Dawkins and Krebs ("Communication occurs when an animal, the actor, does something which appears to be the result of selection to influence the sense organs of another animal, the reactor, so that the actor's behaviour changes to the advantage of the actor", 1978: 282), with those provided by Sebeok ("the discipline within which the science of signs intersects with ethology, devoted to the scientific study of signalling behaviour in and across animal species. The basic assumption of zoosemiotics is that, in the last analysis, all animals are social beings, each species with a characteristic set of communication problems to solve", 1963: 465) and other semioticians, it is clear how, thanks to the semiotic approach, animal information exchange could finally get rid of the rigid stimulus-reaction scheme and achieve a much more significant status.

From that point on, zoosemiotics has enjoyed an increasing popularity among scholars (although not enough to confer it the status of an autonomous field within semiotics, as it will be later discussed). The wide range of topics covered by zoosemiotics, plus its intrinsically interdisciplinary nature, has made this field a rather eclectic one, with incursions in several fields of semiotics, including some apparently-strictly anthropological ones.

Following, but also upgrading, the classification suggested in Martinelli 2007 (32–34), at least two main branches should be distinguished within zoosemiotics, both to be divided, in turn, into two more sub-branches. On the one hand, there is zoosemiotics in the traditional sense, i.e., a discipline dealing with animal semiosis, through the most obvious theoretical tools of semiotics. This branch is named *Ethological Zoosemiotics* (EZ). This field can be divided in a chronological sense, into an *early* current (eEZ) and a *modern* one (mEZ). The former refers to the first stage of zoosemiotics, in which there was no explicit attempt to develop an autonomous paradigm, but rather to use "zoosemiotics" as an umbrella term for gathering different approaches on animal communication. Also, the emphasis on "communication", rather than the broader semiosis, plays a role in defining this early stage. In this phase, zoosemiotics is a discipline largely relying on the Lorenzian ethological school and on the behaviouristic tradition.

In its modern stage, zoosemiotics achieves a few results: first, it operates clearly a transition from having uniquely communication as its research target, to including the entire spectrum of semiosis. Second, it starts developing a paradigm on its own, trying to propose itself as a viable field of inquiry for discussing animal semiosis (and in that sense, it sees the appearance of scholars who explicitly adopt this paradigm, therefore not leaving Sebeok alone with colleagues from other fields). Third, in the majority of the cases, it embraces a cognitive approach, reflecting exactly the type of transition that ethology experienced after the appearance of Griffin 1976. Such transformation, which goes in the direction of a truly semiotic nature (at least according to the traditional definition of "semiotic threshold", which always implies a mental process underlying sign production), emancipates zoosemiotics, as it emancipates ethology, from old mechanistic or semi-mechanistic interpretations of animal behaviour, somehow bringing to final completion some of Darwin's early auspices (not to mention philosophers of the likes of Locke, Porphyry and Hume). It is always difficult to locate exact chronological records of any historical transition, especially when such a transition is the result of organizational demands (as is the proposal of dividing EZ into two historical trends). Therefore, we may gladly welcome the ethologists' effort to spot their own transition in Griffin's work, thus suggesting that, here too, the passage from early to modern EZ occurred sometime in the late 1970's.

The second branch of zoosemiotics, called *anthropological*, in short anthrozoosemiotics (AZ),[2] refers to those studies dealing with the semiotic interaction between human beings and other animals, including those of cultural and/or sociological type. This branch was much anticipated by Sebeok and by the zoologist Heini Hediger. The nature of this interaction has two completely different sides (they should be three, as a matter of fact, but, it shall be shown, one is almost always the natural consequence of another).

The first type is called *communicational* (cAZ). In this context, the human-animal interaction is of a communicative type, i.e., interactive, reciprocal and – with the above-mentioned reservations – intentional. Studies of applied zoosemiotics, such as human-pets or human-cattle interaction, fall under this group, as well as all forms of interspecific communication. In other words, in communicational anthrozoosemiotics, both humans and other animals are semiotic agents, and the study focuses on both parties.

The second sub-category refers to such cases when the non-human animal is a pure source of meaning, an object, rather than a subject, of semiosis. The model is of ecosemiotic type: whereas, indeed, ecosemiotics is the study of human representation of nature, this typology of zoosemiotics deals with the human representation of other animals. This is evidently the case of myths, tales, allegories, but also systematic classifications, such as taxonomy. Now, to be fair, this process corresponds to two different phenomena, signification and representation. On the one hand, the human being *perceives* the non-human animal in a certain manner, and therefore gathers different forms of meaning from it. On the other, this step may be (and is, in most of the cases) followed by an action of representation, in which this perception is shaped, "packaged" and handed over a receiver (virtual or actual, human or not). These two semiotic moments, signification and representation are obviously two different steps, and one (signification) may also occur without the continuation of the other. However, in the economy of this model (which, as said, includes instances from fictional, scientific or everyday discourses), it is very difficult to witness instances where signification and representation operate independently. The mythic representation of, say, the cunning fox is always a consequence of a general perception of that animal as possessing that quality. The (representational) decision of gathering a group of animals under one single species is always the consequence of the (significational) perception of that group as homogeneous under different aspects.

---

[2]Two colleagues, one very young, one very famous – both obviously too busy in showing off their knowledge of Ancient Greek, than in actually grasping the practical side of the question – noticed that the correct shortened formulation should be Anthropozoosemiotics. The result, however, is hardly an economic improvement from Anthropological Zoosemiotics. It may still be one word instead of two, but it is so long that one may easily get lost somewhere in the middle, maybe exactly around the "po" region. This is why, possibly, the social scientists who developed the field of Anthrozoology, decided to skip that syllable too. Semioticians cannot just afford being practical, can they? (And then again, the same famous scholar takes a similar liberty by gladly using the term "Proprioception" in his writings, instead of the etymologically correct, but again impractical, "ProprioREception". Perhaps, when it comes to Latin, he is less demanding).

Table 1.1 Different theoretical trends in zoosemiotics

Zoosemiotics			
Ethological Z		Anthropological Z	
Early EZ	Modern EZ	Communicational AZ	Significational/ representational AZ

For this reason, the name for this type of anthrozoosemiotics will be *significational/representational* (srAZ) (Table 1.1).

One obvious observation, regarding this classification, is that EZ has a close relationship with natural sciences (starting, obviously, from ethology), while AZ is a closer relative of human sciences, especially the so-called anthrozoology and the social sciences, which nowadays show an increasing interest towards animal-related issues. It is thus safer than before to say that zoosemiotics (1) is interdisciplinary, and (2) occupies an intermediary position between natural and human sciences.

Another way to treat this classification is essentially Greimasian (and therefore, among other things, it allows us to dismiss the urban legend according to which it should be impossible to perform any biosemiotic action within non-Peircean structuralist schemes). Besides a distinction between EZ and AZ, indeed, one should also consider the transversal condition of zoosemiotic research in terms of empirical or theoretical approaches. Once again, as the scientific position of zoosemiotics is located somewhere in between biology and humanities, it is not difficult to imagine that the methodological approach or research interest of zoosemioticians may vary from the typical biologist role of the field scholar to that equally typical philosopher role of the speculative thinker.

Naturally, as it is always the case with these categorizations, differences are not clear-cut and the separation of these roles hardly leads to a simply dichotomy. Besides a purely empirical approach, indeed (that is, a direct observation or data collection from the scholar who will eventually analyze those data), one should at least take into account a semi-empirical one, that is, the situation in which the zoosemiotician, although not personally collecting his/her data, relies anyway on (somebody else's) professional findings and offers a purely semiotic interpretation of them, with the support of analytical tools (software, scales, etc.) that normally are part of the field research package. At the same time, also the theoretical zoosemiotician's position is not so sharply defined. The development of a theoretical model based, again, on empirical evidence is clearly a different cup of tea from a mere speculation departing from an abstract intuition. In zoosemiotics, be that a strength or a weakness, there seems to be room for this entire range of possibilities.

Summing up, thus, both EZ and AZ may be investigated by an empirical or theoretical approach. A schematic representation of these four combinations, exactly because they are not just four but virtually endless, is better represented by a Cartesian plan than by the traditional (for semiotics) Greimasian square. This way, for example, the position of purely empirical research will occupy a place much

Table 1.2 Methodological approaches

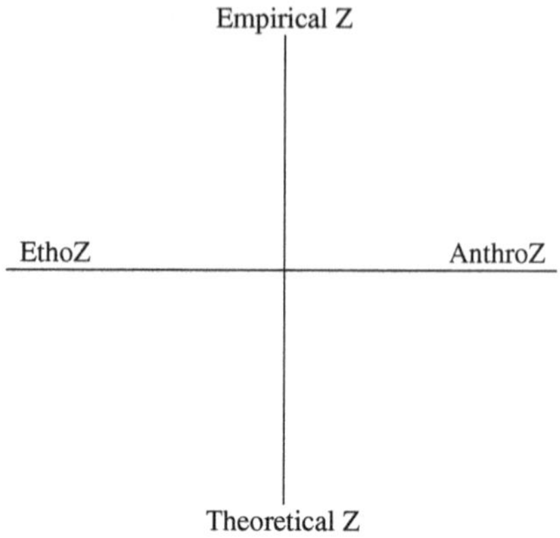

closer to the edge of the empirical-theoretical axis, while the semi-empirical one (depending on its degree of empiricism) will appear more towards the middle of the same axis (Table 1.2). In practice:

A small afterwords to this paragraph. What about language? Is it a zoosemiotic topic or not? And, if yes, what type of zoosemiotic topic is it? In an interview he gave to Susan Petrilli, back in 1987, Sebeok declared the following:

> It is clear that semiotics is interested in two aspects: the study of verbal semiosis (i.e., linguistics) and the study of non-verbal semiosis. However, what most semioticians who are illiterate in biology fail to grasp is that non-verbal semiotics is an enormously wide field, that includes not only the non-verbal human behaviour – which is about 99% of what human beings do – but also an entire, vaste world of millions of animals. In addition, it includes the semiosis of plants and other forms of semiosis that occur within our body, such as the genetic code, the immunological code and other similar mechanisms. Therefore, in terms of pure quantity, it is non-verbal semiosis that largely prevail on the verbal one (translated from Sebeok 1998: 23)

Through these words, one would easily assume that, because language belongs to the remaining 1% of the examples provided by Sebeok, then zoosemiotics is *not* a discipline entitled to discuss language. Surely this is how Sebeok and most semioticians thought. However, the position held in this book is slightly different, and proves once again why this is a "critical" companion. The following observations will serve as points of departure:

(1) It is not clear whether language is a human species-specific feature or not: one has at the same time proofs that other animals do not *use* language, and proofs that they are able to *learn* it, at least to a certain extent. Within such a picture, the criteria for species-specificity become more ambiguous;
(2) Even if language was a human species-specific feature, one cannot use it as an excuse for creating, once again (after Aristotle, Descartes, religions, etc.), qualitative differences between the human animal and other animals. This book is fully supportive of Darwinian theories, and entertains the idea of implementing

them with Uexküllian ones. In both cases, there is no trace of *qualitative* differentiation between humans and other animals. A species-specific trait developes naturally and analogically in one species, from a (cognitive, evolutionary, adaptive) basis shared with more species, so it still bears traces of that common basis. There are zoosemiotic aspects also in language, also in verbal semiosis, and hopefully this book shall give a contribution in this direction;

(3) What exactly language is, is something that scholars are far from agreeing on. Sebeok himself had the rather provocative opinion (which he held firmly, as no provocation at all) that language is *not* a communication device, but primarily a modelling system. Other fields of inquiry would strongly oppose this. And anyway, a common definition of language is far from having been reached;

(4) Among semioticians, however, a rather solid opinion on the definition of language seems to have been established. And, among other things, that opinion seems to be very critical towards those many studies (particularly in primatology and psychology) that seem to have proven that other animals are able to learn human language. The fact that such contrasts exist (the present one, and the one mentioned in point 3) are enough to consider "Language" a critical topic, which therefore deserves a through discussion in this book.

With such premises, a question will arise spontaneously in the reader: is there in humankind something *at all* that is *not* of zoosemiotic interest, if not even language may aspire to be an exclusively anthroposemiotic subject? It is at the same time a very easy and a very difficult question.

On the one hand, one still has the obvious and scientifically unavoidable fact that the human being *is* an animal. Nothing more, nothing less, and nothing else. Everything a human does is something that an animal is doing. So, nothing human *completely* escapes a zoosemiotic insight.

On the other hand, discussing topics like language, or art, or culture, does not mean to discuss what animals species-specifically do with them. And that particularly applies to human beings, who have of course developed an endless number of scientific fields (incuding the semiotic ones) that are specialized in analyzing these activities. Zoosemiotics likes to have a hand in the definition of music, its origins, the components that makes it a biological phenomenon, the notion of performance, the universality of the dance-music association, etc., but cannot (and would not care to) comment on Busoni's transcriptions of Bach, history and evolution of the French horn, differences and similarities between Miles Davis and Chet Baker's versions of *Summertime*, or the use of Pro-Tools in Radiohead's recordings. This is a musical semiotician's job.

Yet, zoosemiotics may (and wants to) discuss what nightingales, or humpback whales, or wolves *species-specifically* do with their music. But this is only because musical semiotics has not developed a branch exclusively devoted to these species. This is why zoomusicology is much more a part of zoosemiotics, rather than of musical semiotics (or musicology tout court). Otherwise, the objection would be just as legitimate as the one raised about humans.

Language, in conclusion, can be either an ethozoosemiotic topic (empirical or theoretical, depending on the instances), when the focus is the investigation upon the possible existence of language in non-human animal species or when one attempts to define language in a zoosemiotic sense, or it can be an anthrozoosemiotic topic of the communicational type (again empirical or theoretical), when the focus is topics such as the experimental programs on interspecific communication.

1.1.2 Zoosemiotics and the Natural Sciences

In 1995, at the Collegium Budapest, where he was Senior Fellow, Thomas Sebeok delivered an important paper (reprinted in Sebeok 2001: 59–73), discussing the initial conditions for a relationship between semiotics and biological siences. He rhetorically wonders what probably most biologists *really* have been wondering about the alleged necessity to let semiotics participate in the natural sciences' discourse: "If one accepts the intrinsic identity of the life science and the sign science, combining at their root into a "natural semiotics" [...] the question still lingers: what is gained thereby?" (Sebeok 1995: 6).

The scope of this paragraph is thus to explore the affinity between zoosemiotics and other natural sciences, and to explore the *role* played by each of them in the construction of the zoosemiotic scientific discourse.

Ethology is inevitably the point of departure, and the major focus, first because the similarity of interests between the two disciplines goes well beyond the differences in methodologies and "philosophy", and second because this connection was already in the agenda of zoosemiotics since its early days. In a 1969 article entitled "Semiotics and ethology" (in Sebeok-Ramsay 1969: 122–161), Sebeok presented the terms of this connection, with the additional service of a very detailed bibliography of, so to speak, "converging" studies. At the time, Sebeok had set important task of showing ethologists (and biologists in general) that the systematics of zoosemiotics was a very effective way for classifying communication, as behavior, and that its theoretical tools had great potentials for analysing it.

As it was already emphasized in the previous paragraph, the early steps of zoosemiotics were mostly characterized by the use of this term as a general way to label the various approaches to animal communication. Not yet having an approach of its own, zoosemiotics was borrowing from the classical ethological school much more than it was lending. The innovations that occurred in ethology during the 1970's, however, determined an advancement that, perhaps by coincidence, perhaps not, had serious repurcussions on zoosemiotics as well. "Cognitive" became a key-word for both fields. The idea that there could be an intermediate stage between a stimulus received by an organism and its behavioral response, completely bypassed by behaviorism, was adopted by both ethologists, in the form of "cognitive processes", and by zoosemioticians, in the familiar form of "interpretation".

If the classical ethology of Lorenz or Tinbergen "was the product of the contemporary behaviourist milieu, and the founders of ethology generally had little positive to say about the possibility of understanding the inner workings of animal minds by scientific methods" (Colin Allen, at host.uniroma3.it/progetti/kant/field/ceth.htm), the new cognitive trend was now devoted to the "the evolutionary and comparative study of nonhuman animal thought processes, consciousness, beliefs, or rationality, and [to] an area in which research is informed by different types of investigations and explanation " (Bekoff 1995: 119). Although some of the contents of this discipline were already anticipated by Darwin and some of his followers, the birth and definition of the term took place only after Donald Griffin's crucial book, *The Question of Animal Awareness* (1976). Griffin had introduced the topic in the following way:

> Ethologists and comparative psychologists have discovered increasing complexities in animal behaviour during the past few decades. [. . .] The flexibility and appropriateness of such behaviour suggest not only that complex processes occur within animal brains, but that these events may have much in common with our own mental experiences. To the extent that this line of thought proves to be valid, it will require modification of currently accepted views of scientists concerning the relationship between animal and human behaviour. Because of the important implications of these developments in ethology, [cognitive ethology] will examine both the pertinent evidence and its general significance in the hope of stimulating renewed interest in, and investigation of, the possibility that mental experiences occur in animals and have important effects on their behaviour. (Griffin 1976: 3–4)

A position like Griffin's, courageous at the times, gave cognitive ethology in the long run a more established and visible role, and a large following, too. Zoosemiotics was one of the fields that took up the challenge. Sebeok (1981) contains already several hints in that direction, while, in more recent times, it is safe to say that most efforts in zoosemiotic research have explicitly embraced this approach, present companion included. Such a development must be considered a significant step forward, and even a crucial presupposition in order for zoosemiotics to exist, at least in the sense of an autonomous semiotics-based discipline. "The complexity of animal communication systems cannot be explained except by assuming that animals do have a mind. What does it mean to have a mind? A first definition may be the following: to have a mind implies at least the capacity of (i) guiding one's own behaviour from the "inside", on the basis of projections not directly connected with what happens outside; and (ii) elaborating and transforming such representations [. . .]" (translated from Cimatti 1998: 9).

Other branches of ethology are in a significant relationship with zoosemiotics. To start with, one must count in all those specialized fields that focus either on a specific portion of animal semiosis (e.g., acoustic signals for bioacoustics, sociality for sociobiology, and so on), or on a given species/family/order (like in the cases of ornithology, cetology, primatology, entomology, etc.). It is crucial for the zoosemiotician to keep up to date with the developments of each of these disciplines, even when (as in the case of bioacoustics) the communion of interests does not correspond to a communion of conclusions (it is safe to say that bioacoustics and zoosemiotics run in the same direction, but on two parallel tracks).

In an ideal world the zoosemiotician may, without fear, answer Sebeok's rhetorical question: "what is gained [by merging the life sciences with the sign sciences]?".

Zoosemiotics, when (1) scientific and (2) up to date, provides a set of theoretical strategies and conceptual bridges that *support* and, with a bit of luck, *improve* the study of animal semiosis (if anything, at least making it clear that "communication" is not the only phenomenon related to information production and reception). Most of all – something that semiotic studies have always been good at – it provide a reliable, flexible-yet-consistent, methodology for framing the different and diverse semiotic phenomena.

The dialogue with natural sciences has not been awfully active, but it has been constant, and has produced remarkable results. It is a mutual relationship that is needed, and wished for, from both sides, not only the zoosemiotic one:

> Interdisciplinary efforts, despite possible pitfalls (. . .), are essential in our quest for knowledge about animal minds. In these joint efforts, open minds and pluralism would also be useful at this stage of the game (. . .). Philosophers need to be clear when they tell us about what they think about animal minds and those who carefully study the behavior of nonhumans need to tell philosophers what we know, what we are able to do, and how we go about doing our research. Although providing alternatives might not be a requirement in thought experiments that conclude that animals do not have beliefs for one or another reason, it would be useful for students of behavior to be presented with some viable alternatives that could be used in their empirical investigations. If it is because philosophers do not have the experience with empirical work that allows them to make realistic suggestions for experimental design, then it would be useful for philosophers to watch ethologists at work (. . .). This experience might allow philosophers to gain a better understanding of what ethology is all about. Even then, it may be the case that ethologists are ill-advised to look to philosophers for a crisp and empirically rigorous definition of intentionality (for example), even if some philosophers promise to provide one (. . .). (Bekoff 1995: 139)

1.1.3 Zoosemiotics and the Human Sciences

The connection with natural sciences is not zoosemiotics' only interdisciplinary task. The truth is, human sciences are nearly as important as the biological ones, at least in the development of specific areas of inquiry of the zoosemiotic field.

As always, everything departs from philosophy, for at least three reasons:

(1) *Historical*: the fact that zoosemiotics was born in 1963 does not mean that the interest in animal semiosis dates back to that year. Questions that are relevant for both the etho- and anthro-zoosemiotic areas have been raised regularly since the dawn of philosophical thinking (a schematic summary will be provided later). Several authors, each in their own way, and within a specific context, have been dealing with semiotic manifestations in non-human animals. Terms like "communication", "sign", "cognition" have of course been employed according to the personal definition of each given philosopher, or according to the definitions dominant in a given period, allowing an extraordinarily heterogeneous picture to emerge, comprised of mere speculation and/or empirical research, anachronistic and/or future-oriented hypotheses, sharply focused and/or allusive arguments.

(2) *Ecological*: in the Chapter 5 to the present companion, issues like the anthropocentrism/non-anthropocentrism debate are discussed. The question of the *approach* to animal-related topics, which (it will be shown) is relevant also at sociological level, was abudantly discussed in philosophy as well.
(3) *Ethical*: also in the Chapter 5, it is suggested that engaging in the ethical implications of zoosemiotic research should not be considered out of place, and may in fact prove useful also at strictly scientific level. There exists, nowadays, an entire philosophical school (with followers like Peter Singer, Tom Regan or Mary Midgley) devoted to the discussion of non-human animals as moral subjects.

Considering that points 2 and 3 of this list are more fully discussed in the Chapter 5, it shall be worthwhile to spend a few more words on the historical aspects of the relationship between zoosemiotics and philosophy, a topic we might easily name "proto-zoosemiotics". It would be an absurd task to attempt to summarize *all* the philosophical contributions to animal semiosis, neither could the selected ones be provided without a lengthy introduction. All that seems reasonable to do, for the purposes of a companion, is to *list* a few of the important and/or influential thinkers of western history, trying to point out the essential data that might prompt the reader into a specific inquiry. The list below is organized according to the following principles:

(1) The list proceeds in chronological order.
(2) When the inscription "About him" appears, it means that the information available was reported by other authors who wrote *about* the given philosopher. This will particularly be a case with ancient philosophers whose work we know only through post-mortem accounts.
(3) The determination of relevance is based upon the following criteria:
    (a) Interest of the given work within the topics dealt with in this companion,
    (b) Interest of the given work within the topics dealt with in zoosemiotics in general,
    (c) Interest of the given work within the topics dealt with in general animal studies,
    (d) Length of the argument (the liberty was taken to omit a few works and philosophers whose mention of animal issues was merely *en passant*, and to consider very short mentions of "low relevance", unless trully important, in the historical and philosophical sense – e.g. Bentham's brief but crucial argument on animal ethics)
(4) This summary will omit mentioning animal scientists with philosophical relevance (like Darwin or Lorenz), or animal right philosophers (like Singer, Regan or Midgley) whose work is *obviously* unavoidable, and therefore does not need to be mentioned in these pages.

Philospher	Work	Parts	Topics	Relevance
Anonymous, various	*Bible*	Genesis, Psalms, Exodus	Human domain over other animals, animal sacrifice for human benefit	Middle
Anaximander (ca. 610–546 b.c.)	About him	–	Human origin from other animal species	Low
Pythagoras (ca. 570–490 b.c.)	About him	–	Cruelty to animals, vegetarianism.	Middle
Heraclitus (ca. 550–480 b.c.)	About him	–	Human irrationality	Low
Empedocles (ca. 492–430 b.c.)	Fragments	104, 117–119, 122	Cruelty to animals, vegetarianism.	Low
Anaxagoras (ca. 492–428 b.c.)	About him	–	Animal intelligence and *logos*	Low
Democritus (ca. 460–370 b.c.)	About him	–	Human origin from other animal species, origins of arts in animals	Middle
Plato (ca. 427–347 b.c.)	*Phaedrus*	248e–249c	Human reincarnation in other animals	Low
	Statesman	XVI, 272b–e, 273a–c	Human-animal coexistence	Low
	Laws	VII, 823–824; XII, 961d	Hunting, Soul and intellect in living beings	Middle
	Timaeus	XII, 41a–d; XXXIII, 76e; XXXIV, 77a–c; XLIV, 90e–92ac	Human-animal biological continuity	High
Aristotle (384–322 b.c.)	*On the soul*	Book II, Chapter III	Human distinction from other animals	Middle
	History of animals	All	Zoology, ethology and taxonomy of animals	High
	Parts of animals	All	Zoology and taxonomy of animals	High
	Movement of animals	All	Principles of motion in animals	High
	Progression of animals	All	Motion and anatomy in animals	High
	Generation of animals	All	Reproduction in animals	High

Philospher	Work	Parts	Topics	Relevance
	Politics	I, 1253–1255	Animal subordination to humans	High
	Nicomachean Ethics	VIII, 11, 1261	Animal subordination to humans	Low
	Metaphysics	I, 980a	Parental care in animals	Middle
Theophrastus (ca. 370–287 b.c.)	About him	–	Human-animal biological continuity, Cruelty to animals, Human rationality and animal irrationality	High
Epicurus (341–270 b.c.)	*Letter to Menoeceus* (as found in Diogenes Laertius' *Lives of Eminent Philosophers*)	–	Vegetarianism	Low
Aenesidemus (ca. second to first century b.c.)	About him (esp. *Sextus Empiricus' Hypoiyposes*)	–	Differences across living beings and their intellectual-expressive abilities.	Middle
Titus Lucretius Carus (ca. 98–54 b.c.)	*De rerum natura*	II, 340–370, 970–973, 986–1012; V, 1035–1090;	Intellect, emotions, dreams and feelings in animals.	Middle
Marcus Tullius Cicero (106–43 b.c.)	*De natura deorum*	II, 153–162	Hierarchy in nature.	Middle
	De legibus	I, 21–27	Superiority of humans over other animals	Middle
Lucius Annaeus Seneca (4 b.c.–65)	*Ad Lucilium epistularum moralium libri*	XX, 121, 124	Self-consciousness and lack of abstract thought in animals	Middle
Gaius Plinius Secundus (24–79)	*Naturalis Historia*	VIII–XI	Zoology, ethology and taxonomy of animals	High
Plutarch (46–127)	*Moralia*	XII, 66–68	Mental faculties in animals, vegetarianism	Middle

Philospher	Work	Parts	Topics	Relevance
Celsus (second century)	Fragments	52,60–62, 65, 74–81, 83–86, 88, 98, 99	Human-animal biological continuity (as atheist argument)	Middle
Claudius Aelianus (ca. 170–235)	*De natura animalium*	All	Animal behavior (mostly anedoctical nature)	Middle
Plotinus (205–270)	*Enneads*	III 1–3	Human reincarnation in animals.	Low
Porphyry of Tyre (233–305)	*On Abstinence from Killing Animals*	All	Animal behavior, intellect, cruelty to animals, vegetarianism.	High
Johannes Scotus Eriugena (ca. 815–877)	*De divisione naturae*	Book III, Chapter 37	Centrality of human beings within Nature	Low
Thomas Aquinas (1221–1274)	*Summa theologiae*	Vol. XV, quest. 65, art. 3; Vol. XVII quest. 64, art. 1	Differences between humans (as rational beings) and animals, legitimacy of cruelty over animals	High
Marsilio Ficino (1433–1499)	*Theologiae platonicae de immortalitate animorum*	XVII, 3	Human being as "god" of the animals	Low
Leonardo da Vinci (1452–1519)	*Bestiary*	All	Moral and emotional qualities of animals (also in allegorical sense)	Middle
Desiderius Erasmus of Rotterdam (1466–1536)	*Adagia*	Dulce bellum inexpertis	Humans, war and animals	Low
Hieronymus Rorarius (1485–1556)	*Quod animalia bruta ratione utantur melius Homine*	All	Animal rationality and morality	High
Girolamo Cardano (1501–1576)	*De rerum varietate*	Chapter XXVII	Animal taxonomy and behavior	Middle
Michel de Montaigne (1533–1592)	*Essais*	Book II, Chapter XI–XII	Human biological continuity with, and moral obligations towards animals	High

Philospher	Work	Parts	Topics	Relevance
Giordano Bruno (1548–1600)	*Cabala del cavallo pegaseo*	Dial. 2, part 1	Anatomy, soul and intellect in animals	Middle
Tommaso Campanella (1568–1639)	*Il senso delle cose e la magia*	Book II, Chapter XXIII, XXV	Presence of memory, discourse and judgment, but lack of rationality in animals; human domination over animals	Middle
Huig van Groot (1583–1645)	*De iure belli ac pacis*	Prolegomena	Lack of rationality, sociality and language in animals	Middle
Thomas Hobbes (1585–1679)	*Elements of Law Natural and Politic*	Part 1, Chapter 5, 22	Lack of curiosity and sense of marvel in animals	Low
	Leviathan	Chapter XXVII	Proto-sociality in animals	Low
René Descartes (1596–1650)	*Discours de la methode plus various letters*	Part V	Animals as machines	High
François de la Rochefoucauld (1613–1680)	*Maximes*	XI	Differences and analogies between humans and animals	Middle
Thomas Willis (1622–1675)	*De anima brutorum*	All	Animals are complex beings provided with soul	Middle
Baruch Spinoza (1632–1677)	*Ethica more geometrico demonstrata*	II, 13; III, 57; IV, 37	Human domination over animals	Low
John Locke (1632–1704)	*An Essay concerning Human Understanding*	Book II	Memory, perception and intellect in animals	High
Ignace Gaston Pardies (1636–1673)	*Discours de la conoissance des bêtes*	All	Soul and knowledge in animals	Middle
Gottfried Wilhelm Leibniz (1646–1716)	*Principles of Nature and Grace, Based on Reason*	All	Rational and logical thought in animals	Middle
Pierre Bayle (1647–1707)	*Dictionnaire historique et critique*	Entry "Beasts (souls of)"	Soul and reasoning in animals (anti-mechanistic stand)	High

Philospher	Work	Parts	Topics	Relevance
Jean Meslier (1664–1729)	*Le testament*	XCI	Soul and thought in animals	Middle
Anthony Ashley Cooper (1671–1713)	*Inquiry concerning Virtue or Merit*	I, II, III	Human domination over animals	Low
Voltaire (1694–1778)	*Dictionnaire philosophique*	Entry "Bêtes"	Against Descartes' conception of the animal-machine	Middle
George-Louis Leclerc de Buffon (1707–1788)	*Histoire naturelle, Histoire naturelle des oiseaux, Histoire naturelle des poissons, Histoire naturelle des quadrupèds ovipares et des serpens*	All	Zoology, ethology and taxonomy of animals	High
Julien Offroy de la Mettrie (1709–1751)	*L'homme machine*	All	Biological continuity between humans and animals	Middle
David Hume (1711–1776)	*An Enquiry concerning Human Understanding*	IX	Reasoning in animals	Middle
	Treatise on Human Nature	Book I, Part III, Section XVI; Book II, Part I, Section XII; Part II, Section XII	Reasoning in animals; moral qualities and feelings in animals	High
Denis Diderot (1713–1784)	*Encyclopédie*	Entry "Animale"	Zoology and ethology of animals	Middle
Claude Yvon (1714–1791)	*Encyclopédie*	Entry " L'ame des bêtes"	Soul and reasoning in animals	Middle
Étienne Bonnot de Condillac (1715–1780)	*Traité des animaux*	All	Animal cognition, and differences between human and animal souls.	High
	An essay on the origin of human knowledge	All	Language, communication, distant space/time semiosis	High

Philospher	Work	Parts	Topics	Relevance
Immanuel Kant (1724–1803)	*Kritik der Urtheilskraft*	II.82	The human being as final goal of nature and creation	High
Jean-Baptiste René Robinet (1735–1820)	*Parallèle de la condition et des facultés de l'homme avec les animaux*	All	Differences and similarities between humans and animals	Middle
Jeremy Bentham (1748–1832)	*The principles of moral and legislation*	Chapter XVII, section 1	Ethical grounds of human-animal relationship	High
Georg Wilhelm Friedrich Hegel (1770–1831)	*Vorlesungen über die Philosophie der Weltgeschischte*	Introduction	Superiority of the human being over animals	Low
Friedrich Wilhelm Joseph von Schelling (1775–1854)	*Erster Entwurf eines Systems der Naturphilosophie*	All	Differences and similarities between humans and animals	Middle
Arthur Schopenauer (1788–1861)	*Über das Fundament der Moral*	II	Against Kant's conception of the human as final goal	Middle
	Parerga und Paralipomena	Chapter I	Ethical grounds of human-animal relationship	High
John Stuart Mill (1806–1873)	*A System of Logic*	Chapter III	Continuity in human and animal logical reasoning	Middle
	Three Essays on Religion	Chapter "Nature"	Biological nature of ethical and unethical qualities	Middle
Friedrich Engels (1820–1895)	*The Part Played by Labor in the Transition from Ape to Man (manuscript published post-mortem in the journal Die Neue Zeit)*	All	Human evolutionary distinction from animals	Middle
Henri Bergson (1859–1941)	*Évolution crèatrice*	Chapter II	Human cognitive distinction from animals	Middle

Philospher	Work	Parts	Topics	Relevance
George Herbert Mead (1863–1931)	*Mind, Self and Society*	All	Human distinction from animals (also in terms of symbolic communication)	Middle
Ernst Cassirer (1874–1945)	*An Essay on Man*	Part I	Human distinction from animals, the concept of *Animal Symbolicum*	High
Theodor W. Adorno (1903–1969) – Max Horkheimer (1895–1973)	*Dialectic of Enlightenment*	Notes and drafts	The human-animal distinction as foundation of human identity	Middle

After philosophy, the area of social sciences is of utmost relevance. Strictly speaking, they are sciences by definition oriented towards the human animal only. However, they play a major role when, once again, the key-word becomes "approach". The history of zoological and ethological sciences (or sciences related to them) is not only a history of what is observed. It is also, and sometimes mostly, a history of *how* something is observed; i.e., what methods are employed, what scientific, philosophical, ethical, or other spirit animate studies on other animals, depending on the times, places and people involved, what taboos are overcome or left in place, and how the results are affected by all these elements. There is a wide area of research, preliminary and mostly propaedeutic to straight zoosemiotics, which accounts for individual and collective mental schemes that human beings project onto other animals. Unfortunately, little accurate research has been done regarding these matters in studies of the animal kingdom. Rather, one finds studies of the interactional dynamics of social types, intergroup relationships, social representations, and so forth. It however makes sense to apply some theories and models to the analysis of the human-other animal relationship (as for instance with the case of intergroup relationship, discussed in the Chapter 5 to this companion). Social sciences might not be interested in non-human animals, but certainly possess the right tools for this kind of analysis (e.g., Tajfel 1981 or Brown 1989).

Finally, as we have seen for natural sciences already, any specialized field that deepens any of the topics applicable to zoo- or anthrozoosemiosis deserves attention. In the case of human sciences, it might seem more difficult to spot those fields, however it is far from rare for researchers in these areas to deal with problems that they only *think* (or claim) are of species-specific human interest, while that usually turns out not to be the case. A good example is anthropological (and ethnological in particular) studies, which, especially at methodological level, show several elements of contiguity with zoosemiotics. Anthropology and zoosemiotics share a similar historical destiny, in the reactions of people (specialists in particular), in that kind of scepticism and hostility on one hand, and naïve and exoticist interest on the other.

The problems that zoosemiotics is supposed to solve in order to "emancipate" non-human animals in the eyes of an anthropocentric public are in principle the same problems anthropology was supposed to solve in order to emancipate non-western cultures in the eyes of a western-centric public. For this reason, research strategies in the two disciplines are very similar – not to mention that anthropologists often have to deal with forms of human civilization with whom it is hardly possible to communicate, and thus have to formulate methods that can make research possible without linguistic interaction. Evidently, and even more so, zoosemioticians face the same problem.

Among the many other specialized fields that discuss topics of potential zoosemiotic interest one might mention at least art research, linguistics, discourse analysis, phenomenology and culturology.

1.1.4 The Importance of the Umwelt Theory

As Jakob von Uexküll has been mentioned several times, it is perhaps time to mention how instrumental his theories have been in the development of zoosemiotics. In the attempt to understand what is the most suitable general methodological framework for a discipline of this type (in other words, how a human researcher should theoretically approach other animal species), it turns out that, among the many possible solutions, scholars developed models roughly classifiable in three main categories of approach: *gradualism*, *discontinuity* and *pluralism* (or *Umwelt theory*).

By *gradualism* one means a generically Darwinian approach. The idea is that of an evolutionary adaptive continuum in which, depending on the contexts, a given animal species displays the most refined version of a given physical, behavioural or other trait, and in which, position after position, the characteristics of the diverse animals (and fungi and plant species) are less and less complex and refined, although, in most cases, adequate for ensuring the survival of the species in question. From a semiotic perspective (and also from many other perspectives), the gradualistic approach ensures a position of solid dominance for the human being. For instance, language is in this framework considered a unique and gradual structure, which finds its maximum development in the *Homo sapiens*. This means that signs and signals emitted by other animals may easily be considered "language", but their apparently lower complexity, the lack of elements present in the human one and other such differences, are considered as manifestations of a comparatively inferior development. To mention but one example, typical gradualistic attitudes are recognisable in those who consider birdsongs as proto-musical, and who more generally maintain that the origins and rudiments of art can be traced to several animal species. For instance, Hamilton and Marler take a gradualistic approach when they declare: "we must also bear in mind the possibility that some aspects of song variation [in birds] are a manifestation of some kind of primordial exercise in aesthetics" (Hamilton and Marler 1966: 446).

The word *Discontinuity* refers to an attitude that is generally sceptical of, if not hostile towards, the hypothesis that other living species can be compared to humans on a qualitative basis. The typical approach here is to emphasise indeed a "discontinuity" in the evolution of human beings, in comparison with all other animals. In other words, a sort of autonomous and peculiar development started at some point in the human evolutionary course, such that every behavioural element articulated from then on constituted an exclusively human characteristic. An example of discontinuity is the opinion that language is a typically human phenomenon, which has nothing to do with signs and signals emitted by other animals. Such manifestations may *sound/look like*, but definitely cannot *be* language. The long, at times pathetic, history of the human attempts to define the human identity, *on the basis of* a qualitative separation with the rest of the Animal Kingdom (up to the anti-scientific assumption, from the part of most religions, to consider humans as not animals at all), reveals the full symptomatology of the discontinuity approach (the human as the rational animal, then as social animal, then as the cultural animal, then as the symbolic animal, etc.).

Finally, the *pluralistic* approach (based on the concept of Umwelt, developed by Uexküll already at the beginning of the twentieth century – see Uexküll 1899–1940), starts by the assumption that the environment where an organism lives does not only consist in the actual environmental niche, but this latter is merely a physical portion of a bigger, not purely physical "environment" (Umwelt, indeed), which is perceivable and meaningful in its entirety only from the perspective of that particular organism:

> Let us consider, for example, the stem of a blooming meadow-flower and ask ourselves which roles are assigned to it in the following four Umwelts:
>
> (1) In the Umwelt of a girl picking flowers, who gathers herself a bunch of colorful flowers that she uses to adorn her bodice;
> (2) In the Umwelt of an ant, which uses the regular design of the stem surface as the ideal path in order to reach its food-area in the flowerpetals;
> (3) In the Umwelt of a cicada-larva, which bores into the sap-paths of the stem and uses it to extract the sap in order to construct the liquid walls of its airy house;
> (4) In the Umwelt of a cow, which grasps the stems and the flowers in order to push them into its wide mouth and utilizes them as fodder.
>
> According to the Umwelt-stage on which it appears, the identical flower stem at times plays the role of an ornament, sometimes the role of a path, sometimes the role of an extraction-point, and finally the role of a morsel of food. [...]
>
> The same components that are subjected to a certain building-plan (Bauplan) in the flower stem are torn asunder into four different Umwelts and are integrated, with the same certainty, into various new buildingplans (Baupldne). Each component of an organic or inorganic object, on appearing in the role of a meaning-carrier on the life-stage of an animal subject, has been brought into contact with a "complement", so to speak, in the body of the subject that becomes the meaning-utilizer. (Uexküll 1982: 29–30)

The golden zoosemiotic implications of the Umwelt theory are three:

(1) What looks/sounds/appears like a incoherent/incomprehensible/stupid behavioural pattern in a given species, depends in reality on the fact that the animal in question values the same situation according to very different perceptual criteria than a human being (or than another species, for the matter).

(2) In order to understand non-human animal semiosis, one first needs to investigate how a given animal/species organises its own experience, i.e., what is pertinent to it and what is not.
(3) Something interesting or pertinent for a non-human animal may not be perceived by humans (or any other species) at all, and vice versa.

Rather erroneously, the term "Umwelt" has often been confused with that of "environmental niche", or in other cases with "habitat", and in the most inattentive cases, with "environment". It is evident, though, that Umwelt does not designate a touchable and tangible category, but rather an array of subjective and perceptive elements.

More specifically, Uexküll considers Umwelt to be the result of two main elements: the *Merkwelt*, i.e., the specific perceptive field of a given organism, and the *Wirkwelt*, i.e., the field of actual interaction, the operational dimension of the same organism. Perceptual and operational factors contribute to form a specific Umwelt, which is exclusive for each species, and – proceeding by levels, and establishing adequate proportions – for each community, individual, class, family and so forth (Fig. 1.1).

To conceive animals in the light of Umwelt theory means at the same time to acknowledge biological, zoological and species-specific traits. It means, in other words, to take into account the biological foundations of certain behavioural patterns, and the autonomous and peculiar developments of other ones. Common bases and specific development thus must both be taken into account.

Umwelt theory is a third way for zoosemiotics, but in fact it is mostly an upgrade to Darwinian Gradualism. By adding room for subjectivity and pluralism to the interpretation of life, Uexküll's reflections prevent Darwinian theorists from turning the *scala naturae* of gradualism (with its implications of a kind of "domination" of certain species over others) into a simple, slightly softer, variant of discontinuity. In this sense, the Umwelt theory might even be considered as truer to Darwin's original findings and interpretations.

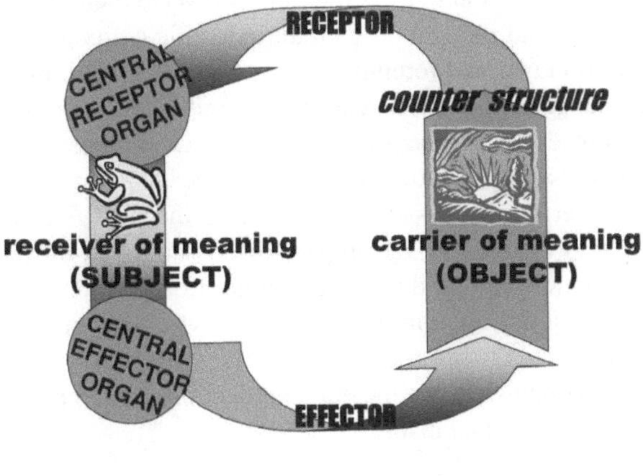

Fig. 1.1 The umwelt circle

Semiosis cannot be conceived as a unique continuum, simply divided by degrees (although, as any other thing, it actually evolves by means of selection). To locate a given element on one level instead of on another, implies the understanding of where exactly that element should be considered pertinent (e.g., is symbolic use of communication an exclusive matter of human beings? Or great apes? Or primates? Or mammals? Or animals? Or living beings?) and also where (i.e., at which point) certain traits can be analyzed in their specific autonomy. In this sense, the Umwelt theory is not dissimilar to how Giorgio Prodi envisioned the coexistence of continuous and discontinuous traits in a living organism. There cannot be reciprocal exclusion between continuity and discontinuity, because

> The progressive complication of given functions may turn, at a certain point and with certain conditions, into a "critical" restructuring [. . .]: the characteristics of human language can be explained with gradually progressive skills in the interpretation of meaning, which – at a certain point – reached a certain threshold, interact among themselves forming a completely novel situation, which is anyway still explainable with a gradual evolution of its single components (translated from Prodi 1983: 180)

Semiosis is the result of an interaction between a subject and an object, between a structure and a counter-structure, between a receptor and a carrier of meaning. These two parts are in constant and reciprocal informational exchange. In fact, the exchange itself is the real generator of any semiotic phenomenon, since the latter would simply not exist if the subject was not affected by it and did not affect it. Any zoosemiotic research, from pheromones to whale songs, should take into account such a conception, otherwise it risks perverting the essence of the phenomenon of semiosis itself.

1.2 Critical Issues in the Past, Present and Future of Zoosemiotics

In this section, the focus shall be on the most delicate aspects of the zoosemiotic study, in relation to its paradigm, its diverse methodological and theoretical characteristics and its relation with the scientific community. Such a discussion calls into question first and foremost the nature of the relationship between zoosemiotics and its supposedly closest relative biosemiotics. It is a rather interesting relationship, that reveals a number of strengths and weaknesses of the idea itself of approaching natural sciences with a semiotic attitude.

After this discussion, that is, after the next two paragraphs, a few words will be said concerning the impact of zoosemiotics in the scientific world: here, too, one notices pros and cons, and some strong criticism will not spared on the way the zoosemiotic community manages its own environment and its relations with external ones.

In conclusion, the inevitable reflections on the perspectives and potentials of the discipline will be made.

1.2.1 The Turbulent Relationship Between Biosemiotics and Biology

Besides the ones already quoted in the paragraph "What is zoosemiotics?", another definition of biosemiotics worthwhile reading is that of Kalevi Kull:

> Biosemiotics can be defined as the science of signs in living systems. A principal and distinctive characteristic of semiotic biology lays in the understanding that in living, entities do not interact like mechanical bodies, but rather as messages, the pieces of text. This means that the whole determinism is of another type. [. . .] The phenomena of recognition, memory, categorization, mimicry, learning, communication are thus among those of interest for biosemiotic research, together with the analysis of the application of the tools and notions of semiotics (text, translation, interpretation, semiosis, types of sign, meaning) in the biological realm (Kull 1999: 386)

Apart from those notions that have already emerged from the previous definitions, Kull is here suggesting that the introduction of semiotics in biological studies is not only worthwhile in the direction of a new theoretical paradigm (e.g., the centrality of semiosis rather than morphology), but also in strictly methodological terms: Nature, in other words, can be read as a text, it can be interpreted, it has a meaning, and so forth.

So, is biosemiotics, after all, about just "everything"?

When a perspective of this type comes to mind, it is difficult not to feel a bit dizzy, and that is why in this paragraph it is urgent to point out the dangers of, so to say, indulging too much in that promised land of answers to all questions that sometimes semiotics seems to display (or, to be more precise, that many semioticians seem to envision). It is a risk that other scientific communities have already perceived, and this is perhaps why the reputation of semiotics is not so high within the academic world at the moment, after having enjoyed a golden age of respect and admiration back in the 1960's and 1970's (i.e., when semiotics did *not* claim to have such an omnicomprehensive range of interests). It is therefore important, on this occasion, to address colleagues from other fields (from which semioticians have only to learn) and from the semiotic environment itself, in a clear statement about the position (i.e., potentials *and* limitations) assigned to zoosemiotics within the realm of animal studies. Given that the author is in the peculiar and ambivalent position of (a) often having to explain (if not defend) his own zoosemiotic research to entire squads of sceptic (or simply inquisitive) colleagues, and (b) having himself been pretty critical towards biosemiotics on more than one occasion (with an emotional participation that denotes most of all care and affection), reflecting upon the issue will certainly not be a banal task.

The theoretical foundations of the whole biosemiotics have been exposed to a significant amount of criticism, from both semiotic insiders and outsiders (typically from other biological sciences). Before discussing these criticisms from a strictly semiotic perspective, let us shortly review the contacts between biosemiotics and other fields of life sciences, mostly "uneasy" ones. According to Hoffmeyer,

> 20th century life sciences have been characterised by two major trends. One trend is molecular and genetic reductionism. This trend is well known and need no further comment.

> Beginning as an undercurrent to this trend, however, another much less noticed but in the long run just as important trend has gradually been unfolding: The semiotisation of nature. (Hoffmeyer, 1997: 355)

It can safely be said that the second trend mentioned by Hoffmeyer was the one that was welcomed most reluctantly by biologists. The history of modern biosemiotics has also been a history of contrasts and subtle fights between semioticians and scholars belonging to other fields. To be fair, it should be said that this was due both to a conservative attitude of the traditional biological sciences, typically hostile to theoretical/methodological changes in their field, and to a certain occasional arrogance from biosemioticians, firmly convinced of the undisputable superiority of semiotics over other disciplines. To mention one example, Konrad Lorenz was definitely inspired by the work of Uexküll, and certainly ethology can be credited with having an intimately semiotic nature (its main branches being animal communication and sociobiology). But pushing that link as far as to say, like Sebeok did, that ethology is 'hardly more than a special case of diachronic semiotics' (Sebeok, 1976: 156) is a bit reductive towards a field which not only may easily have a synchronic approach too, but – more importantly – focuses its investigation on topics that all in all are not of great semiotic interest at all.

Another interesting contact between biosemiotics and other sciences happened in 1953, after Watson-Crick's modelling of DNA and consequent understanding of the genetic code. This event marked the beginning of a theoretical process that led to the understanding of the importance of endosemiotic research, along with the exosemiotic one. Moreover, Roman Jakobson, in 1973, underlined the several similarities between genetic code and human language, in terms of dynamics and articulation, but still his claim remained unheard among biologists, who – with very few exceptions – resisted the possibility even of applying semiotic terminology to their work.

It was interesting, in this respect, that Eugene F. Yates had nevertheless noticed that a semiotic-related vocabulary was anyway very much in use within the field of biochemistry (Yates 1985). Terms like recognition, messaging, signalling, etc. were just as ordinary as any other (metaphorical or not) expression used to describe biological processes. Yates took a sample of 60 articles for his investigation, and noticed that almost 50% of them were titled with semiotic-friendly words and expressions.

Nevertheless, if biologists are interested in that kind of terminology, that is mostly due to their interest in information theory, rather than semiotics. Information theory, a mathematics-based paradigm, presupposes that information is an entity that can be measured objectively. Predictably, such a stand cannot be accepted by semioticians, not without reasons:

> The tacit assumption behind the idea of biological information seems to be that such information is the same sort of thing as "mathematical" information, i. e. an objectively existing property of so-called informational molecules such as DNA, RNA or protein. Thus for instance the famous "central dogma" formulated by Francis Crick holds that information is always passed from DNA to RNA and from RNA to protein, never the other way around. Information, then, is something which can be moved or transported.

> [However], when biologists and physicists talk about information, they talk about different kinds of things. While information as understood by physicists has no connection to values, relevance or purpose, biologist think about information in a much more everyday language sense, and in fact biological information always serves a purpose in the system, if nothing else it at least serves to promote survival. The point is that biological information is inseparable from its context, it has to be interpreted in order to work. For example, if we discuss genetic information it should be noted, that contrary to the general image raised in textbooks there is no simple relation between the DNA coded messages and the construction of the organism, whether single celled or multi-cellular [...].
>
> DNA does not contain the key to its own interpretation. In a way the molecule is hermetic. In the prototype case of sexually reproducing organisms only the fertilised egg "knows" how to interpret it, i.e., to use its text as a manual containing the necessary instructions for producing the organism [...]. The interpretant of the DNA message is buried in the cytoskeleton of the fertilised egg (and the growing embryo), which again is the product of history, i.e., of the billions of molecular habits having been acquired through the evolution of the eukaryotic cell [...] in general and the successive phylogenetic history of the species in particular. (It took evolution two billion years to produce this marvellous entity, the eukaryotic cell. Having accomplished this deed evolution spent only one and a half billion years on producing all the rest). (Hoffmeyer 1997: 357–358)

Another discussion between biosemiotics and other fields revolves around the notion of "values", or "usefuleness". Arguably, traditional biological sciences tend to underrate the usefulness of a given object as a relevant part in the understanding of life processes. Values are activities of many kinds, from eating to sleeping, from reproducing to moving. According to Sharov,

> Usefulness is not a quality but a relation between an object and user. But at a closer look, a user is nothing but a collection of useful objects. Organs are tools that are used by an organism for performing specific functions, but there is nothing in the organism besides organs. Thus, the user is just a set of relations between useful parts. Obviously, not all kinds of relations can be considered useful. Some relations may destroy the system. Relations are useful only if they preserve and augment the same relations in the future, i.e., if these relations are self-reproducing. This idea was first formulated by Pattee (1982, 1995) and was called "semantic closure". Semantic closure is a new criterion for autonomy (or wholeness) of systems. A set of elements connected by relations is autonomous only if it is semantically closed, i.e., it reproduces itself in the future and defines its identity in the process of self-production. The value of each component or relation in an autonomous system corresponds to its contribution to the ability (or probability) of the system to reproduce itself (Sharov 1998: e-text)

For instance, the reproductive *value* of an organism is related to how much it contributes (therefore, how much it is *useful*) to the growth of its entire population. The mechanism that maximizes the value of simple organisms at each step in the life cycle is probably natural selection. However, in higher animals, an estimation of the values *before* natural selection occurs, is possible. A value can apply both to an entire organisms or to its single parts.

Now, why should biologists use semiotic terminology in instances like this? Why should a biologist prefer the expression "semantic closure" to that of "self-reproduction"?

> "Self-reproduction" seems to be a convenient term that does not have uncertainties associated with signs or semantics. But this simplicity is illusive; self-reproduction includes the

> word "self" which comes from the field of semiotics rather than physics or biology. In the process of self-reproduction, an organism defines itself; in other words, self is what is preserved in the process of self-reproduction. Self-reproduction is simultaneously a process of self-measurement, self-interpretation, and communication from parents to offsprings. [...] Normal communication requires that signs have positive values both for a producer and receiver (Sharov 1992). An organism spends its resources to produce a sign only if the sign has value, i.e., it increases the rate of self-reproduction either directly or indirectly. In the same way, the receiver never interprets a message unless it expects to increase its fitness after interpretation. Here I mean expectation in a broad sense including evolutionary (unconscious) expectation. Only in higher animals and humans expectation becomes conscious. But in some cases, the value of signs may be negative. For example, some predators may intercept signals produced by their prey. In this case, the value of a sign is negative for the producer. Other predators may emulate signs that attract their prey. In this case, the value of a sign is negative for the receiver. But negative values are not normal. If a sign has negative value too often, then organisms will simply avoid using it. (Sharov 1998: e-text)

Another uneasy contact between biosemiotics and biological sciences concerns the challenge launched against neo-Darwinism in the field of evolutionary biology, by the so-called *infodynamics* (after Brooks and Wiley 1986, Salthe 1993 and others). Infodynamics "subsumes thermodynamics and information theory, essentially animating the latter by means of the former" (Salthe 1993: 6).

Still Hoffmeyer claims that the idea is intrinsically semiotic:

> The general idea as originally suggested by Dan Brooks and Ed Wiley is that information capacity (disorder) increases spontaneously in developing systems, being produced along with physical entropy as the system grows and differentiates. Since such self-organisation is a prevalent property of our universe, natural selection should not be seen as the dominating force of evolution, but rather as playing the more modest role of pruning down the novelty that is constantly and autonomously being generated by the requirements of the second law of thermodynamics. [...] I have discussed the surprising correspondence between these ideas and the "cosmogonic philosophy" of Charles Sanders Peirce (Hoffmeyer 1997: 359).

The discussion remained lively also in fields like artificial life, where again biosemiotics had (and has) something to say that would call into question the traditional paradigms of biological sciences. At present, the situation is still unclear, and although more and more biologists are accepting and promoting biosemiotic approaches, it is also true that in most of the cases this promotion occurs through channels that are – so to say – slightly less "official" than those used by other disciplines. The academic attitude towards a semiotically-inclined biologist is still that of the black sheep among the white "normal" ones, although it is undeniable that, slowly but steadily, the number of black sheep is increasing.

Perhaps, as a conclusion to this short review, it is worthwhile to quote an entire passage from Barbieri 2008, which possibly summarises, in the most efficient and least demagogical way, the main points of the fracture between biosemiotics and biology:

> Modern biology has not accepted [...] that the existence of the genetic code implies that every cell is a semiotic system. And this is no accident. The rejection of the semiotic

nature of life has been, and continues to be, extremely widespread because it is the logical consequence of at least three concepts that lie at the very heart of modern biology.

(1) The first is the model that describes the cell as a duality of genotype and phenotype, i.e., as a biological computer where genes provide the software and proteins the hardware. The crucial point is that a computer has codes but is not a semiotic system because its codes come from a "codemaker", which is outside it. This makes it legitimate to say that cells too can have a code without being semiotic systems. All we need, for that conclusion, is the idea that the genetic code was assembled by natural selection, i.e., by a codemaker that is outside the cell just as the human mind is outside the computer.
(2) The second basic concept is physicalism, the doctrine that everything in life, including signs and codes, is ultimately reducible to physical quantities. This implies that the genetic code is not a "real" code but a linguistic expression that biologists have adopted simply because it was intuitively appealing. Deep down, according to this view, the genetic code is but a metaphor because all its features must be completely accounted for by physical quantities.
(3) The third basic concept of modern biology is the belief that every biological novelty has been brought into existence by natural selection. The codes, be they organic or mental, are outstanding phenomena, but as long as they are not a mechanism of evolution, they do not account for anything fundamentally new. This conclusion is reinforced by the fact that the genetic code appeared at the origin of life, whereas the codes of culture arrived almost four billion years later. They came into being respectively at the beginning and at the end of life's history and are considered, therefore, as utterly exceptional phenomena, not as ordinary biological processes.

The genotype–phenotype model, physicalism, and natural selection are the three pillars of modern biology, and they are totally alien to the idea that semiosis is fundamental to life. This idea, therefore, can become part of biology only if we prove that all the above concepts can be replaced by more general ones. That is what biosemiotics is really about. It is about a new biological paradigm that gives us (1) a new model of the cell, (2) a real alternative to physicalism, and (3) a new mechanism of evolution. These are the great novelties of biosemiotics (Barbieri 2008: 578).

However, as anticipated, this is not the end of the story. A considerable amount of criticism comes also from the insiders of the semiotic community (although, perhaps, not necessarily insiders of *bio*semiotics), and – possibly even more than in the case of criticism from outside – there are good reasons behind all this. Mostly, for the purposes of the present companion, one has to reflect on the role that zoosemiotics plays *inside* biosemiotics. That is, how much is zoosemiotics *inside* biosemiotics?

First and foremost. Consider an essay collection of, let us say, twenty articles, or a symposium of twenty papers. How many articles or papers out of twenty should one scholar be interested in, in order to buy the book or participate to the congress? And how many s/he should have some competence about, in order to contribute to that book, or giving a presentation to that congress?

Let us go by degrees. Let us say the person in question is a comics artist, and she is hoping to become the next Charles Schultz. To a strict degree, she is of course interested in articles or papers like "Stylistical traits and evolution in the Peanuts, from the origins to nowadays", or "The role of Charlie Brown in American Society". To a slightly larger degree, also essays on Stan Lee or Walt Disney are fine. And, of course, our friend would not mind learning something about media, satire, newspapers. She *might* be happy also with something about drawing in general. Maybe painting too. Sculpture? Well, actually... Architecture? Ouch, not really... Interior design? Snoopy, help!

So, would our friend buy a book where 5 articles are about LeCorbusier, 6 about ceramics decoration, 4 about Baroque churches, 3 about French impressionism, and 2 about Superman? Probably not. Still, this is the status of a zoosemiotician, inside the biosemiotic circle.

The interests covered in the present companion include the human-animal semiotic relationship, the concept of "animal", animal aesthetics, play behaviour, deceiving behaviour, several ethical and methodological issues, the cultural representation of the animal, Interspecific communication, the concept of language, zoomusicology, mimicry, cognitive issues, ethological issues, etc. It really does not look like a restricted range of interests. There are some 1,250,000 animal species on this planet, and they *all* do many interesting things with signs. Yet, on the occasion of a biosemiotic congress, symposium or session, it is already half a miracle if one, sometimes two, scholars speak of any of those topics.

As members of the biosemiotic communitity, a zoosemiotician can be sent articles to review, and they are called things like *The Physics and Metaphysics of Biosemiotics*; *Biosemiotics as a Mode of Thermodynamics in Second Person Description*; *Beyond bioinformatics: can similarity be measured in the digital world?* What is a zoosemiotician supposed to know about these things? After all, a paper on Snoopy is still more pertinent for him/her.

Of course there is profound and absolute respect for these topics, and even more for the people who deal with them. It may be difficult to follow what they talk about, but it all looks very clever. No, the problem is different: is it acceptable to feel so lost in this huge galaxy of topics? Biosemiotics seems to deal with all kinds of things, and all sizes: from the infinitely small (DNA, for one) to the infinitely big (the Chosmos). And all that stays in the middle: cells, molecules, plants, animals, ecosystems, light, virtuality, you name it. All seems possible. Why? Because BIOS is a Greek word for life. Now, is there anything a scholar can talk that is not about *life*? In taxonomy, before reaching Life, one needs approximately 9 steps: Species, Genus, Family, Order, Class, Phylum, Kingdom, Domain, and finally Life. And human being, the subject of a massive number of semioticians, is just a species, i.e., the first step.

That raises the question: What is exactly biosemiotics? Is there such a thing as *one* biosemiotics? After all, Sebeok had anticipated it, there is cytosemiotics, phytosemiotics, zoosemiotics, micosemiotics, endosemiotics (shall chosmosemiotics be added?). Etcetera. Is it really so that talking about mushrooms has to happen in the same context where the next paper is about subjectivity, the one after that is about wild geese, and the previous was about the origins of the universe? Why not a paper on Paul McCartney, then? He is alive, he is vegeterian and he wrote "Blackbird".

It appears that it is very hard to find anything that is not life, but it is exactly when one realizes that something is about everything that it becomes about nothing. Are all these topics belonging to the same discipline? If yes, what is the advantage?

Just having more people in a conference (if that is really the case, because there hardly seems to be a reason why a specialist on any biological topic should find so attractive a place where hardly anybody speaks about his/her things)?

And what are the costs? Is it healthy for a discipline to be so generic? How can a zoosemiotician give helpful feedback to the micosemiotician colleague? And how can s/he say something useful in the context of – let us say – whale songs? Of course there would be interesting discussions anyway: but how deep, how precise, and how precious – that is a bit tricky.

The explicit choice of the present book to address issues that are exclusively zoosemiotic (without suggesting an extension outside the kingdom Animalia), is also to be interpreted as a suggestion for the near future of biosemiotics. That is, to decide whether this label wants to be un umbrella term, so that everybody is welcome, former Beatles included, or a discipline with a strong – and strong often means limited – paradigm. If the latter option is chosen, as in this book, as far as zoosemiotics is concerned, then one needs to trace a path that goes towards choice, distinction, and specialization.

This is called for also because, and it is probably the *main* fear (or misunderstanding), not too rarely biosemiotics is experiencing something similar to a metaphysical drift, or – to put it mildly – an empirical support to the interpretation of given life processes does not always seem to be a priority. It is beyond the scopes of this companion to go into details, but certainly the insiders of the biosemiotic community know perfectly what these details are, and the resulting picture should be put in relation with the not-amazingly-positive reputation that biosemiotics has in other scientific fields.

1.2.2 The Ever Present Cartesian Dualism. Part I: Razors and Cryptosemioticians

But let us not underrate the McCartney example as "just a joke". Any biosemiotician, biologist, or else, knows perfectly that a paper on a pop-rock singer-songwriter would hardly be accepted at a biosemiotics congress, or in similar contexts. Now, Sir Paul McCartney and the vastness of topics just mentioned are two sides of the same problem. The problem is that when biosemiotics looks at its left, it sees no limits whatsoever, while, when it looks right, it sees the Pillars of Hercules. Biosemiotics sees its analytical potentials as unlimited, but only (if "only" is the word) when they do not interfere with the untouchable dualism Nature-Culture: Nature allowed, Culture not. Culture is not a living process, apparently. It has no biological basis, and has no part in detecting the meaning of life. Yet, it would not be so difficult to prove otherwise. Both rethorically and scientifically. Not to mention that scholars who like Peirce so much (as biosemioticians do) should be anxious to support the notion of synechism. But, thankfully, at least human culture is left out.

It is difficult not to interpret this choice not as the result of a pondered scientific decision, but more as an ideological one. In an environment, the semiotic one, dominated by scholars who deal with human matters, biosemiotics came to cover just everything-that-is-not-human. Which is a definition by negation, not by affirmation. Take humans out, and what remains is biosemiotic.

Zoosemiotics, thus, gets to be placed exactly in the middle of this tug of war: *pulled in* by a community of people who mostly study life-related topics *outside* the animal kingdom, and *pushed out* by a community who study the animal species *Homo sapiens*.

Regretfully, this is not the kind of specialization called for. Because if the point is to accept once again, although in disguised form, and pretending to do something else, the Aristotelian/Cartesian dualism, then it becomes hard for a zoosemiotician to identify him/herself with this community. Regardless of how much against this perspective one might be ideologically, the thing is that most of the research interests of zoosemiotics fall into the category of cultural processes. Or where else should reasonably fit such topics as zoomusicology, the human-animal relationship or the notion of language?

Semioticians like Prof. Kristian Bankov, from New Bulgarian University, became so annoyed by this unlimited extension of the biosemiotic field, to actually conceive, in 2005, a sort of Ockam's Razor (ironically baptized Bankov's Razor by Bankov himself) to revise, among other things, "(1) the philosophic grounds of the biosemiotic discourse" and "(2) the scientific output of biosemiotics" (Martinelli-Bankov 2008: 399):

> Concerning the first point, one must distinguish between several interrelated discursive orientations within the field of biosemiotics. (1) A strong ontological pretense in biosemiotics, a philosophically-grounded claim on how things hang together, what the real essence of nature and life is, etc. "Life is coextensive with semiosis" is characteristic of that kind of reflections. (2) A huge part in biosemiotics, dedicated to a kind of history of ideas, relevant to its predecessors or discovery of biosemiotic ideas in "unexpected" thinkers and scientists. That includes contributions of systematization of the contemporary advances in the field. (3) A descriptive, strictly scientific, orientation, concerning facts in the Universe and the biosphere: Big Bang, cells, molecules, plants, bacteria, animal behavior, etc. The above-mentioned theoretical mimicry can be found mostly in those contributions. (4) Last but not least, the contribution on current socio-political issues like ecological consciousness, "biomoral" and others, where biosemioticians are expert voices in the public space.
>
> Considering this variety of discursive orientations, it is difficult to establish any clear identity for biosemiotics. After hearing that biosemiotics is "the study of signs, of communication, and of information in living organisms" (as stated in the Oxford Dictionary of Biochemistry and Molecular Biology) and even more: "the scientific study of biosemiosis" (Emmeche et al. 2002: 26), an unprejudiced observer would imagine people with an experimental agenda. Kalevi Kull claims that "Biosemiotics means biology" (2002: 332). My expectations were similar, but when I started reading books and papers I found much more philosophy than laboratory/field work. Somehow, the identity of the discipline was to be established more as an interpretive than as a productive approach. And I think that more or less all scholars in biosemiotics would agree on the precept, present in any extended definition, that biosemiotics offers the existing biological theories a kind of generalizing view which they lack and they need. (Martinelli-Bankov 2008: 399–400)

Although controversial in many aspects, the notion had the great merit of shaking the coconut tree of biosemiotics, forcing some insiders of the field to seriously reflect upon the scientific role of biosemiotics, its weaknesses and its strengths.

There has been quite a hype around the Razor in the 2–3 years following its enunciation, within the semiotic circle, or – better – within the *bio*semiotic circle, i.e., the circle of those semioticians who move around the same area of knowledge exchange. That circle meets often in Finland, particularly in Imatra, in Tartu, in Copenhagen and in other European places in the occasion of the so-called Gatherings in Biosemiotics.

In one the main points of his argument, Bankov notes that a certain "scientific humility" that he expects from biosemiotics is, "not present in the majority of biosemiotic papers I have read" (Martinelli-Bankov 2008: 401). It is difficult not to agree, although, probably, for reasons that are not necessarily the ones indicated by Bankov (see the paragraph "In defence of Biosemiotics"). A certain arrogance has indeed been displayed in quite a few occasions, when it comes to both the foundations of the discipline, and its theoretical – so to say – heritage. The above-mentioned metaphysical drift is already something that speaks for itself, but let us see other instances.

Firstly, there is the so-frequently experienced presumption of many biosemioticians to have finally discovered the ultimate answer to all problems. One can see them, happy and proud, reminding us how the biosemiotic revolution has turned the obsolete knowledge upside down, for better and for good. The thing is, it may also be secretly thought that biosemiotics offers a more consistent view of natural phenomena (although one should not be so confident about it), and a semiotic interpretation of them may also be fully accepted. But, a biosemiotician does not have the reputation to say that openly, this reputation having completely to be achieved. And in fact, even if s/he did have it, it would still be an unpleasant attitude.[3] Certainly biosemiotics has to face prejudices, incomprehension, conservative and reactionary feedback. But boldness cannot be an appropriate response.

Another example of this arrogance lies on the luckily-just-occasional tendency of refusing, or at least underrating, the contributions to semiotics provided by those semioticians who were never *bio*semioticians, or who anyway refer to completely different fields of inquiry. The obvious instance is the strong stand in favour of the Peircean sign paradigm, to the detriment of the Saussurean and Greimasian ones. This companion tries not to fall into the trap, and models belonging to different semiotic traditions are democratically applied in the attempt of making zoosemiotics a trully semiotic field of inquiry.

Also, from time to time, this rejection of extra-biosemiotic approaches may become a momentary lack of memory. A couple of years ago, the author of this companion was asked to review an article on modelling systems, written by a very famous semiotician. The article certainly provided a comprehensive account of the state of the art of the theory on Modeling Systems, very efficiently summarising the main issues related to the concept. The reviewer was naturally in favour of its publication. However, on one aspect he found the author surprisingly quiet, and

[3]Of course, scholars in other disciplines, including natural sciences, often display a similar arrogance, in support, or celebration, of their own paradigms. However, committing a fault just because others do the same, does not really make that fault easier to excuse.

that was the total – total! – lack of mention to the enormous contribution that the Tartu-Moscow semiotic school in general, and Yuri Lotman in particular, made to the concept. To maintain, as the author of the article did, that, "modelling systems theory *blossomed* forth as a comprehensive theoretical and methodological framework *only* in the pivotal work of Thomas A. Sebeok" (emphasis added) completely neglected the contribution of the Tartu-Moscow school, which is even more ironic when one thinks that Sebeok himself repeatedly expressed his debt to Yuri Lotman on this account. Is the point here that all semioticians are equal, but biosemioticians are more equal than the others?

The picture totally changes when the war against non-biosemioticians ceases being a *civil* war inside semiotics and displays the characteristics of an international conflict among different fields of knowledge. There, biosemiotics (as in fact, the whole of semiotics) assumes a completely opposite attitude, much more eager for inclusion than exclusion. Dozens of scholars coming from all kinds of sciences are acknowledged as crucial in the historical development of semiotics. With one particular caveat: they all *become* semioticians. They are all called "cryptosemioticians", sometimes "protosemitocians", even when any concept vaguely resembling semiotics (let alone the terminology itself) was clearly outside their scopes. their reflections only have to fit to the semiotic agenda, and they turn into people who, *if only they knew*, would have passionately joined the semioticians' community. Illustrous victims of this process are, among others, Aristotle, John Locke, Augustine, Thomas Aquinas and Jakob von Uexküll.

A similar attitude goes of course hand in hand with the post-modern program of scientific unification. When this attempt is, so to say, well-tempered (and it is what is attempted in this companion), the benefits are certainly more numerous than the risks of trivialization of the scientific discourse. However, such risk is constantly around the corner, and it takes an extremely careful approach to avoid it. In this respect, mystifying the scientific identity of a scholar for the purpose of having him/her in the semiotic *militia*, then hinting that, after all, that unified paradigm has always existed and, by the way, it has existed under the semiotic umbrella, *as it can clearly be noticed by the fact that all these people were in fact semioticians*, well, that does not seem the best way to avoid trivialization.

1.2.3 The Ever Present Cartesian Dualism. Part II: Semiotic Animals and Metasemiosis

Turning the discussion on its head, there is another form of *horror vacui*. If on the one hand, biosemiotics identifies the concept of "life" with nearly everything, and on the other hand it can be noticed that the human subject is carefully kept outside the discussion (meaning that so far no biosemiotic paper/article – not that the author knows of, at least – focused on human-related topics). Is *Homo sapiens* not a living form? As mentioned, the motivations behind this differentiation appear to be more ideological than methodological, and in this paragraph this point will be illustrated by discussing the arguably best example of this differentiation within scientific discourse, that is, the already mentioned history of definitions of "humanity", as instruments for qualitative (discontinuous) distinction from the rest of the animals. It is a highly educational history, with a distinctive, and altogether amusing, pattern. The goal: founding human identity by establishing a radical difference with all the other animals. The procedure: launching (sometimes in good faith, sometimes not) a manifesto-statement on the specific and special diversity of the human animal. The methodology: trial and error. As soon as the statement is dismissed by a scientific counterproof, another statement, concerning another supposedly species-specific human characteristic will be issued. Alternatively, if the characteristic in question is perceived as possessing strong potentials of human exclusivity, there is the option of narrowing the definition of the characteristic itself, until, fatally, other animals are excluded. Of course, in that case, several human communities or subjects are excluded too, but this is what in military jargon is called "friendly fire". And "war", or better "struggle", is an appropriate word to describe this process: a struggle for affirming the "ingroup" identity, in a specific historic moment where the only morally-acceptable discrimination can be operated on the outgroup "animals".[4]

There is no pretension here to reconstruct a philological chronology of such statements, but at least the following should be mentioned:

- The rational animal
- The symbolic animal (or *Animal symbolicum*)
- The playful animal (or *Homo ludens*)
- The moral animal
- The laughing animal (or *Homo ridens*)
- The spiritual animal
- The social animal
- The cultural animal
- The linguistic animal (or *Homo loquens*)
- The abstract animal
- The signifying animal (or *Homo significans*)
- The political animal (or *Homo politicus*)
- The tool-maker animal (or *Homo faber*)

Some of these definitions stem from authoritative thinkers, who, in a certain moment of philosophical history, reflected on the characteristics of human nature (like Aristotle with the notion of rational animal, or Cassirer with the symbolic animal). Some others derive from recent studies in sociology, psychology or else, and some others, finally, are of commonsensical or superstitious/religious nature.

---

[4]On intergroup dynamics, see at least Tajfel 1981 and Brown 1989. The topic is also discussed in part IV of this companion.

Now, as resulting from different investigations (mostly in natural sciences, but occasionally also in humanities, including semiotics[5]), all these supposedly distinctive features of humanity were challenged and generally invalidated exactly in their uniqueness. Any of these definitions would last a certain amount of time as the "ultimate sign of human distinction", and then the likes of Darwin, von Frisch, Griffin, Goodall, Sebeok and others would raise their hand and go like "well, actually, we found out that also animals can do this".

It is rather tempting to picture an imaginary "International Committee for Human Identity" gathering in the occasion of any of these discoveries in animal studies, and putting in their agenda the need for finding another "ultimate sign of distinction". Rational animal? Failed. Cultural animal? Failed. What about Symbolic animal? Did we try this? Not yet? Good. Approved. Launch the "symbolic animal" campaign.

Alternatively, the strategy adopted, as mentioned, is that of narrowing the concepts, when it turns out that the previous definition was loose enough for allowing other animals in. It happened with "culture", "language", "mind", "abstraction", "art", and other critical topics, whose definition and conceptualization are in constant progress, and therefore very liable to be manipulated (i.e., in most cases, restricted). Commonly, this procedure is performed with the good old dogmatism of the *petitio principii*, an all-time favourite among different religions. Quite simply, inside the definition and without providing additional explanation, we say things like, "culture is the integrated pattern of *human* knowledge, belief, and behavior that depends upon the capacity for learning and transmitting knowledge to succeeding generations" (as appearing in the Webster dictionary). Why learning capacities and knowledge transmission (extremely common phenomena in most animals) should be exclusively human remains a mystery, but this way we avoid the confrontation: we set the dogma by default, good night and good luck.

In addition, another problem may arise, what was here called "friendly fire": more often than not such restrictions end up excluding from the picture also several human communities or human subjects, creating a discrimination that, *in this historical moment*, is definitely not acceptable. Some of the most recurrent definitions of "music" are for instance so circumscribed that not only humpback whales or nightingales may be resentful, but also the majority of human peoples around the globe, plus illustrious musicians like John Cage or Karl-Heinz Stockhausen.

The main problem remains anyway one: this vital urge, from the part of human beings, to (1) establish their identity on the basis of claimed radical differences with other animals, and (2) setting this separation in form of qualitative dualism, as in human beings on one side and all-other-animals, indistinctively, on the other. The former being basically an anti-Darwinian stand, the latter being anti-Darwinian, anti-Uexküllian, anti-scientific, and – for the matter – anti-semiotic. The rhetorical dynamics of human exclusiveness in terms of dualism are discussed at length in

[5] It was for instance Sebeok himself to dismiss once and for all the idea that only human communication can be of symbolic type.

the paragraphs on anthropocentrism (in the Chapter 5), so the question will not be further explored here.

What is most important, at this stage, is the question, Where does semiotics come into the picture here? Approximately, in all stages of the process. That is, there has always been, in one moment or another of semiotic history, (a) campaigns in favour of the uniquely human application of semiotic studies (the so-called semiology), or (b) attempts to contribute to those processes of progressive restriction of given concepts (semiotics had a go on at least "language", "symbol", "culture", "abstraction" and "signification"), or finally (c) creation (or participation to the creation) of new labels of human distinction (the linguistic animal, for one).

The last, specifically semiotic (and it could not be otherwise, given its name), creation offered by the International Committee is the concept of the "Semiotic Animal". A colleague and friend like Prof. John Deely, the main promoter of this concept, subtitles it "a post-modern definition of *human being*". As he recalls:

> The earliest record we have of the expression "semiotic animal" in English dates to Deely 1990,[6] in Italian to Petrilli 1998, but both these usages find a considerably earlier ancestor in the test of Hausdorff 1897, written under the *nom de plume* "Paul Mongré": [...] "The human being is a semiotic animal" [...] There is, however, this considerable difference in the original German coinage and the later coinages in English and Italian: Hausdorff explains his expression in the terms that would by the mid-twentieth century be associated with the glottocentric minor tradition of semiotics commonly called "semiology" (Deely 2005: 25–26)

By this, Deely means that Hausdorff was an *ante litteram* supporter of the heavily anthropocentric view of semiotics that dominated the scene through the whole twentieth century until roughly the 1980's: the linguistic-semiological tradition of Saussure. In this sense, talking about a "semiotic animal" means nothing less than affirming, as Hausdorff did, that the "human nature lies in the fact that, instead of expressing naturally his needs and wants, [man] has developed a language of signs which is conventional, symbolic, and only indirectly understandable" (Deely 2005: 26, translated from Hausdorff 1897: 7).

On the contrary, the semiotic tradition followed by Deely, that red line that connects – among others – Peirce, Morris and Sebeok, assumes the almost opposite stand of "opening" semiotics to the multiple instances of non-human signification, communication and representation, "seeing cultural creation itself as a natural extension of the activities of the semiotic animal according to what is proper to it as part of nature" (Deely 2005: 26). In that sense, there is never, in Deely's theories, the denial of the animal nature of human beings, and in fact (at least according to his own words in various conferences and talks), the notion of semiotic animal is apparently conceived for overcoming the human-animal dualism and creating that continuity-discontinuity dialogue envisioned by Prodi (1983: 180):

> The anthropos as semiotic animal is an interpretant of semiosis in nature and culture alike, that can only be because the ideas of this animal, in their function as signs, are not limited

[6] That is, the first edition of Deely 2009, in this companion's bibliography.

> to either order, but have rather [. . .] the universe in its totality [. . .] as their object. (Deely 2005; 27)

And, more to the point:

> At the heart of the difference between the human Umwelt and the Umwelt of other cognitive organisms is the "idea" in this specifically semiotic sense: the relationship itself that constitutes signification is grasped in its proper being at once imperceptible and distinguishable both from a given signified and from a given sign-vehicle – and therefore as detechable from any given vehicle and attachable to any other vehicle, as well as directable to some other object, or to the same object only, in its new attachement. (Deely 2009: 89)

What does this mean in practice? It means that the very species-specific characteristic of the human being is the capability of emancipating its semiosis from the constrictions of its own Umwelt and (a) access other organisms' Umwelten (also, but not only, at purely speculative level), and (b) creating possible worlds, i.e., Umwelten that are distant in time, in space, or that are not even existing at all, but are created through language (there is no such a thing like mermaids, yet there is such a thing like mermaids, at the very moment at least one person linguistically produces a concept, or an idea, of them). Such abilities, Deely firmly points out, do *not* make the human being *something else* than an animal, yet they establish the ground for human biological identity. The identity being: the human beings, unlike other species, are *semiotic* animals.

John Deely is one of the firmest supporters of zoosemiotics, and a most credible "heir" of Thomas Sebeok, of whom he was also an intimate friend. In that sense, the last thing one might suspect is that he is anthropocentrically-biased. In this respect, he would *not* qualify for a membership in this fictional International Committee of Human Identity. The fact that this time the sign of human distinction is (a) limited to a tiny particular in the entire spectrum of possibilities previously explored,[7] and (b) expressly formulated in such a way that it is made unmistakably clear that the human being *is* an animal in all respects, must be certainly regarded as the fairest step taken so far in this special category of speculations.

The present observations indeed are of purely methodological nature, and partly call into question the research ethics of this type of intellectual enterprises. In this respect, they must be read as general observations, which only use Deely's formulation as an excuse, and not at all as the main target.[8] With this in mind, there is still a feeling that a number of questions are left unanswered, ambiguous and possibly biased.

First and foremost. The motivation. Why is it so important, for a semiotician (and for a scholar in general), to establish a clear sign of qualitative distinction with

[7]There is no doubt that being able to create possible worlds is a much milder form of distinction than, e.g., being "social" or "spiritual", or "rational", all being traits filled with countless implications at different levels of cognition and behavior.

[8]And this is said also as one more demonstration of the author's esteem to Deely as a scholar, which has nothing to do with the few critical remarks advanced in this text. It must also be repeated that the "semiotic animal" is the last of the problems in this special category: the reason why it is worthwhile to discuss it is its primarily and specifically semiotic nature.

other animals. The question is not rhetorical, and in fact neither ethical (although it may appear so), but mainly historical and anthropological. There seems to be a need, the already called "vital urge", from the part of *some* human beings to define themselves, as an ingroup possessing certain characteristics, in relation to a bigger outgroup which does not possess them. In this specific historical moment, the other animals are identified as that outgroup. It was not always like this. Many forget that the original *Scala Naturae* conceived by Aristotle (from which the notion of rational animal is based), is not only an expression of human-animal dualism. The truth is, that was not a dualism at all. It was a comparison between a single category of privileged human beings (those who were also male and free) over three main categories of unprivileged: women, slaves and, yes, other animals. If a dualism must be established in the Aristotelian framework, thus, its definition should rather be borrowed from Peace Studies jargon: it was an oppressor-oppressed dualism, in Dussel terms, or a centre-periphery one, in Galtung's.

In general, the shaping of the ingroup identity was, until very recently, not necessarily characterized by the urge of defining "humanity" *in toto*. It could take all sorts of disguises: male people, free people, white people, western people, sane people, rich people, Christian (or Muslim, or Jewish, etc.) people, heterosexual people, and so forth. What happened (in different times and places of the twentieth century) is that most (to a certain extent, all) of these forms of discrimination finally and thankfully came to an end, at least on a generally-accepted ethical level. There is no ground, anymore, for anybody to say that men are rational and women are not, without being taken as a chauvinist pig, or as a cheap joker. Whoever advances any post-nazi claim that people of Caucasian ethnicity are more intelligent than, say, Afro-Americans, raises immediately general irritation, and it also takes a special effort to consider a similar statement just a bad joke.

Given this picture, what is left to a poor intellectual who still wants to exercise his/her anthropological urge to be part of a small elite of elected individuals? Racism, chauvinism, classism, eurocentrism, and all the rest is gone. Vanished in that annoying black hole called civilization. How to reconcile the primary need for discrimination with the modern pressure of a civilized and tolerant outfit? The answer is what philosopher Peter Singer named *Speciesism*, the genesis of which he explains exactly in the terms here illustrated:

> In recent years a number of oppressed groups have campaigned vigorously for equality. The classic instance is the Black Liberation movement, which demands an end to the prejudice and discrimination that has made blacks second-class citizens. The immediate appeal of the black liberation movement and its initial, if limited, success made it a model for other oppressed groups to follow. We became familiar with liberation movements for Spanish-Americans, gay people, and a variety of other minorities. When a majority group – women – began their campaign, some thought we had come to the end of the road. Discrimination on the basis of sex, it has been said, is the last universally accepted form of discrimination, practiced without secrecy or pretense even in those liberal circles that have long prided themselves on their freedom from prejudice against racial minorities.
> One should always be wary of talking of "the last remaining form of discrimination." If we have learnt anything from the liberation movements, we should have learnt how difficult it

is to be aware of latent prejudice in our attitudes to particular groups until this prejudice is forcefully pointed out. (Singer 1989: 148)

Speciesism, *so far*, is the "last remaining form of discrimination". Considering the encouragingly increasing number of animal welfare legislations in different countries that are extending the basic human rights to other Great Apes, that is, to species belonging to our same family *Hominidae*, it is fair to forecast that the next sensation, in terms of discrimination trends, will be called "familyism". And this is not necessarily a joke.

There is still ground for speciesist statements without being politically incorrect. There is no formal program for a "he or she" formula, as applied to non-human animals, and the same expression "non-human animals" that is carefully used in this companion, is easily replaceable by "animals", achieving the same result, in people's understanding, of referring to "all other animals except the human one".

So, a first question to the International Committee for Human Identity, is "why?". Why in general, but also why in the specific context of scientific communities who have (or should have) fully accepted the Darwinian notion of evolutionary continuum. It is difficult to grasp the coherence here, and inevitably one is reminded of Richard Dawkins' warnings on the great difficulty of complete emancipation from religious beliefs, even in a secular context.

Just In case it is not yet very clear, there is here an absolute – again: absolute – support for the Darwinian Revolution. The theory of evolution, despite the peculiar comments of some biosemioticians who find it "anti-semiotic" (and it would be nice to learn why), is at present – from both the philosophical and the empirical point of view – the most reliable scientific theory for explaining life on this planet. Among the millions important implications of this revolution, there is also a firm warning not to discuss *anymore* the differences across species in terms of *qualitative* distinctions. There is no such point in which an organism radically *stops* being/doing something and *starts* being/doing something else. Everything occurring in life is a natural continuation of a process in continuous development, with traces detectable in different living forms, and with aspects that are only *quantitatively* developed in certain forms rather than others. However, and that is crucial here, *no more* are scholars allowed to place a threshold at any point of evolution and affirm that "this is it, from now on it is a completely different matter, whose features concern only this species, and not all the others". Also, and this is the lesson instead learned from Uexküll, it is time to quit this *human and-all-other-animals* dualism once and for all. Human beings are a species, not the conceptual counterpart of the Animal Kingdom.

This is certainly one of the sensitive areas within semiotic discussion, something that quite a few semioticians do not seem to grasp entirely, probably because they think that semiotics, with its own tools and procedures, is alone able to overcome certain problems. The Peircean process of endless semiosis is awfully charming, and it is understandable that a few scholars are tempted to turn it into straight philosophy of science. But then, these scholars should not be surprised if they end up being labelled as rather arrogant, and by consequence the whole semiotics is regarded suspiciously. This is why, besides a need to support Darwin against fancy retro-futuristic notions (i.e., those notions that, while attempting something new, are again restating pre-Darwinian anthropocentric ideas), another statement is also called for: when semiotics aims to discuss science, it ought to be scientific. Period. If statements like a supposed qualitative diversity of the human being are issued (or, on the other side, the presence of interpretation processes in plants' semiosis), proofs should be provided, not only speculation.

Which kind of leads to the second question: "Have we not learned the lesson yet?". The human intellectual history, as shown, is full of these repeated attempts to create a "special" ingroup, and a specific care, in the last century, was taken to establish this uniqueness to the detriment of other animal species. Now, all these attempts failed. Long ago, or recently. Honourably (like the "symbolic animal" or the "abstract animal", which admittedly took accurate research and speculation to be finally dismissed), or miserably (like the "social animal" or the "cultural animal", which clashed not only against basic scientific inquiries, but also against lay-people and children observations). Now, what a "reasonable animal" would do, in order not to look like a "pathetic animal", is to behave like a "careful animal", and avoid advancing more hypothesis by using the same old dualistic framework. There is vast ground for building the human identity (again, provided people cannot really help establishing one) on a fair Darwinian level, replacing the digital with the analogical, the concept of differentiation with that of specialization. And, most of all, without making the whole enterprise look like a Freudian phallic pulsion to show the world that "we do it better" or "we have it bigger".

Plus, apart from scientific arguments, the main reason why it should be time to stop doing this is that it is, simply, very childish.

Unfortunately and evidently, this is yet not the case. Which leads to the suggestion that the best definition for the human being is that of the "stubborn", or the "frustrated" animal. This animal still seeks for a dualistic, qualitative confrontation. Because we humans cannot just accept the idea that we are animals. Or, if we surrender to this scientific evidence, we still need to prove that we are special animals, but not special to the extent that every species is special (see the Umwelt theory). No, we want to be special in a special way. Second-degree special. Once again: all animals are equal, but some are more equal than others. Orwell did not know how *extensively* he was right.

Third point. So, dualism it is, once again. The question: if we humans are semiotic animals, what type of animals are the other animals? Non-semiotic animals, unsemiotic animals, semiological animals? This does not explicitly appear in Deely's or Petrilli's writings, but it was cleared during a friendly chat between John Deely and Dario Martinelli, in June 2009, in the occasion of the International Semiotic Institute annual congress in Imatra (Finland). Non-human animals, it appears, are semiotic animals, with an *s*. Therefore, the capability of creating possible worlds is what distinguishes semiotic from semiotic cognition.

So, clearly, a new semiotic threshold is now established, is it not? There used to be one at the times of semiology, and one of the main achievements of the Peircean school was exactly to put into question this threshold up to lowering it at a virtual minimum ("The universe is perfused of signs". Anything goes). With the notion of "semiotic animal" (which curiously comes exactly from proudly Peircean scholars) the issue of the threshold is addressed once again. Not only: it is set to an extremely high point of cognition. Nothing less than the possible worlds. Semiotic are those animals that manage to create and access possible worlds, semiotic are those that cannot. Regrettably, it does not seem a great improvement from the semiological threshold, mostly based on intentionality, which achieved the same identical result: humans above, other animals below.[9] Not to mention that at this point it is probably needed to diversify the discipline formerly known as semiotics into two fields: the logocentric (or metalogocentric) "semiotics", and the huge (99% of semiosis, Sebeok used to say) "semiosics", that is, a field that investigates all semiosis processes, except those that lead to the creation of possible worlds. But, no doubt, Deely for one cannot possibly want this.

However, let us analyze the concept a bit further. *Sic stantis rebus*, the upper level of the semiotic threshold should be considered as corresponding to the realm of abstract imagination. Or should it? Of course, that is not the end of the story: a crucial precondition to this form of imagination is the ability (this also attributed to humans only, in the "semiotic animal" formulation) of using signs being aware that they are signs. Only with this *metasemiosis*, as Petrilli[10] calls it (1998: 8), the human animal becomes aware that "there are signs", and that these signs can be detached from any association and manipulated to create endless new associations, even imaginary ones. It is metasemiosis that makes mermaids (and unicorns, angels, God, Neverland, Spider Man and Dr. Octopus) possible. That, it is claimed, occurs only among humans.

Now, let this point be the first in the checklist. To tie inevitably metasemiosis with possible worlds, one should assume that metasemiosis *only* produces possible worlds. Because, if it produces something else, then it is not just alien Umwelten that animals are not able to access: they would fail to access also all the other outcomes of metasemiosis, as metasemiosis, too, it is claimed, is a species-specific human phenomenon. There goes the point. The awareness that "there are signs" is a precondition for a million of other cognitive abilities. Deception, for one ("semiotics studies everything that can be used to lie" is the famous quote from Umberto Eco). Then Play. Symbolic representation. Aesthetic semiosis, Ritualization. To a certain extent, also interspecific communication. They all imply (simple or complex) forms of sign manipulation that are possible *only* by being aware that what is being used *is a sign*, and not the actual entity it refers to. And, looking closely, one

[9]Incidentally, in this companion the classical threshold is accepted and adopted, exactly to leave no doubt that, even at that level, non-human cognition perfectly qualifies for the "semiotic", with a t, denomination.

[10]As with John Deely, also in the case of Susan Petrilli the author cannot but state his profound admiration for the work of this outstanding scholar. The reservations expressed here concern a very minimal part of Petrilli's work, and luckily there has been an occasion to express them personally to her, so none of what is written here should come as a surprise.

might even argue that codes themselves cannot be established without an at least vague understanding of the nature of signs.

It is not clear whether Petrilli was aware of the "size" of metasemiosis, and, if so, whether she actually intended that non-human animals are indeed incapable of all this. What is certain is that there is no shadow of doubt, at the present state of *all* animal-related research fields (including zoosemiotics), that non-human animals are unmistakeably capable of all these behavioral patterns: there are differences, of course, across species, and it would be very interesting to finally discuss *those*, rather than posing the problem in dualistic way, but nobody here seems to be interested in that, as the outgroup "animals" is systematically treated as one big cauldron.

Now, once we have understood that this idea of "metasemiosis" is quite frankly out of range, let us return to one of its (many) implications: the possible worlds. Would it be right to assert that no animal except the human one is able to cognitively access alien Umwelten, particularly the imaginary ones? It is a quite interesting question, with several possible answers:

(1) One could for instance follow the logics of Peircean abduction, and state immediately that "As the unavoidable precondition for possible worlds cognition is metasemiosis", and "As metasemiosis is empirically existent in many non-human animals", *then there are good reasons to think* that many non-human animals are able to cognitively construct possible worlds. This reply alone will hardly satisfy the readers, but it is a reasonable point already. Language is not the only condition for metasemiosis: of course it is not. If metasemiosis is equalled with possible world cognition, there is a *de facto* denial that the latter is produced exclusively by language. And, in fact, there are very good reasons to think so;
(2) The concept itself of accessing alien Umwelten may be in fact a contradiction in principle. Taking Uexküll's theories very faithfully, one shall deduce that if we, members of the Umwelt A, manage to cognitively access the Umwelt B, then we cannot really talk about different Umwelten anymore, because the occurrence of this very process would prove that the Umwelt B is simply part of the Umwelt A, therefore not "alien", and in fact not Umwelt, either. When we talk about *things*, in a way, they are no longer *things*, but already *objects*, or at least philosophical *objects.*[11]
(3) In any case, even accepting the language's capacity to create strong dialectical, philosophical and even rhetorical configurations of alien Umwelten, a few

[11]*Now there is a great difference between an object and a thing, however confusedly the two notions are made to play in popular culture. For while the notion of thing is the notion of what is what it is regardless of whether it be known or not, the notion of object is hardly that. An object, to be an object, requires a relation to a knower, in and through which relation the object as apprehended exists as terminus. A sign warning of "bridge out" may be a lie, but the thing in question, even in such a case, is no less objective than in the case where the sign warns of a" true situation".* (Deely 2000: 18).

questions remain unanswered. For instance, dialectics, philosophy and rhetorics do not give *full* access to another Umwelt, even though they certainly allow a good deal of reflections and some conclusions about it. This is important to remember, especially at an ethical level: sometimes, indeed, we take too much for granted that we understand of other animals *all* that is to be understood. To make a specific zoosemiotic example, we may understand, speculate and talk about echolocation, but a complete, senso-motorial, psycho-physiological access to a process like echolocation (at least so far) is not possible, and that makes a huge difference (especially when we think of the dominant role that echolocation have in the semiosis of many cetaceans);

(4) That is not all. We should not be so enthusiastic about our invention of myths, tales and superheroes, at least not to the extent of founding a new definition of the human being upon it, because all the imaginary entities produced by language are *not* entirely untied to our perception. They might not exist, but they are a combination of (mostly visual) perception-bound elements (a horse plus a horn for a unicorn, a human female plus a fish for a mermaid, a zoo- or anthropomorphic outfit for most gods, etc

We know already from narratology (see at least Genette 1972, Searle 1975 and Bruner 1986) that possible worlds are forms of (a) paratisism and (b) limitation of the real empirical world. We know from logics (see at least Lewis 1986, Herrick 1999, Divers 2002) that their ontological status in relation to reality is far from independence, and in particular their modal status completely relies on this relation (the six propositions – true, false, possible, contingent, necessarily true and impossible – are no less than hermeneutic variations applied on empirical reality). We finally know from Freud (1899) and later Jung and the whole psychoanalysis[12] that the "true" abstract imagination, i.e., dreams (those we cannot even logically reconstruct after waking up because they escape the cerebral centres for thinking activity, which indeed do not function during dreams), belongs to a realm, the unconscious, which is neither linguistic nor para/pseudo/proto/pre-linguistic: it has simply nothing to do with language, but with cerebral centres for sensorial and nervous activities. A little too "biological" for being a "ultimate sign of human distinction".

(5) Let us anyway stay on this particular (perception-bound) case of abstract imagination: the possible worlds constructed thanks to the rhetorical-narrative properties of linguistic semiosis (hoping that it is clear that this is an entirely different matter from the trully intangible possible worlds investigated in psychoanalysis). The question is: what elements do we have to be able say that these features of human semiosis are *qualitatively* different from other animals? If many Interspecific Communication Experiments (ICE) turned out to be a successful attempt to teach human language to other animals, as they did, in mostly everybody's opinion except semioticians', and if language belongs to the human

[12]With the possible exception of Lacan, who never gave the impression of exactly knowing what he was doing with psychoanalysis.

Umwelt, then it follows automatically that those language-trained animals *did* access the human Umwelt, in fact they accessed what is commonly regarded as the quintessence of it. An extensive mastering of such a powerful device as language, as possessed by humans, certainly allows a drastic *increase* (note: "increase" is a term that does not imply qualitative distinctions) of this type of semiosis, therefore the more refined and articulated our linguistic knowledge, the deeper our cognitive abilities in this respect.

However, the outcomes of those experiments seriously call into question our exclusivity in this respect, and – at the same time – reinforces the hypothesis that this specific form of metasemiosis is a language-specific feature. Which makes a subtle, yet dramatic, difference. Because these features now mark a difference between language and other communication and modelling systems, *rather than* between humans and other animals. If other animals were able to learn language, at least to some extent, that categorically means that it is in their cognitive potential to learn it.

And, by the way, as commented on part III of this companion, it is always surprising to notice that ICE are never considered in their phylogenetic aspect. It has been said that apes learning language reach the cognitive abilities of a young human child. The assessment might be accurate, but, for a change, is anthropocentrically unfair. The first – very rudimental – traces of a language presumably appeared in the genus *Homo* (4–5 millions of years ago), and took millions of years to take recognizable shape: the anatomycal and neurological features of a language-inclined human being developed gradually and slowly, and with a good amount of random events. Now, compared to this immense time, it is a bit too much to demand perfection from the first generation of non-human animals who have been introduced to human language (we are talking about less than one century). Alex, Koko, Kanzi and the others have shown that it is in the potential of many species to learn a sign system of that kind: if we were already expecting *Planet of the Apes*-type of scenarios, then clearly we did not consider this aspect.

(6) One point is however missing in the previous argument. The mentioned ICE provide evidences that other animals are semiotically able to access only alien *existing* Umwelten, not imaginary ones. None of what has been remarked suggests actively that non-human animals are able to invent (let alone conceptualize) an imaginary entity. In this very detail, the notion of "Semiotic animal" may be accurate. No trace of imaginary possible worlds has been found in non-human cognition).

No, better: no trace of imaginary possible worlds *appeared* in non-human cognition. It is a crucial correction, because in order not to *find* something, we first need to look for it. Did we check whether non-human animals have their mermaids and gods? Of course not: how could we? How is it possible to detect a number of signs in a given animal, to separate the apparently meaningful from the apparently meaningless signs, then to analyze the apparently meaningless ones, and finally to understand that some of them are not in fact meaningless but refer to an imaginary entity? Only imaginary entities like Superman can possibly do that.

To my knowledge, the closest we got to a systematic study of animal imagination are the following cases:

(1) Robert Mitchell's edited collection *Pretending and Imagination in Animals and Children* (2002), which however tends to present a definition of imagination that is pre-linguistic and closely related to the idea of deception and play.
(2) Studies like Correia-Dickinson-Clayton 2007, Clayton-Bussey-Dickinson 2003, Emery-Clayton 2008a, Mulcahy-Call 2006, and Raby et al. 2007, on alternative future-scenarios in the cognition of species like jays and apes. Once again, we might here object that, although "alternative", those scenarios are always perception-bound, so we do not really solve the "semiotic animal" problem (or perhaps, as it is maintained here, we simply provide another proof that it is *not yet* possible to investigate anything else than this).
(3) Studies like Shanahan 2006 and Shanahan-Baars 2005, where it is argued (through the adoption of a model of information flow from global workspace theory) that the neurological prerequisites for consciousness, emotion and imagination are typical of the mammalian brain. This model – it has been suggested – can be applied to other animals too (Emery-Clayton 2008b: 135). We are extremely far from semiotics, in these cases, but then again semiotics has not the appropriate tools for investigating these issues on an empirical basis.

Even with those animals that learned human language there is very little we can do. It is known that some of them are often caught performing signs at random, for their own entertainment or in front of their trainers. Since this is a completely speculative environment, why not launching the hypothesis that some of those random signs may actually be the result of their own imagination? Maybe the likes of Washoe, Kanzi and Koko are accessing imaginary Umwelten, and even describing them to us. It may sound like a ridiculous hypothesis, and it probably is, but once we are in the field of speculations, all logically-acceptable hypotheses have a similar ontological status.

To be precise, there are even more authoritative speculations on this subejct. Jane Goodall, during her very long and well-known experience with chimpanzees in their habitat, described a group performance that she called a "rain dance" (Goodall 1971: 54). Naturally, as Sebeok (1981: 219) remarks, the meaning of this behavioural pattern is not at all clear. However, what seems to be certain is that chimps dislike rain, thus their dance is possibly not a way – as it is with humans – to invoke water. Is it then to make it stop? It is not known: a display of disappointment is probably the simplest option. But one cannot exclude a certain rough degree of spirituality (i.e., the claim that the very act of rain-dancing has some kind of influence on the weather conditions).

Darwin, too, was convinced that traces of belief in supernatural were present in other animals as well:

> [M]y dog, a full-grown and very sensible animal, was lying on the lawn during a hot and still day; but at a little distance a slight breeze occasionally moved an open parasol, which would have been wholly disregarded by the dog, had any one stood near it. As it was, every time that the parasol slightly moved, the dog growled fiercely and barked. He must, I think, have reasoned to himself in a rapid and unconscious manner, that movement without any apparent cause indicated the presence of some strange living agent, and no stranger had a right to be on his territory. (Darwin 1871: 67)

And, along with this, Darwin would also push the notion as far as to assert that also religious devotion had a basis in the non-human realm:

> The feeling of religious devotion is a highly complex one, consisting of love, complete submission to an exalted and mysterious superior, a strong sense of dependence, fear, reverence, gratitude, hope for the future, and perhaps other elements. No being could experience so complex an emotion until advanced in his intellectual and moral faculties to at least a moderately high level. Nevertheless we see some distant approach to this state of mind, in the deep love of a dog for his master, associated with complete submission, some fear, and perhaps other feelings. The behaviour of a dog when returning to his master after an absence, and, as I may add, of a monkey to his beloved keeper, is widely different from that towards their fellows. In the latter case the transports of joy appear to be somewhat less, and the sense of equality is shewn in every action. (Darwin 1871: 68)

Again: all these are mere speculations. But so is the hypothesis that only humans have this kind of imagination. It is a theory constructed on two points: (a) the fact that humans clearly have this ability; and (b) the fact that no similar ability *appeared* in other animals. Period. This allows for a fair speculation on the subject, but should not reasonably allow to *affirm* that non-human animals, *therefore*, are not capable of imagination. Otherwise, we witness the same type of argumentation of those people speculating that there are no other intelligent living forms except those inhabiting the planet Earth. The argument is: we have not seen any, so far (or, worse, nobody showed up yet), therefore they do not exist. A far cry from providing evidence of the absence of such living forms. especially when we consider how little of outer space we have so far been able to explore.

Which creates an interesting parallel with the affirmations on human qualitative distinction within the Animal Kingdom: such affirmations are indeed, in most of the cases, not supported by solid knowledge of animal behavior. In that sense, speculation for speculation, one prefers to trust Jane Goodall or Charles Darwin, who at least spent a life among non-human animals, so they might have a clue of what they are talking about.

All in all, in fact, this is the thing that puzzles the most in this whole enterprise: the lack of specific competences. Nearly everybody in semiotics sooner or later has a go at non-human animals and on the human qualitative difference with them, yet nearly nobody had a specific training in animal studies to make them entitled to issue such important statements. With almost the sole exception of Sebeok himself, essays on human uniqueness within the animal kingdom contain an impressive amount of references related to human knowledge, humanity and human behavior, and at the same time they display an astoundingly poor number of references to animal studies. To speak of what humans do and are, as compared to what other animals do not and are not, apparently requires only a solid knowledge of what humans do and are.

No care is required to actually check whether it is *true* that other animals do not or are not.

Defining the semiotic animal (Deely 2005) is a 95 page-long book, which, despite its brevity, still manages to contain no less than 134 bibliographical references. It may read even as excessive, but in fact it is rather typical, considering John Deely's impressive knowledge on philosophy and semiotics. There is, however, an unpleasant surprise: in a book that discusses so extensively the uniqueness of humankind over the rest of the animal kingdom, one would expect that, if not half, at least one third of the references would actually come from the enormous catalogue of studies in ethology, zoology, sociobiology, primatology, and so forth. Exactly because, if one says that Kenyan people, *unlike Belgians*, are such and such, it would be normal to expect that a comparative study of *both* Kenya and Belgium was performed, and that a respectable balance in the sources concerning both countries was displayed in the bibliography. Well, apparently, references to Belgium are not necessary, because Deely 2005 contains no more than two (two!) references from animal studies: von Frisch's classic on bee-dance (1967) and a lesser known article on balloon flies (Kessel 1955). Two out of 134. Not exactly a fair contest (leaving aside the fact that the most recent reference dates to 38 years before the publication of Deely's book).

And this is just one example of many. "Bodies, Signs and Values in Global Communication", one of the manifesto-articles in which Augusto Ponzio and Susan Petrilli announce the birth of Semioethics to the world, has a very strong incipit, for which no whatsoever evidence or reference is provided:

> Similarly to nonhuman animals being-communication in human animals presupposes the construction of worldviews on the basis of a species-specific modeling device. However, while modeling and being-communication identify in nonhuman animals, in human beings they do not. The specific modeling device in humans, also called language, allows for interpretations, evaluations and responses. Humans are endowed with a capacity for metasemiosis or "semiotics" which presupposes language which is a species-specific primary modeling device. Syntax, deconstruction and reconstruction, the engenderment of infinite possible worlds, the capacity for "semiotics" thus understood, therefore, for evaluation, the assumption of responsibility, inventiveness, creativity and the capacity for planning are all prerogatives of language. (Ponzio-Petrilli 2008: 113)

In this case, we have a 34 page long article, with 58 references. How many references shall we think it takes in order to Cartesianly state that only humans are allowed for interpretations, evaluations, responses, syntax, deconstruction, reconstruction, assumption of responsibility, inventiveness, creativity and planning (and to have in the article such paragraphs as "Signs of humanity, humanity of signs")? Whatever figure the reader may come up with, the answer is: less. Indeed, the answer is zero.[13]

As to the reasons for this peculiar phenomenon only guesses can be made: maybe there is some kind of stereotypical *topos* in human sciences not to treat the animal subject seriously (a topos probably encoouraged by the extensive use of animal characters in kids' programs); maybe animals is one of those topics that everybody likes

---

[13] Let alone the fact that in general, in all these works, there is not an even vague, *en passant*, trace of the studies mentioned earlier, like Mitchell 2002, Goodall 1971 or Darwin himself.

to discuss even when an acceptable background is lacking (probably because they are one of those things with which we are in constant and multiple relation anyway); maybe – once again – the vital urge of defining human identity makes people rather tolerant towards their own prejudices. Maybe something else. Maybe, in this specific case, semiotics enjoys being an exclusive club where only semioticians and cryptosemioticians (i.e., many philosophers) are allowed the right of speech,[14] while – for instance – zoologists, ethologists, sociobiologists are not. It is a fact, anyway, that the animal subject holds a record for being one of the most trivialized topics in the whole sphere of (at least) humanities.

A last remark, before drawing some conclusions. At this point of the discussion, there might be a temptation, from members or supporters of the International Committee for Human Identity, to change the cards on the table. Once the option of the semiotic-narrative possible worlds is removed, why not trying the "actual" possible worlds, that is, those related with Freudian studies on dreams and unconscious? What about dreams? Do animals dream? We can be sure that, in the hands of some human scientist (maybe a semiotician), this topic would quickly become the new "ultimate sign of human distinction": after all, as shown, it seems it is enough to affirm that animals do not dream. Who needs references?

If, however, the scholars in question would care to check some specialists' literature (that is, in this case, neurology and other brain sciences), and particularly some recent ones, they would find that the dreaming activity of non-human animals has been proved, in all degrees of complexity (i.e., far beyond the stereotype that cats simply dream of catching mice), and in different research programs. Interesting readings, in this respect, mighy be Smith C. (1995), Siapas-Wilson (1998), Poe et al. (2000) and Louie-Wilson (2001).

Having pointed this out, what is left is a very simple observation, perhaps a plea. There is nothing wrong, nor abnormal, in wishing to define the identity of the category one feels s/he belongs to. The formation of ingroup-outgroup dynamics, we learn from social sciences, are the result of a rather basic need to cope with the complexity of reality (see Allport (1954) and Chapter 5 to this companion). What is heavily biased is the second step of this process, i.e., constructing this identity while actively deconstructing the outgroup's identity (and consequently placing the ingroup in a dominant position). In an ideal "live and let live" type of world, the outgroup's identity is by no means a concern for the ingroup. The identity of the ingroup is not less consistent, if the issue of the outgroup is *not* addressed.[15]

A person may characterize him/herself as many things, let us say: Jean-Paul by name, French by nationality, teacher by profession, single by marital status, rugby player and jazz listener by hobby. A bunch of characteristics of this type are certainly

[14]This would also explain the fact that in both Deely 2005 and Ponzio-Petrilli 2008 the information regarding non-human semiosis are either produced by the authors themselves, or borrowed second-hand from the interpretations and re-interpretations of Sebeok, who at least took the trouble to check from direct sources.

[15]In fact, in the author's opinion, the ideal world is not of the "live and let live" type, but rather based on the "I care" model. This part, however, belongs to ethical reflections discussed somewhere else in this companion.

very relevant for this Jean-Paul to define his identity: it is obvious, indeed, that these and other conditions/activities of his make him a distinctive individual, clearly recognizable in many circumstances, and so on. That is, Jean-Paul has clearly an identity. Now, in no point of this process, which led to the definition of somebody's identity, there was a need to outdo one or more persons who are *not* Jean-Paul: our French friend does not need to add to his profile that he is such and such, *unlike* Pedro, William or Mariangela. This step adds nothing to the process. If Jean-Paul takes that step, it is not identity he is after: it is domination, it is discrimination. And if, on top of that, Jean-Paul compares himself to Pedro or Mariangela, without really knowing them, then, after domination and discrimination, one also has to write stereotype and prejudice. All of a sudden, Jean-Paul's love for rugby and jazz becomes very marginal compared to his strong identity trait of being chauvinist, ignorant and arrogant.

If scholars understand this, they might possibly undersand also that the entire enterprise of human identity definition does not carry the compulsory requirement of additional comments on other animals. Particularly not if these scholars do not possess the necessary knowledge for making such comments (this way also ending up putting in question significant achievements of modern science, like the Darwinian Revolution). A discussion on human sociality, rationality, culture, language, etc. remains extremely interesting and important, if proposed by sociologists, psychologists, culturologists, semioticians, etc. Let the other animals be taken care of by those who have devoted to their study the necessary amount of time and energy.

1.2.4 In Defence of Biosemiotics

After creating some distances from what can be considered "excesses" in the theoretical program of biosemiotics (and semiotics in general), a few words should be also spent in defence of aspects for which, on the contrary, biosemiotics is unfairly criticized.

As Bankov's Razor was mentioned as a semiotic insider's criticism of biosemiotics, a more extended picture of those remarks should be now offered, emphasizing also the parts that may be considered erroneously addressed. For instance, while agreeing on Bankov's plea for more humility within biosemiotics, it was also pointed out that such agreement bears motivations that are not necessarily the ones indicated by the Bulgarian semiotician. He says:

> This presumed superiority of biosemiotics on conventional biology and mechanistic science is out of place. Biosemiotics is totally dependent not only on the existing data achieved within certain sciences. It is entirely dependent on the very scientific worldview, on its articulations and categorizations of the surrounding world, on the biological terminology, on the theory of evolution and the like. Biosemiotics would be nothing without them, whereas they, the conventionally affirmed disciplines, have existed, exist and will exist totally unaware of biosemiotics. (Martinelli-Bankov 2008: 401)

Although one can hardly deny that there is a certain form of dependence, in the way described above, it is rather hard to see this as an exclusive problem of biosemiotics. All disciplines depend in some form from those that previously dealt with the same issues. Did not musical semiotics depend on musicology and music theory? Did not classical semiology depend on linguistics (up to the point that many scholars we happily called semiologists or semioticians, like Saussure, Hjelmslev, Jakobson were in fact linguists)? Of course they did, except that, exactly like biosemiotics, they were able to develop an identity of their own, autonomous enough to enable us to consider them "something else". It is acceptable to say that the autonomy of semiology from linguistics, when it comes to specific analysis, is generally superior to that of biosemiotics from biology, but then again, semiology is older. Give biosemiotics some time. And still, let us always keep in mind that everytime a teacher introduce "semiotics" to his/her undergraduate students, s/he is in fact introducing them mostly to linguistics and structural linguistics. Signifier, signified, paradigms, syntagms, pertinence, commutation, referentiality... All these are entries of a linguistic dictionary.

Biosemiotics may borrow from biology by paraphrasing it in semiotic terms, but at least it has the decency of "paraphrasing". Semiology seems to be more interested in plain robbery. More importantly, however, Bankov's statement seems to ignore the fact that biosemiotics *did* propose a consistent number of novel theories and methodologies, maybe even too many, and these have just nothing to do with those "other" sciences, starting from the principle itself that animates the entire biosemiotic idea, i.e., identifying the biosphere with the semiosphere, with all the thousands consequences and the shift of mentality implied. And that, we may like it or not, is something that biologists never thought of (another matter is of course to establish whether that suggestion was correct or not, but obviously this goes beyond the scope of this book).

Nothing appears to be "borrowed from biology" in the theories of Barbieri, Kull, Hoffmeyer, Weber, Emmeche, Markos, and many others, if not in sporadic cases, and with a legitimate right of overlapping with other fields every now and then. If, let us say, Barbieri argues about the possibility of interpreting the cell as a semiotic system, he obviously needs to refer to a number of notions that are already existing in other fields of natural sciences, but this is not a problem at all, because it is the *thesis* that needs to be genuinely semiotic, i.e. that a cell can be interpreted as a semiotic system, not the whole thing (what is a cell, biologically speaking, is certainly not something that requires semiotic research or terminology). Football features a lot of characteristics borrowed from other sports: like in other sports (born long before football) the players run, are organized in two opposite teams, are ruled by referees, are encouraged by supporters, etc. But it is difficult to imagine a Carl Lewis' razor claiming that football is *borrowing* from athletics. And that is because the "theses" of football (kicking the ball with feet, allowing hands to only one player per team, etc.) are intrinsically football*esque*.

And then, possibly, the terminology should be semiotic too, because the essay happens to be addressed to the semiotic community, therefore it is reasonable to use the language of that community, even when it implies re-baptizing a few concepts. It is still a fairer option than the robbery exercised by semiologists of the formalists, the Prague school and all the others. Once, François-Bernard Mâche, the founder of zoomusicology, said that this discipline aims to speak of animal "music" *without* the inverted commas. Well, one can safely say that biosemiotics, to say the least, has removed the inverted commas from biological discourses about "codes", "signs", "communication" and the like.

In fact, sometimes the arguments of biosemioticians are so semiotic, that one may happen not to agree with them on that very basis. Sometimes (and this book confirms it), a semiotician needs to *defend* biology against the over-semioticization that some colleague undertake. To mention one, it still hard to understand why Darwin was so wrong and so anti-semiotic (as a few biosemioticians maintain).

Naturally, an entirely different matter is to browse through all the biosemiotic contributions, and put in a separate box (1) those that are probably not necessary, as they *only* reformulate with different terminology what has already been known for a while; and (2) those that risk exposing biosemiotics to an excessive amount of criticism, given their purely speculative basis, which sounds a bit out of tune in a discipline that after all aims more to improve biology than philosophy. But two factors can be comforting: firstly, the majority of the scholars are still interested in biosemiotics in terms of empirically-based research. If anything, that is a *silent* majority, less glamorous than the "fancier" metaphysical minority. Secondly, the epistemological debate on biosemiotics is still very open, and has already produced consistent criticism to certain speculative excesses (again, Barbieri's distinction between code-based and sign-based biosemiotics can be mentioned here).

However, let us dwell a bit more on the debate empirical-speculative, as no impression is meant to be given of despising the philosophical side of biosemiotics. Bankov says:

> Of course, one could say that every scientific revolution was preceded by a drastic paradigm shift, but while we are waiting this auspicious event, it would be better, for the sake of biosemiotics, to adopt a line of empirical inquiry; a line which certainly exists, but is dominated by other discursive orientations. Besides the standards of validations of the contributions among the scientific community, there are also commonly shared resistances of the natural world; observable phenomena, which resist to fit to any existing hypothesis. Abduction and experimentation are the ways to struggle with the resistance of the natural world, and the validity of any new paradigm consists in its "fresh" lead for new abductions and the ensuing confirmations. Poets and ontologists are those who propose possible worlds, whilst ignoring the judgements of facts. (Martinelli-Bankov 2008: 401)

To start with, the author of this book is a philosophically-oriented scholar, too, with a background in humanities, although he never felt that he was a metaphysician because of this. He might not perform direct observations, but he normally relies upon other scholars' professional observations. His zoosemiotic actions consist in a theoretical interpretation (and/or construction) of empirical work. This still means speculation, whether we like it or not. To use Bankov's terms (Martinelli-Bankov 2008: 402–403), discourse validation of this type tends towards a loose extremity.

Is this a problem in principle? It seems that Prof. Bankov is resentful at biosemiotics for exactly this: taking this loose extremity, precisely on those issues where, on the contrary, a "rigid" option is called for. Now, besides those cases that are already criticized in this book, generalization should be avoided, and one should never forget that biosemiotics, among other things, wants to issue an important statement to the scientific community. That is (in Bankov's terms): are we sure that the whole biological discourse requires only rigid-extremity validations? Is biology only about yes/no measurements? The funny thing is, even the sacred laboratory/field scientists that Bankov proudly defends against the abstract biosemioticians, have never been so strict as the Razor likes to portrait them. Donald Griffin, the one who dealt the fatal blow to obsolete mechanistic conceptions about animal behaviour (the first blows being already dealt right after Descartes), was a pure laboratory/field scientist. A zoologist, not even an ethologist. Yet, someone who dared to be "loose".

Moreover, it is not like empirical yes/no observations are normally autonomous processes. For what we know, there is no single mathematic or physical theory that did not have an even vague precedent in philosophical speculation, or that is not still followed (and often improved) by it. Bankov (justly) talks about a monopoly of scientificity over truth. Fine, but that should not mean stubbornness of the former over the existing perception of the latter. Even those scientific findings postulated "after Enlightenment" with strictly empirical methods were not only preceded, but often challenged and overcome by "better", or more accurate, reflections (first of all), observations and experiments.

The way fractal geometry challenged and overcame the Euclidean one followed faithfully this sequence. It is fair to say that the name "fractals" came out after Mandelbrot's work, but where would Mandelbrot be without Poincaré, Klein, Fatou, etc.? And where would they be without Weierstrass? But most of all, where would everybody here be without Mr. Leibniz, and his mere, razor-sensitive, and in fact pre-enlightenment, speculations on recursive self-similarity? Would we still be able to enjoy the pleasant redundance of Koch's snowflake without Leibniz' loose extremity (and sorry for the double meaning)?

Another little problem found in Bankov's argumentation[16] is his conviction that biosemiotics is promoting a "gradual transition scenario", as opposed to his favoured either/or (yes/no) discourse validation. Well, in fact, biosemiotics is *not* promoting that kind of Darwinian approach (in fact actively rejected by many biosemioticians), which is still a dualistic one (although now relying upon a more/less axis, rather than a either/or one, as the Descartes/Razor connection instead suggests). The scenario usually proposed is a pluralistic one, heavily based on Uexküll's theories, and finally stresses on bio- and semio-diversity as a result of life's *complexity*, not simplification.

This leads to a key-point: the theoretical unification that biosemioticians are in principle promoting (or so they like to believe) is not meant as an "all together now", "cosmic solution", "let us all be friends" interpretation. On the contrary. It urges to reflect upon the fact that life contains many more inter-relations than we

[16]Of course, the present argumentation is using Prof. Bankov only as a synecdoche for more general issues. Kristian is a great scholar and a friend, for whom one can have nothing less than deep esteem.

like to think. The X-vs-Y solutions that "normal" semiotics hints (be that Nature vs Culture, Body vs Mind, Saussure vs Peirce...) are the actual over-simplifications: it is when we divide the world in two that we are being superficial. If putting Nature in one box and Culture in another seems to "normal" semioticians the solution to superficiality, then we are really in trouble. Unifying the two (as entities and as concepts) means the exact opposite of reducing them from two to one. It means saying that it is unacceptable to treat them separately, because too many and too complex are the relations between the two. We *cannot* analyze any cultural phenomenon as completely untied from natural contexts. There is a lot to learn about human culture from Paola the Iguana (Bankov's pet, which he himself mentions as an example of his arguments), not only from the likes of Bergson.

Having said that, who said that biosemiotics is not able (or does not aim) to be accurate and "rigid" (this time, in a positive sense) in its formulations? It is clear that Bankov mistook a specific trend in biosemiotics for the whole of it (and regardless how right or wrong he may be on that particular trend). Marcello Barbieri, to mention one, is a passionate supporter of a so-called "Scientific biosemiotics":

> [. . .] a scientific biosemiotics is within our reach, but [. . .] we need to use precise definitions and testable models in this as in any other field of science. The fact that such a simple conclusion has been criticized is neither surprising nor upsetting. [. . .] Biosemiotics is much more than the union of biology and semiotics. It is the long-awaited reconciliation between the two cultures, and this is an issue that strikes very deep, no doubt about that.
> At the end of the day, however, scientific biosemiotics is merely the attempt to find out the truth about semiosis with the imperfect tools of science. How semiosis came into being, how it evolved during the history of life and how it eventually gave origin to language and culture. That very culture that today we use to look back, to reconstruct what happened and to understand what made us. Personally I find that the best description of scientific biosemiotics was given not by a scientist but by a poet like T.S. Eliot: "The end of all our exploring will be to arrive where we started and to know the place for the first time" (Barbieri 2009, forthcoming).

In conclusion, although in the previous paragraph it is clearly stated that notions such as the Bankov's Razor have several merits, it must also be said that Bankov shot (or shaved) towards the wrong target. Particularly the inclusion of zoosemiotics in the problem seems a rather inaccurate, and *very* outdated, mistake. A circumscribed addressing of the razor to, for instance, those biosemiotic approaches that resolve in pure speculation, unsupported by empirical research (or *discussion around* empirical research), and move in the rather vague area of pansemiotism, might have been a legitimate concern, significantly reducing the number of counter-arguments – and, in fact, Bankov *was* going in that direction, when he said he "would imagine people with experimental agenda".

Why not focus on that rather than wanting a *tabula rasa*? Scientific biosemiotics, and particularly zoosemiotics, have a full right to exist, within the semiotic panorama, and in the general scientific dialogue. To deny this means, roughly, to deny a good century and a half of good rigid *and* loose investigations in both human and natural sciences.

1.2.5 The Zoosemiotic Program

Perhaps in the light of these controversies, perhaps not, it is fair to stress that the importance of zoosemiotics as a discipline, or even as a simple idea, has not walked hand in hand with its success, within the category of semioticians and other scholars. One cannot really say that the scientific environment has been invaded by zoosemioticians who are anxious to be acknowledged as such. On the one hand, less than 50 years of age is still not enough to entitle anyone to such statements, neither should we forget that semioticians are still complaining that institutions are refractory to officially "accept" semiotics, the whole of it. A few considerations are worthy to be mentioned in any case:

(1) In terms of intensity, the spreading of zoosemiotics is not awfully encouraging, especially if compared it to equally (or even more) recent fields within semiotic research, such as musical semiotics or biosemiotics. The publication of an explicitly zoosemiotic text is a much rarer event than that of a musical semiotic or biosemiotic one.
(2) As compared to other branches of semiotic research, zoosemiotics can hardly be considered a specialisation on its own. In other words, if it is not very difficult to encounter comments or topics of zoosemiotic concern, it is, however, rare to encounter self-styled zoosemioticians: rather, they might either belong to different disciplines dealing with the same issues (quite often the case of ethology, as the likes of Marc Bekoff illustrate), or deal with zoosemiotic issues only in exceptional cases, their specialisation (and academic identity) being of quite different type (it is the case of John Deely or Susan Petrilli). This is another reason why
(3) Zoosemiotics has not yet achieved a scientific autonomy. If a musical or media semiotician is rarely confused with a musicologist or a mass-mediologist, zoosemioticians often seem on the threshold of an identity crises (ethologist? Biologist? Zoologist?).

An interesting support for such considerations can be provided by Internet research, through the use of the so-called search engines, like the ultra-famous www.google.com. We can try to type some key-words like "zoosemiotics", "zoösemiotics" (variation proposed by John Deely, in order to "force" the reader to pronounce the prefix "zoo-" in the same fashion as "zoology" rather than in the word "zoo"), "zoosemiotic" (as adjective), "zoosemiotician" (as profession), "zoosemiotica" (translation of the term in Spanish and Italian), and "zoosemiotique (French translation). Here are the results as of 17th of June 2009:

Zoosemiotics: 9,120
Zoösemiotics: 170
Zoosemiotic: 9,070
Zoosemiotic: 764
Zoosemiotician: 65
Zoosemiotica: 1,430
Zoosémiotique: 686

In order to handle some comparative material, we can then search the terms "semiotics", "ethology", "biosemiotics" and "biology", plus the respective adjective form ("semiotic", "ethological", "biosemiotic" and "biological") and nouns indicating the profession ("semiotician", "ethologist", biosemiotician", "biologist") on the same search engine (which is normally considered the most authoritative as far as scientific-academic purposes are concerned). The reasons for the terminological choices should be quite clear: semiotics is the mother-discipline of zoosemiotics, biosemiotics is the closest relative, ethology is a discipline zoosemiotics is often referred to, and biology is finally an example of an extremely well known discipline, in comparison to which one can figure out the notoriety of zoosemiotics in the relative sense (Table 1.3). Here are the results, as of 17th of June, 2009:

This data certainly calls for a few reflections. First of all, it appears that zoosemiotics is quite definitely the least known among the disciplines mentioned. Biosemiotics alone is about three times more recurrent in websites. Secondly, the adjective form "zoosemiotic" is very rarely found. Even more impressive is the recurrency of the term "zoosemiotician". Google found only sixty-five matches. In other words, the label "zoosemiotician", as applied to anyone, is a rather exceptional event. This is probably the most striking result, although, in this sense, Biosemiotics seems to experience a similar problem.

Further considerations need to be made about the contents of the websites found, also because they unveil two major limits of Internet-based surveys. In the first place, it is mostly the last and the second last generations of scholars that make use of the Internet. A zoosemiotics-related research on the Internet ends up providing information on the last 10–15 years only of zoosemiotics, not on its entire existence. No wonder that Sebeok, who should be supposed to be a massive presence, occupies after all a marginal position within the Internet world.[17]

Secondly, countries that are more eager to use Internet devices obviously figure more prominently. Scandinavian, Baltic and Anglo-Saxon countries are more often represented than New-Latin, Asian or African ones. By consequence, scholars living or working in the former areas are more often displayed.

The 9,120 matches for "zoosemiotics" are made even less consistent by those sites in which the term is used in generic, indirect or even metaphoric fashion

Table 1.3 Google entries for semiotics, ethology, biosemiotics and biology

	Discipline	Adjective	Profession
Semiotics	3,260,000	1,130,000	48,000
Ethology	731,000	173,000	70,200
Biosemiotics	30,500	6,750	196
Biology	112,000,000	78,300,000	8,350,000

[17]Most of the time, Sebeok only appears as a bibliographical reference of authors who made use of texts like *Perspectives in **zoosemiotics*** or *Talking with animals: **zoosemiotics** explained.*

(instances that, for the purposes of this research, shall be called "interference"). The term indeed also appears in the following contexts:

(1) Articles or essays concerning ecosemiotics and/or biosemiotics. Here, zoosemiotics is (correctly) quoted as part of the above-mentioned disciplines, but there is no specific treatment of the matter;
(2) Articles or essays concerning Sebeok's life. Here, zoosemiotics is referred to as a main specialisation/innovation of the American semiotician, but – once again – the term is just mentioned within a list;
(3) Occasional messages posted in discussion groups. Here, the term is just marginally mentioned for the purposes of totally different contexts;
(4) Sites that – scientifically speaking – have nothing to do with zoosemiotics, but rather employ the term as an effective metaphor (such as a multimedia installation named exactly "zoosemiotics", and a photographic exhibition of animal tracks, both events found during the Google search).

The nature of this datum, however, is softened by the fact that these are common problems to any of the terms used in this survey. Particularly, "semiotics" and "biology" are words often employed in purely metaphorical contexts.

Moreover, if it is not very difficult to encounter comments or topics of zoosemiotic concern, it is, however, rare to encounter self-styled zoosemioticians: rather, they might either belong to different disciplines dealing with the same issues (quite often the case with ethology, as the likes of Marc Bekoff illustrate), or deal with zoosemiotic issues only in exceptional cases, their specialization (and academic identity) being of quite different type (it is the case of John Deely or Susan Petrilli). This is another reason why zoosemiotics has not yet achieved a scientific autonomy.

It is quite significant that Winfred Nöth, in writing the chapter for zoosemiotics in his precious *Handbook of Semiotics*, ends up referring quite exclusively to scholars belonging to other disciplines. Exceptions are evidently Sebeok, W. John Smith and – to a fair extent – Günter Tembrock (who is, again, not a semiotician, but, yes, in this case it is fair to say that he is extremely semiotic in both his subject matter and approach).

Therefore, where are, who are the zoosemioticians? In this companion there has been an attempt to provide an entry for each of them, still keeping in mind that the notion of "pure zoosemiotician" is almost an utopia.

So, finally, where is zoosemiotics going, and what will be its main concerns in the near future? What can be noticed at the moment are at least the following seven points:

(1) Zoosemiotics, along with other semiotic fields, is acquiring more and more an ethically-minded approach. When one thinks of the establishment and the rapid spreading of such theoretical projects as Semioethics (proposed by the scholars of Bari University) or Existential Semiotics (proposed by Eero Tarasti), it becomes clear that semiotics has probably emancipated itself from the role of a purely descriptive field of inquiry, and it aims to an increasingly relevant prescriptive paradigm. Zoosemiotics seems to be willing to follow a similar route,

often putting a special emphasis on questions related to animal rights and welfare (also, from a strictly formal point of view, e.g., by encouraging the use of such expressions as "non-human animals", or "other animals", in place of the apparently demagogic "animals"). The prediction is that this attitude will encounter more and more favour, thus going hand in hand with the general, institutional and scientific, increase of attention towards these issues;

(2) More generally, the aspects related to the human-other animal relationship, as analyzed through a semiotic interface (what is called here anthrozoosemiotics), are an increasingly popular interest, among zoosemioticians, and it is quite safe to affirm that the two areas (ethological, that is the traditional one, and anthropological) are at present occupying almost two equally-consistent places;

(3) What has been here called cognitive approach, i.e., the anti-mechanistic and anti-behaviouristic paradigm, is enjoying increasing consensus among zoosemioticians. Most of the current generation of semioticians interested in animal semiosis seems to agree on the existence of a very active mental life in all animal species (each with their own sources and species-specific limitations), that underlies any semiotic action, from the most complex to the simplest one (one may easily compare, to mention a couple of cases, the works by Maran and Turovski with previous ones focusing on mimicry and comfort behaviour, to realise how accurately this approach has been developing). If anything, what changes among zoosemioticians is the methodological motivation: for some, this paradigm seem to be the natural continuation of what is happening already in other animal-related studies (ethology being the most relevant case); for others, the reason is intrinsically semiotic, and relates to the nowadays clear prevalence of Peircean semiotics over the structuralist (semiological, rather than semiotic) tradition;

(4) With all the due difficulties and contradictions, zoosemioticians seem to prefer nowadays dealing with the most critical topics available in the field, namely those that tend to question the human uniqueness in performing given behavioural patterns or possessing given features. Culture, aesthetics, symbolic signalling, and – most of all – language, are all traits whose human species-specificness has been sooner or later questioned from a zoosemiotic perspective. In some cases, aesthetics especially, there seem to be no more doubts, among semioticians, that categories of this type can be – if not easily – justly applied to the semiotic behaviour of other animal species. In some others, language primarily, the question remains open, and the discussion sharp. It is to be predicted that in the future, zoosemioticians will focus more and more often on these issues, also in the light of the new findings coming from empirical sciences;

(5) Little by little, and with no hurry of any sort, zoosemioticians are trying to explore different paths from the ones proposed by Sebeok, whose shadow is sometimes so big that one could be tempted to identify zoosemiotics exclusively with its founder. Although nobody attempts to deny the (justly) unavoidable importance that the Hungarian-born scholar holds in this field, a few cases exist where scholars are either following other approaches, or even daring to question

some of his assumptions as not awfully accurate. To interpret it psychoanalytically, such occurrence might be a timid yet clear sign of emancipation: it is the young kid who turns to an age when s/he starts seeing his/her father as not that undisputable hero that s/he used to think he was. In the future, it will be seen whether the adult age will bring even more departures, or alternatively a (total or partial) restoration of the traditional paradigm;

(6) In any case, a firm, neat emancipation of zoosemiotics from other fields of semiotics is yet to be achieved. Zoosemioticians are still those strange animals that venture either into Biosemiotics congresses (where they might also feel at home, but it turns out to be a huge house that comprises scholars in plants, micro-organisms, genetics, fungi, not to mention increasingly fashionable approaches on the meaning of life itself), or into human science gatherings where they, brave and lonely, try to remind everybody that issues like architecture, music, culture, etc. are of zoosemiotic concern too, and that an analysis on the paintings of the elephant Siri is not semiotically less worthwhile than a paper about Van Gogh. Zoosemiotic congresses and symposia are being organized here and there, now and again, but we are still far from referring to these events with words like "tradition" or "regularity". Other than to a lack of people, which is still *the* issue, the problem also seems to be related to a lack of organisation. So far, zoosemioticians seem to prefer working on their own, rather than enhancing and encouraging interaction. When the possibility of interaction is envisioned, it is not rare that personal views and egos prevail on the necessity of developing a unified paradigm (let alone a "school");

(7) In particular, the apparent ease with which zoosemioticians are happy to be identified as just a special case of biosemioticians is rather tricky to interpret. On the one hand, it is true that being part of a larger community increases the chances of exposure, and – in the specific case – contributes to empower the biosemiotic project, therefore – among other things – improving a condition from which zoosemiotics itself fully benefits from. On the other hand, however, in doing so, zoosemioticians encourage a strongly anthropocentric equilibrium within semiotics that they for first (and together with all other biosemioticians) should reject and fight against, i.e., the implication that all of the nature-related fields should be concentrated in one (no matter how big) single pot, while all cultural areas of semiotics have a right to enjoy a space of their own. When one, for instance, thinks that a single human body consists of about 25 trillions cells, a number which – alone – is 2,000 times more than the entire human population on this planet (plus, all these cells have direct or indirect connections with each other through more than one modality), it becomes clear that an area like cytosemiotics is at least as entitled as – say – literary semiotics to claim exclusive property of some land.

In conclusion, one may safely say that the big challenge for zoosemiotics, in its next future, is the search for an affirmation of its own identity. It is certainly a discipline with a robust theoretical (methodological in particular) apparatus, but with too few followers who would be convinced that following this path is any more worthwhile (or sometimes any different) than the ones proposed by such disciplines like ethology or zoology. If biosemiotics, social semiotics, musical semiotics, and several other fields were able to convince a fair number of biologists, sociologists and musicologists that the semiotic approach does actually add something to their own study, ethologists, zoologists, sociobiologists and other categories have so far found nothing particularly different or charming in zoosemiotics, if not in the above-mentioned few cases.

The question is, Did these scholars ever have a chance to find out? In other words, how often zoosemioticians were able to expose zoosemiotics to colleagues from other fields? The answer, it must be feared, is that these occasions were very few, and, among those few, most of them did not really help, as they either ended up in strong polemics (the most famous instance being Sebeok's harsh rejection of interspecific communication scholars, despite repeated invitations from some of them, like Sue Savage-Rumbaugh, to actually visit their centre and see them working), or in that – unfortunately not rare – presumptuous attitude of semioticians to consider semiotics the ultimate carrier of scientific truth, with the implication that the scholars involved in similar topics, but according to different frameworks, are merely wasting their time.

More humility, but most of all a better organisation and coordination, will certainly lead zoosemiotics to occupy the place it deserves within the scientific panorama.

Chapter 2
Ethological Zoosemiotics

2.1 General Systematics of EZ

Ethological zoosemiotics, as seen in Chapter 1, is the primary and the most classical field of inquiry for zoosemiotics, i.e., the one that actually deals with the semiosic processes occurring among animals. It was also mentioned that the field can be divided into an *early* stage (eEZ) and a *modern* one (mEZ). In its early phase, zoosemiotics was mostly conceived as an umbrella term for gathering different approaches on animal communication. And "communication" was still the key-word, as in this stage there was a clear emphasis on this specific process, rather than the broader semiosis. Classical ethology and the then very trendy behaviouristic schools, finally, provide the scientific framework for (and document the uncertain semiotic identity of) this type of EZ.

Sometime in the late 1970s (or thereabouts), zoosemiotics enters its modern age, this transition being marked by a few important innovations:

1. Zoosemioticians extend their interests to the entire spectrum of semiosis.
2. They start treating zoosemiotics as a discipline of its own, with its own methodologies, its own fields of interest and its own jargon. This is also possible because
3. cognitive approaches appear, producing something similar to what the publication of Griffin (1976) had produced in ethology. Indeed, through this transformation, zoosemiotics develops a strictly semiotic paradigm (as the traditional definitions of "semiotic threshold" always imply a mental process underlying sign production). These changes, experienced by all branches of animal studies, bring to a completion what Darwin (and the likes of Locke, Porphyry and Hume) had already envisioned.
4. In a general sense, zoosemiotics becomes more aware of itself, and scholars start systematizing the discipline in different respects. It is in the 1970s that we witness the first attempts to trace a history of zoosemiotics (Tembrock 1971, Sebeok 1979); and it is still in the 1970s that we find a first annotated guide to zoosemiotic (or, mostly, zoosemiotic-friendly) literature (Sebeok 1972).

D. Martinelli, *A Critical Companion to Zoosemiotics*, Biosemiotics 5,
DOI 10.1007/978-90-481-9249-6_2, © Springer Science+Business Media B.V. 2010

Systematising EZ should thus be the starting point. First and foremost, as already mentioned, the target of EZ relies on all semiosic phenomena occurring among animals: *signification*, occurring when the receiver is the only subject taking part in the semiosis, and a true sender is missing; *representation*, occurring when the sender is the only semiosic subject, and *communication*, occurring when sender and receiver both take part in the semiosic phenomenon. The sign production and exchange among animals can be *proprioceptive*, *intraspecific* or *interspecific*. The first is the case when sender and receiver are the same subject. In other words, the animal is sort of talking to itself – a typical case being the so-called *echolocation*: animals like bats (see Fig. 2.1) or dolphins can receive information about the surrounding environment by emitting ultrasonic sounds in the direction of the object they are aiming at. By hitting the object, the sounds come back to the source and give the animals the information they need.

Semiosis is intraspecific when sender and receiver (now two different subjects) belong to the same species: this is probably the most common instance of sign exchange, in all kinds of organisms. However, to belong to the same species does not necessarily mean to fully understand each other. Several species (most notably, human beings) are socially organized in communities that have very few features in common, and can hardly understand each other. Similar cases are also quite common in non-human semiosis, and they are known as "dialects".

Finally, one speaks of interspecific semiosis when sender and receiver belong to different species, family, orders, etc. The instances are not so few, and cover a large number of relations (prey-predator, parasitism, territory competition, etc.)

Another crucial distinction in zoosemiotics, as in all other branches of semiotics, needs to be made between *syntactics*, *semantics* and *pragmatics*, these terms being introduced by Charles Morris. A syntactic investigation concerns the sign repertoires of living systems in themselves, their structure, their spatial-temporal features, the way they form meaningful combinations. In Sebeok's words, zoosyntactics "deals with combinations of signs abstracted from their specific signification or their ecological setting" (Sebeok 1990: 43). The key-question, therefore, is "What does it look like?". For instance, when studying and classifying the songs of the sprosser nightingale *Luscinia luscinia*, Olavi Sotavalta (1956: 7–8) found that they are usually organized into six main parts:

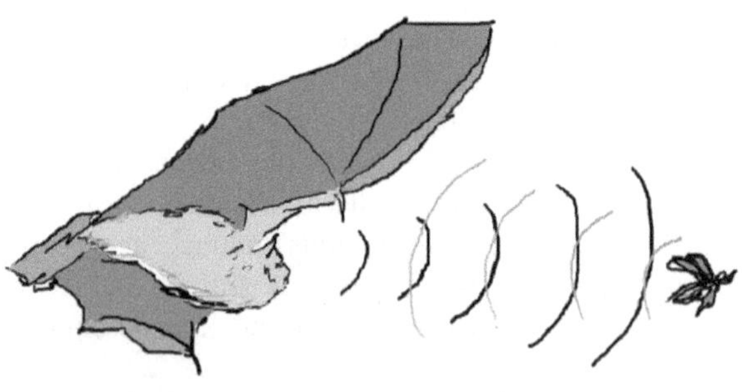

Fig. 2.1 Echolocation in bats

a. Introductory: a starting sound repeated twice in a basic duple metre, and another sound a half-tone lower, still in duple time, but more complex, which links to the second part.
b. Antecedent: this part is optional, and the bird may every now and then skip it, proceeding directly to the next part. When present, the antecedent consists of only one, relatively low-pitched, repeated sound.
c. Characteristic: this is one of the main and most melodious parts. It consists of "successive single notes of different pitch, or of two-note chords of a certain interval. The rhythm can be duple or triple, in eighth- or sixteenth-notes" (Sotavalta 1956: 7). Rhythmic patterns constantly change (no less than 10 or 11 times), and the melodic line contains quite large intervals.
d. Postcedent: it normally consists of only one relatively low-pitched, "flute-like" repeated sound, but the rhythmic pattern may change (it is usually duple, but sometimes triple).
e. Finale: "The finals showed a remarkable uniformity in tonality and partly also in rhythm. As a rule, the finals were chirpy, xylophone-like, repeated chords; the commonest final was the minor sixth F# – D, or the same inverted, embellished with auxiliary notes and sometimes with acciaccatura. Tonally different finals were very rare" (Sotavalta 1956: 7). Rhythmical patterns are usually duple.
f. Cadence: it usually consists of a simple "tambourine rattling" in duple time, but it can also be more complex. Sometimes it is absent. Occasionally, and usually between any two successive sections, the so-called bridges and links may appear. Links are short, single or double whistles of high pitch; bridges are made up of more than one link, interspersed with low-pitched notes, shakes or clicks. Rhythmical patterns are variable.

This type of classification is of the syntactic type, and it is extremely useful not only for identifying formal functions, but also for detecting meanings and functions.

Zoosemantics "is devoted to the signification of signs, and must take account of the context referred to by the source and apprehensible by the destination; this is the least well understood dimension of animal communication studies" (Sebeok 1990: 43). An investigation of the semantic type focuses on the meaning of a given message, or combination of messages. The key-question is "What does it mean?". Sebeok is right in saying that, altogether, this is the part we master the least in the understanding of other animals' semiosis, yet the catalogue (the dictionary, one might say) of animal signs to which a rather accurate meaning was attributed is rather encouraging. We may just think of the rich variety of signs in a dog's ethogram (the different tail positions, the role of territorial markers, the so-called play-bows, etc.) which do not seem, at least in general, to present big mysteries for human knowledge anymore.

Finally, "Zoopragmatics may be said to deal with the origin of signs in the source, or sender, the propagation of signs through a medium, or channel, and the effect of signs on the destination, or receiver" (Sebeok 1990: 43). A pragmatic investigation, thus, deals with the biological aspects of sign functioning, including the processes and the determinants of propagation of the signs. The key-question is "How does it work?". One of the most impressive examples of animal communication is certainly the dance (or language) of the honey-bee. By means of this dance, one or more bees inform the hive about the presence of food, nesting sites or water sources in the more or less immediate vicinity. A pragmatic form of inquiry focuses the attention on the many interesting dynamics that constitute the bee-dance: the fact that it adopts at the same time tactile, olfactory and acoustic channels; the fact that it communicates data such as source type, distance, direction and type of route, and the fact that it uses the sun as a compass, and it contains elements of iconicity, indexicality and symbolicness. In this companion's glossary a thorough description is provided of this impressive phenomenon that Karl von Frisch (the main scholar behind its discovery) did not hesitate to call "language".

The field of zoosemiotic research can also be divided into *pure*, *descriptive* and *applied*. By pure zoosemiotics, it is meant the study of theoretical models in zoosemiosis (like, for instance, the study of mimicry). In descriptive zoosemiotics, the focus is on individual sign systems (for instance, the sign repertoire of a single animal species). Finally, in applied zoosemiotics, one seeks an actual application of zoosemiotic studies to specific human activities (like biotechnologies, breeding, etc.).

A theoretical systematization of zoosemiotics also involves the study of the channels involved in zoosemiosis. By channel, it is meant the sensory mode used to transmit a message. It would be a mistake to think that these sensory modes correspond only to the five human senses. Zoosemioticians classify at least seven channels: the *olfactory*, the *gustatory* (these two normally considered under the more general *chemical channel*), the *tactile*, the *thermic*, the *electric*, (these three normally grouped into the general *tactile channel*, as they present a clear similarity in mediation – they all require physical contact), the *acoustic*, and the *visual*. Depending on the research context, an analysis of the channels may include their possible classifications, their constitutive parts, the occurrences they operate in, the species making use of them, and their characteristics. To take but one example (definitions of all channels are available in the glossary of this companion), the acoustic channel represents a rather familiar territory for most species, including the human one, for whom it is second only to the visual channel (by far the main mediator of perception and experience for the human species). Acoustic signs are produced both vocally (i.e., through a vocal apparatus) and nonvocally (e.g., by hitting parts of the body or external objects). The acoustic channel has a few important advantages: broadcast transmission (we can communicate with someone even if we do not see them, like in the darkness), directional reception, rapid fading (which may sound like a disadvantage only, but it is not, as it makes a fast continuation and response possible in the communication), immediate feedback, and a very high degree of specialization (although applicable to many channels, language developed first of all as a vocal-acoustic sign system). The *rapid fading* is of course also a minus, in that it does not allow a message to last in time, and that is why many animals always have chemical communication as a support for the acoustic one.

Finally, it may be worth discussing whether the type of semiosis that originates from *magnetoception* (that is, the ability to detect a magnetic field to perceive directions and locations, as widely used by birds in their migrations, but also by certain insects, reptiles, crustaceans and cartilaginous fish) is a *channel* in all respects. Leaving aside the general difficulties in studying the topic altogether, what appears so far is the capacity, from the mentioned species, to make sense *out of* magnetic fields, and not to *produce* sense with them (i.e., there seem to be only a magneto-*reception* and not a magneto-*representation*, as instead occurring in all other channels). However, research on this topic is still at too a preliminary stage either to open or close any *door* of interpretation.

2.2 Typologies and Functions of the Signs

A sign, in the classic Augustinian formulation, is an *aliquid pro aliquo*, something that stands for something else. If the goal of semiotics is the study of semiosis, a sign is, quite simply, the *currency* of semiosis, what is being used for the latter to occur. To study semiosis means to study how signs are constructed, organized, assembled, distinguished, interpreted, codified, and so forth.

In order to investigate the different typologies of signs involved in animal semiosis, choice has been made in this companion to compromise between two renowned classifications: the famous Peircean distinction between Icons, Indexes (or Indices) and Symbols, and a slightly less known distinction, provided by Sebeok (e.g., in 1994), between *Signals* (mechanical or conventional signs meant to elicit a reaction from the receiver); *Symptoms* (automatic, non-arbitrary signs linking the sign itself to the entity it refers to); *icons*; *indexes*; *symbols* (the last three roughly equivalent to Peirce's classic formulation); and *names* (signs of personal identity and class extension).

As it happens, both the notions of signal and symptom are either too generic (signal), or too similar to other typologies (an index can be *also* a symptom, and a symptom is *always* an index), to pass the test of time, and indeed it is no coincidence that these two sign typologies are rarely used by semioticians (and anyway, incomparably less than Peirce's triad). The category "names", however, may be still preserved for the purposes of this companion, because it becomes "critical" when applied to non-human animals, and because this type of sign has a general property of defining the identity of its object in an almost unique and exclusive way, and in that sense it presents features that makes it, if not a category of its own, a *very special* case of "symbol". Related with symbolicness, finally, is also the notion of "syntax", which is also critical enough to deserve a discussion of its own.

2.2.1 Indexes and Icons

Despite the fact that, in the Peircean system, the *icon* is the sign-object relation connected to *firstness* (i.e., the mode of being of what is without reference to any subject or object), the analysis in this paragraph shall start from indexes, simply because this is the typology of sign that the vast majority of animal semiosis is concerned with. The significance of indexicality, within zoosemiotic study, is to be considered more central than in the case of anthroposemiotics, particularly its aspects related to human culture, whose logocentric nature predictably manifests a greater interest in symbols and icons. In a sense, one may go as far as to say that one of the main side-effects of the production of an artificial and/or cultural sign system is exactly its emancipation from indexicality, in favor of iconicity and (most of all) symbolicness.

A great number of markers, symptoms, indicators and signals produced through any of the available channels are in one way or another indexical, from the tail-wagging of a dog (index of its excitement) to the wheel display of a peacock (index of its health conditions). An inattentive scholar might infer that this high recurrence of indexicality is an indication that mental processes do not play a major role in animal semiosis, as, after all, an index is most often a natural reflex, an instinctive reaction, a physiological display. However, this conclusion would be as inaccurate as saying that a human display of joy as the smile is alien to any cognitive implication, ignoring that elaborate path that goes from joy to smiling and that includes the *reasons* for joy, the specific *causes* of a smile, the frequent presence of *receivers* of the "smile" message, the different *degrees* of display (a smile can be emphasized, hidden, or else, depending on the situation and on the subject), and others, that obviously require the type of mediation that a mind can provide. The same applies to a similar non-human manifestation of joy as the tail-wagging of a dog, who has the same cognitive "problems" to solve, and normally provides an analogous range of solutions.

What instead is true is that indexicality, although requiring mental processes in most cases, does not necessarily requires complex and *multi-layered* mental processes, and that is primarily because an index tends to require temporal presence and spatial continuity, and its encoding is strongly bound to the *hic et nunc* of semiosis. The decoding may still happen later or further (e.g., some chemical sign may be carried by the sender for days and kilometres), but, what we might call the "packaging" of the sign is an immediate process that does not require too much cognitive articulation.

Finally, it is not entirely true to say, as Nöth does, that the "degree of indexicality in animal semiosis is definitely higher than in human semiosis" (Nöth 1990: 163), not at least if we interpret this statement to the letter (i.e.: non-human animals use *more* indexes than humans), perhaps thinking that human language was able to *replace* indexes with more efficient symbolic descriptions. What has happened, and what possibly Nöth himself meant, is that language added a series of communicative and cognitive elements *on top* of the existing ones, not *in place of* them. And indeed, natural semiosis is still intact in human beings, and goes on roughly to the same extent as it used to before language appeared. What has crucially changed is the decoding stage: precisely because human language is admittedly a very efficient strategy for informing (or deceiving) each other about several topics and conditions, the human being learned to pay attention mostly to that, losing – or better: stiffening up – his/her abilities in non-verbal communication. In other words: there is not a single pheromone less spread by a person who is sexually attracted by another person, but it is likely that the two people concerned will negotiate this attraction by means of words rather than armpit smelling.

Iconicity is slightly less recurrent than indexicality in animal semiosis, but still very easy to find. The many fascinating forms of mimicry (extensively studied in semiotics by Timo Maran, in, a.o., 2005, 2007a,b, 2008a) are normally the first category of examples mentioned. In a more general sense, a great deal of signs related to deception is of iconic type, as the often mentioned case of sham behaviour (discussed later in this part, in the paragraph "On deception", or, a.o., in Hailman 1977: 179–83).

Mimicry and imitation do not in any case serve exclusively deceptive purposes: the enormous area of birds' vocal imitations, therefore, qualify in most of the cases for iconicity (and quite often for symbolicness, too). Other examples of iconic signs are recurrent in play (for instance, when a canid intends to mock-fight, s/he will not rarely offer an iconic representation of someone sleepy, or dead – therefore defenceless – by lying on his/her back), in ritualized social behaviour, in aesthetic signs, and in several other situations.

Before concluding this paragraph, a brief comment, once again related to the issue of an animal's Umwelt. It was hopefully made clear (through such examples as vocal imitation) that, despite its name, an "icon" is not only a sign typology that concerns the visual channel. An icon is not just a sign that *looks like* its object: it may also *smell like*, *taste like*, *sound like*, etc. Obviously, though, the fact alone that a word like "icon" was chosen to represent this type of sign says a lot about how central the visual channel is in human semiosis. This may lead to slightly inaccurate approaches to other animals' semiotic processes, especially in those species which give other channels a perceptive priority. An example of these inaccuracies comes from the many language-training experiments conducted by psychologists and primatologists (and often commented upon by semioticians). On those occasions where the research subject (normally a chimpanzee or a gorilla) was practically required to display its iconic competencies, both researchers and commentators failed in giving the necessary importance to the way the perceptual field of a given specimen (therefore, probably, of its entire species) affects the actual reception of a linguistic sign. The fact that the famous chimpanzee Washoe, once she had learned the American Sign Language symbol for *flower*, took to name that way everything with a strong smell is quite interesting in this respect (Gardner, Gardner, and Van Cantfort 1989: 82). As humans, we tend to establish an immediate connection between object and representamen on the basis of their visual relation. By contrast, the first thing that caught Washoe's attention was the smell of the object, i.e., she paid more attention to the olfactory channel, in this case. As a consequence, she had a reason to iconically associate the sign "flower" to the semantic category of "smelling objects", rather than look-alike flower-shaped ones. Cigars and aftershaves were thus as much *flowers* as roses and orchids, while the non-smelling Van Gogh picture was obviously something else. If we forget for a moment the human way of categorising reality, we must admit that this way of reasoning is by all means legitimate, and – naturally – intimately semiotic.

2.2.2 *Symbols, Syntax and Names*

The concept of symbolicness has been for many years what in the introduction was defined as "the ultimate sign of human distinction", that is, a concept (and a consequent mental/interactive capacity) of which only humans were claimed to be capable. As with many other anthropocentric barriers, this one too met its unavoidable fate. As symbolicness is primarily a semiotic concept, it is no wonder that it has been up to a semiotician, Thomas Sebeok, to set the record straight:

> The fondly cherished mythic characterization of man, adhered to by E. Cassirer's epigones and many others, as a unique, animal-Symbolicum can be sustained only if the definition of 'symbol' is impermissibly ensnared with the concept of natural language, which G.G.Simpson quite aptly characterized as "the most diagnostic single trait of man." By every other definition — invoking the principe of arbitrariness, the idea of a conventional link between a signifier and its denotata, Peirce's 'imputed character', or the notion of an intentional class for the designatum — animals demonstrably employ symbols. (Sebeok 1990: 42)

In order to prove this point, Sebeok deliberately chose two examples from the world of insects, that is, from those animals with supposedly a very low and rigid degree of reasoning. The first one refers to the dipterans of the family *Empididae,* one of those that practice cannibalism: in order not to risk being eaten, the male, before the copulation, offers the female an empty balloon.

> The evolutionary origins, that is, the increasing ritualization, of this gesture have been unravelled, step by step, by biologists, but this story is irrelevant in a synchronic perspective: the fact remains that the gift of the balloon features a wholly arbitrary symbol, the transfer of which merely reduces the probability that the male himself will fall prey to his female partner. (Sebeok 1990: 42–3)

As for the second case, Sebeok offers a very interesting interpretation of the celebrated dance of the honey bee:

> It is common knowledge that if its food source is farther away than 100 m., the bee's tail-wagging dance conveys, among other bits of information, the direction of the goal, the sun being used as a reference point. Now if the bee dances on a horizontal surface, von Frisch tells us that "the direction of a waggling run points directly to the goal", that is to say the display is indexic (the rhythm, incidentally, depicts the distance iconically, since the farther away the goal, the fewer cycles of the dance occur in a given period). If, however, the dance takes place on a vertical comb surface — as is the case, normally, in the dark hive — then "the dancer transposes the solar angle into the gravitation angle", according to von Frisch. In other words, if a vertical honeycomb is involved, when an angle with respect to gravity is substituted as the orientation cue, the indexical aspect of the display attenuates to the extent that, temporarily, its symbolic aspect comes to rank predominant. (Sebeok 1990: 43)

As a matter of fact, then, one can argue that also in normal (horizontal) conditions the dance is symbolic, because it anyway employs a communicative strategy that is based on a model. Thus, if this model was simply indexical, besides not being a model at all, it would have changed its configuration, at the changing of the contextual conditions (from horizontal to vertical position). As it seems to be symbolic, it becomes understandable that the model (which is by nature conventional and abstract, and therefore resistant to contextual variations) is not affected by changes of sort. To clarify this aspect, we may think of the positioning of an arrow, in a street, for signifying the direction "forward". Now, as we all know, a sign for forward (which is most often bi-dimensional) may appear in at least three configurations: (1) pointing forward, in a horizontal position, as for instance when the arrow is painted on the asfalt of the road; (2) still pointing forward, but in a vertical position, as it may appear on a street sign, on the right side of the street; and finally, more significantly (3) pointing up, still on a street sign, when it appears poisitioned in front of the driver. Now, not only do configurations 1 and 2 not provoke any significant changes in the behavior of the driver, but neither does configuration 3, that is, the driver does not start flying with his/her car. This is possible when the appearance of the arrow is perceived neither as indexical nor as iconic, but simply as the result of a convention, untied to any physical connection (resemblance or consenquence) with the object represented. And it is all the three configuration which have a symbolic value, not just the third one. It would be too laborious, even for a human mind, to separate the notion of indexicality, iconicity and symbolicness as applied to the same sign. Much more economic is to establish one single convention (still a convention, but at least "one"), and accord one's behavior on that basis.

For ethologists, who are obviously less demanding when it comes to defining what symbolic semiosis is (as the concept is less central in their interests), it is the entire bee-dance (which they in fact often call bee-*language*) that presents features of symbolicness. Among the evidence they bring in support of this view, a rather interesting example is the geographical variability in the significance of the various signs. Different communities of the same species *Apis mellifica* may for instance use the very same articulation of the distance-sign (that is, the time employed to cover the middle axis of the 8 figure) to represent completely different distances. The same amount of time will signify 5 m for an Egyptian *Apis mellifica*, 25 for an Italian, and 75 for the predictably more efficient German bee (see Mainardi 1992: 455).

In addition to this, ethologists have two more favourite examples of symbolicness in animal semiosis. One is, of course, the singing activity of many bird species, which present features of cultural learning and development, geographic dialectal variation, and – most of all – conventional and abstract sign-object codification, that makes it hard to consider these songs as merely composed of indexes and/or icons. The other example, perhaps more interesting as it is less obvious, is the system of alarm cries developed by species like the ground-squirrels of the genus *Spermophilus*, the prairie dogs *Cynomys*, the monkeys *Cercopithecus* and others, to inform their cospecific of the arrival of a predator. These species emit differentiated cries depending on the type of predator, and it is highly specialized in the case of the vocalizations of the Vervet Monkeys *Cercopithecus aethiops* (see Fig. 2.2) and of the Greater Spot-nosed Monkeys (*Cercopithecus nictitans*). When one vervet monkey sees a predator, s/he produces a vocalization that alerts the rest of the group. This type of alarm is of three different kinds, and depends on the specific kind of predator in the vicinity. A loud barking call is given for leopards (therefore, it means "leopard", or whatever the monkeys call that animal), a short, double syllable cough stands for eagles, and a "chutter" "sound stands for snakes". The calls are

Fig. 2.2 A vervet monkey launching a cry of alarm

not similar to the sounds that those predators utter or produce, they are therefore not of the iconic type, and the response of other monkeys to a given call is appropriate for escaping the corresponding predator (i.e., when the leopard call is heard, the monkeys run to the trees; when the eagle call is heard, they look up into the air and seek shelter; and when the snake call is heard they stand up on two legs and look in the grass). In many communities, in addition, specific cries were also noted for two more possible predators, which evidently the vervet monkeys perceive as rather similar to each other: baboons and human beings (no offence meant, hopefully).[1]

Once again, it is the strong dialectal variation (plus the fact that the cries are learned, and it takes a while before the young specimens provide an adequate response to them), to persuade scholars of the symbolic character of this sign-catalogue. At a distance of few kilometres, a leopard is signalled with a completely different cry. One may argue that the dialect variation does not exclude a priori the possibility that these cries remain of indexical or iconic type. However, the notion can be accepted only if fairly traded with the hypothesis that most of human language, too, is therefore only indexical or iconic, *despite the existance of different languages and dialects*. If we take the example of honomatopoeias, it is clear that the creation of such words departs from iconic logics. Yet, words describing the same acoustic phenomenon can be completely different, depending on the language, as in the case of animal vocalizations (a dog goes "Woof" in the UK *and* the USA *and* Australia, three geographically separated places, yet s/he goes "Bau" in a country, Italy, which is much closer to the UK than any of the other English-speaking countries).

Or, alternatively, we could accept the hypothesis that no sign is actually 100% symbolic, as a link (even a remote one) with a physical entity is sooner or later detectable: even street signs themselves, which are normally taken as the most

[1] For a thorough description of vervet monkeys' calls, see Cheney and Seyfarth (1990).

extreme example of symbolicness, show different degrees of this relation (an arrow remains after all iconic, to some extents; the use of the red colour for interdictions and dangers is easily related to the recurrence of this colour for blood and poisonous substances in nature, etc.).

A last important criterion for detecting the symbolic character in a sign was established by Peter Marler (1984), and concerns the possibility of discerning a sign from its function. A sign is a true symbol when in principle it can arbitrarily be used in a different context, or provoke a different reaction. As Marler himself puts it (1984: 354), precisely with reference to the monkeys' alarm calls, "If an alarm call is truly a referential symbol it should be potentially dissociable from acts of fleeing". W. John Smith comments on this criterion, and applies it to the very case suggested by Marler:

> By dissociable he meant that an individual uttering a call would not always flee. In fact, individuals do not always flee after calling — not just in the case of vervets but in ground squirrels and, indeed, in any well-studied case of which I am aware. Fleeing is simply one of several options that become predictable from such a call. Other options may include pausing to monitor or continuing with a previous activity that was interrupted by calling; the particular conditions of each event determine whether escape, monitoring, or some other alternative follows calling. No single class of behavior occurs as the exclusive and inevitable correlate of a signal. (Smith 1991: 216)

But there is yet another aspect in the semiosis of these animals that probably demolishes many of the assumptions we had, as humans, about out own language, and that is syntax. In general, after Marler (1977), scholars distinguish between *Phonological Syntax* (that is, a combination of sounds that, taken alone, do not necessarily have a specific meaning, or at least not a qualitatively different one from the combination), and *Lexical Syntax* (that is, a combination of sounds that are also meaningful as single units, and that also mean "something else"). Predictably, no particular effort is required for a language scholar to find examples of phonological syntax in non-human animals. For instance, in Robinson (1984) and in Mitani and Marler (1989) the phenomenon was studied respectively in wedge-capped capuchins (*Cebus olivaceus*) and in male gibbons. The difficult task is to find evidences for lexical syntax.

Now, the Greater Spot-nosed Monkeys have two main alarm sounds: something like "pyow", which designates a leopard, and something like "hack", which stands for "eagle". The use of these sounds pragmatically proceeds in the same vein as with vervet monkeys. Recent research, however, has revealed an interesting phenomenon with regard to the Greater Spot-nosed (Arnold-Zuberbühler 2006). They form sequences with these sounds (sort of "pyow pyow hack hack"), sequences whose meaning is not simply "A leopard and an eagle" (or two leopards and two eagles), but means something entirely different, that is, "Let us move to another place". This is clearly a case of lexical syntax, where existing meaningful units are combined in order to produce a new meaning, that is not simply the sum of its constituents. Keeping up with the hypothesis that languages are never *fully* symbolic, one might guess that this new meaning developed from an indexical formulation of the type "eagles everywhere, leopards everywhere", which syllogistically turned into a kind of "as dangers are everywhere, this is definitely not a safe place", and refined into the more practical "let's go somewhere else". This type of interpretation can reasonably be applied to all languages including the human one, and can easily explain why, for instance, we went from the red of blood and poison to the red of warnings and interdictions in street signs.

The last category of signs suggested by Sebeok is that of the "names", defined by the great Hungarian-American semiotician in the following way:

> A name is a form that identifies a human being (Alexander, Sarah, etc.), or, by connotative extension, an animal, an object (such as a commercial product), or event (such as a hurricane). A name has both indexical and symbolic properties: it is partly an indexical form because it identifies a person and, usually, points to his/her ethnic origin; it is partly a symbolic form because, like any word, it is a product of conventionalized representational practices. Less often, named are coined iconically: Trivial but instructive examples of this can be seen in names given typically to household animals-Tuff, Pooh-Pooh, etc. (Sebeok and Danesi 2000: 27)

Now, nevermind the initial unjustified attribution of the notion of names to humans only (nothing new, really), and let the focus be on the properties of the "name" sign. Given such properties, can one advance the hypothesis that non-human animals name themselves/each other? Once again, the answer seems to be "yes", at least when it comes to birds and cetaceans, the two categories that have received the most attention in this respect. Very interesting is the case of the Bottlenose dolphins (*Tursiops truncatus*), to which specific attention has been devoted in a number of investigations, particularly the seminal Janik et al. (2006). Janik and his colleagues found that dolphins possess individually distinctive signature whistles that are used to maintain group cohesion. While identification signals are present in many other species, dolphins' names have the characteristic of being influenced by vocal learning. By using a sample of 14 dolphins at Sarasota Bay, Florida, the researchers recorded and then played the signature whistles to the animals, who recognized their names, and did so (in 9 cases out of 14) also when they heard an artificially-synthesized version of the whistle. The conclusions of the research team is that dolphins extract identity information from signature whistles, even after a complete artificialization of the signal, where all voice features are removed from the signal and only the time-frequency relation is preserved (which corresponds more or less to a complete stylization of a colored painting into a simple pencil contour drawing), showing that dolphins seem to be a very rare case of animals, besides humans, that transmit identity information regardless of the caller's voice or location:

> This study demonstrates individual identity information being encoded independent of the signaler's voice or location in the natural communication system of a nonhuman animal. We found not only that the frequency modulation pattern of signature whistles carries sufficient information for individual discrimination but that this information is used by the receiver to identify individuals. (Janik et al. 2006: 8295)

Great care is taken to point out that recognition is different from simple discrimination:

> animals recognize each other's whistles individually rather than just discriminate between them. We define recognition as perceiving something to be identical with something previously known. Discrimination can use but does not require such previous knowledge if it is solely based on the comparison of distinctive features. (Janik et al. 2006: 8295)

Finally, it shall be interesting to point out that the choice of the personal name, which is up to the carrier him/herself and occurs during the first year of life, is the result of a combined imitation of several other names the young dolphin is exposed to – more or less as if a woman choses the name Lina because her uncle is named Linus, her friend is called Tina, and her teacher is called Linda.

One might conclude these reflections with the following question: Are non-human animals' semiotic systems comparable to human language? This issue shall be explored in the next part of this companion, as the answer lies mostly[2] in a form of human-other animal interaction, the interspecific communication experiments, that belong to the realm of anthropological zoosemiotics. At this stage, however, keeping up with the spirit of this whole book, it may be anticipated that the question is once again badly formulated. The investigation should not focus on whether language constitutes a qualitatively specific feature of the human being, because it may be assumed already that it is not, and the evidence provided in this paragraph demonstrates already that three of the characteristics generally perceived as being exclusive of the human language (symbolicness, syntax and naming) are far from being so (plus, in Chapter 1 the notion of language-produced possible worlds is also discussed).

What should be more interesting, and possibly methodologically fair to discuss, is how language developed as a fundamental human *need*. That is to say, the capacities (cognitive and practical) to acquire language are not inaccessible to other species, however the human being has been and continues to be the species who appears to be most interested in pursuing it, and – as a consequence – is best disposed to use it. In this sense, it may be suspected, the role of the constitution of the human vocal apparatus in the acquisition of language seems to be highly underrated, in favour of the idea that the entity that is most responsible for such a process is the human brain. This, and others, will be the main focus of the discussion on language provided in Chapter 3.

2.2.3 The Six (or Seven?) Functions of Communication

Another important way to classify semiosis in animals is through its *functions*. As already proved by Cimatti (1998: 59–105), the classic theoretical model proposed by Roman Jakobson for classifying the functions of communication can be easily applied in zoosemiotic contexts. The model was originally conceived for the

[2] Not only, anyway. Language in non-human animals can be studied also in a purely ethological dimension, and that has been particularly recurrent in the field of avian communication: "there is ample justification for speaking of bird utterances as language" (Armstrong 1963: 27).

specific area of semiosis called "communication", and "communication" will be the main point of reference here. Yet, an extension to other fields of semiosis is easily possible, for at least four of the six functions described below (the expressive, the referential, the metalinguistic and the aesthetic), do not explicitly refer to a sender-receiver interaction.

As widely known Jakobson (1960: 353–7) classifies six main functions within a communication system, each of which concerned with a particular element of the semiotic process: *expressive* (referred to the *sender*), *conative* (referred to the *receiver*), *phatic* (referred to the establishment of *contact* between sender and receiver), *referential* (referring to the *context* of the message), *metalinguistic* (referring to the *code* used in a given communication act), and *aesthetic* or *poetic* (referring to the form of the *message*). None these functions are not to be considered in isolation or as mutually exclusive. A message is usually the intersection between two or more functions, even if one of them is often dominant and more evident.

By *expressive* is meant the use of communication in which the most relevant part is the display of the emotional state and the identity of the sender. All kinds of messages more or less display some feelings or emotions (a datum that can be easily confirmed by anybody who enjoys the company of a cat or a dog). Rogers-Kaplan, for instance (1998: 86), report the case of Australian magpies *Cracticus tibicen* performing their whole repertoire of songs throughout the entire year. For this species, singing is a ritualized way to give instructions about food, migration, territories, etc. Yet at the same time, and according to intensity, speed, loudness and other parameters, these songs are also an indicator of anxiety, joy, alarm, enthusiasm, attention or other emotional states.

By *conative* is meant the use of communication in which the sender uses the message in order to make the receiver have a certain reaction and act consequently. Conative functions stand out clearly in all mating and all territory signs, i.e, the bulk of the sign-repertoires of practically all species are basically of a conative type. The sender wants the receiver to do something (mate with him/her, leave his/her territory, etc), no matter how ritualized (and aesthetic) such signs may be. Conative are usually also those signs used for deceptive purposes, and in general all those aimed to some form of manipulation of the receiver. It may be curious to notice that the traditional definitions of animal communication, as provided by classical ethology, tended to emphasize only the conative aspect of sign exchange: "Communication between animals involves the giving off by one individual of some chemical or physical, that, on being received by another, influences its behaviour" (Frings and Frings 1964: 3); "Animal communication evokes a change of behaviour in another individual"(Cullen 1972: 103); or "Communication occurs when an animal, the actor, does something which appears to be the result of selection to influence the sense organs of another animal, the reactor, so that the actor's behaviour changes to the advantage of the actor"(Dawkins and Krebs 1978: 283).

The term *Phatic* refers to the use of communication mainly aimed to establish and/or keep contact between sender(s) and receiver(s). Such a function is probably best illustrated by the ducks' repeated "I am here, where are you?" famously described by Konrad Lorenz. A more articulated example concerns the wolves, *canis*

Fig. 2.3 Howling wolves – An example of phatic semiosis

lupus, and other *canidae*. After hunting and rallying together as a pack, wolves act very excited and joyfully howl together (Fig. 2.3). The meanings of this practice are of celebrative type: celebrating the success of the hunt (i.e., the presence of food), the wolves' safety, and the joy of reuniting after some time of separation and difficulties (see Lopez 1978/1999: 55). In addition, howls serve the purpose of *practically* rallying the wolves within a pack, before and after hunting, and of preventing packmates from wandering off on their own. This range of meanings somehow covers the entire spectrum of a message's phatic function (and of course, of the conative, aesthetic and expressive ones, too, but it was anticipated that it is only very rarely that a function appears alone). In howling one finds the simplest (the establishment of a contact) and the most complex (the strengthening of comradeship) extents of phatic communication. In particular, the act of celebrating the reunion of the pack is a most revealing example of how close cultural, ritual, aesthetic (musical) and phatic elements, gathered in a single behavioural pattern, can be.

The fourth function of communication systems is the *referential* one. In this case, signs refer to the context, to the surrounding environment, i.e., to everything except the sender of the message (and sometime even to him/her). Referential signs are thus descriptive, and may refer to places, times, people, sensations and so on. "It's 9 o'clock", "Franz Kafka was born in 1883", and "It's summertime, and the living is easy" are all referential messages. In this case, too, there are at least two paradigmatic examples that were abundantly described in this companion, the dance of the honey bees and the alarm calls of vervet monkeys, which certainly do not need further illustration at this point. It can also be added that several displays of an animal's health, identity, social status, etc., share an expressive and a referential dimension. Finally, several birds use acoustic communication in a referential fashion for showing migration routes. A few other, like the Herring Gull *Larus argentatus* and the Greater Honeyguide *Indicator indicator* may use sounds also to indicate sources of food.

The fifth type of communication function is *metalinguistic*. In this case, signs are used to refer to other signs, as occurs in playing, deception, or several forms of ritualization. In playing, particularly, the adoption of metalinsuistic signs is crucial, in that several forms of play are nothing but unserious imitations of "real life" situations, most of which rather dangerous, as hunting or fighting. The signs of aggression, thus (running towards the receiver, displaying anger, etc.), must always be accompanied by other signs, which refer to the former, indicating that all seemingly serious actions taken from that point on, must be interpreted only in a playful way.

Another very interesting example comes from the vast catalogue of interspecific or intergroup imitations (particularly through singing). The imitation of another species' or group's singing style is an iconic (and generally referential) sign of that style, but at the same time another important intention is again playful: the animal finds it pleasant and challenging to imitate another specimen. In other cases, imitation serves very precise functions, one of which is related with territoriality: by imitating a potential enemy belonging to another species (including predators), the imitator pretends to belong to the species imitated, and practically "says" that the given territory is already occupied by a member of the potential enemy's species. Finally, imitation in singing may also recall the – very metalinguistic – use of columns in medieval cathedrals. Not only do they hold the church up, but variously styled columns coming from different places, placed throughout the cathedral, would serve ostentatiously, as a sign of richness. As the mating season approaches, the male of the Superb Lyrebird *Menura superba*, whose usual mating task is to impress the female, adorns his singing with a huge set of short imitations of different species and objects (including dog whines and doorbells). The rule is *the richer the better*, thus signs made for totally different reasons (like the doorbell) are here used metalinguistically as a display of the creativity and musical skills of the Lyrebird.

Finally, now singing has been mentioned, it is also clear that several signs and sign-repertoires related with mating (and not only with mating), are of *aesthetic* type, that is, they concentrate on the message itself, elaborating on its representation and form. Architectural, pictorial, kinaesthetic and musical signs (to use the classification adopted in Sebeok 1981) always carry a (normally predominant) aesthetic dimension (this aspect will be deepened further on in this Chapter 2 of the Companion).

As a bonus, it may be worthwhile to mention a seventh function of communication, elaborated and proposed by Felice Cimatti (1998: 102–105): the *cognitive* one:

> To speak about a Cognitive Function of language means to deal with how an animal's inner and outer behavior change in consequence of the possession itself of a language. In particular, we are interested in establishing how the way of thinking of an animal changes according to the fact that it is able to semiotically interact with other specimens (Cimatti 1998: 102).

The example brought by Cimatti is once again that of the monkeys' alarm calls. When a danger appears, the monkey's mind is probably activated in two ways: on the one hand, the mental representation of the perceived object (what kind of danger, how close, how fast ...); on the other hand the activation of a signal, i.e., that specific alarm call that stands for that specific danger. Here one notices already a "change" in the semiotic animal: if s/he did not possess a "language", his/her reaction would have been of another type (namely, most probably, escaping from the danger).

In addition, a second characteristic of the cognitive function is that the call does not only act in a heteroceptive way (i.e., as a referential sign describing the danger and as a conative sign for the monkey's community), but also proprioceptively, in that it increases the self-control skills of the animal (who is now *expressing* his/her emotional status and *self-conatively* guiding his/her own behavior). In other words, the typical semiotic description of language as a communicative system *and* a modelling system finds here a perfect match: not only the mental complexity of an animal functions as *source* for a rich sign-repertoire The process works the other way round too: the richer and more complex the "semiogram" (i.e., the complete sign catalogue of an individual), the more sophisticated the mental life. The active participation to semiosis *opens up* for further – and increasingly articulated – cognitive abilities.

2.3 The Mind/Body Dualism and the Problem of the Point of View

"Consciousness is what makes the mind-body problem really intractable". This is the very beginning of the famous Thomas Nagel's article "What is it like to be a bat?" (1974: 435), which became a seminal point of reference for various studies on the classic problem of mind/body dualism. In this paragraph, an attempt will be made to turn the issue upside down? What if the question was: "The mind-body problem is what makes consciousness (and many other issues) really intractable"?

Usually, when the question is not posed in scientific terms, we are content to accept that mind and matter coexist independently of each other. But when we get a bit more serious, we find that the issue is extremely complicated. To start with, if mind and matter are mutually independent – how come that we think "I'll go to sleep" and our body actually goes to sleep? How come that our thoughts, that are totally different from a physical event or entity, do affect our actions?

Moreover, when we try to discuss it within the area of non-human animal studies, the problem tends fatally to enlarge, and involves the big (huge) question of the possibility of adopting an almost objective perspective towards phenomena that are by nature subjective. In other words, Can we interpret mind through matter? One of the best-known arguments against this possibility points out that "every subjective phenomenon is essentially connected with a single point of view, and it seems inevitable that an objective, physical theory will abandon that point of view" (Nagel 1974: 436). Although this point is ontologically undeniable, and it also serves as an ethical reminder of human arrogance, it should be now explored whether such a position is totally or only partially acceptable. Particularly when it comes to zoosemiotic research, where – obviously – the problem is of the utmost relevance. In detail, the principal questions are:

- Are certain types of methodological relativisms always productive for scientific research?
- Is a distinction between objective and subjective perspective to be shared a priori? Or maybe the dualism mind/body this distinction presupposes is exactly what should be demonstrated?
- Can certain human cognitive models be used as tools for answering such questions in the realm of other animals?
- What are the roles of speculative sciences on the one hand and the empiric ones on the other?

Zoosemiotics has not answered these questions in a systematic way, but it has dealt with a number of topics where the mind/body issue is relevant, generally showing a position that is in favour of the continuity mind-body, i.e., against the idea of considering it a dualism.

2.3.1 Etic Versus Emic

Zoosemiotics has very little to do with admiring a dog who brings the newspaper to the master every morning, and considering him/her clever simply for that reason. Zoosemiotics is rather concerned with thinking that dogs possess an intelligence that works in their own Umwelt and that they are able to project their cognition while relating to the environment, including us, other animal species or other dogs. The point is not only our disposition, as humans, to consider dogs as actively involved in semiotic processes, but to see if such a "thing" exists *within* the dog's mind. Is him/her a semiotic animal, or are we simply imposing upon him/her a semiotic dimension? In fact, such a question is not easy even when studying humans (particularly certain communities or certain groups). How will this be possible with other animals? Everything we say is based on human criteria, which we are arbitrarily applying to other animals, since nobody could ever say if animals really think in the ways we think they do.

Borrowing from linguistics (and particularly from Pike 1954), these opposite perspectives (so to say, the active and the passive one) have been named *etic* and *emic*. *Etic* stems from "phonetics", the study of linguistic sounds without regard to their significance in a language; and *Emic* comes from "phonemics", the study of speech sounds that are meaningful in a language. In brief, by the emic perspective linguists mean the *insider*'s or *native*'s interpretation of or reasons for his or her customs and/or beliefs; i.e., what things mean to the *members of a society*. By the etic perspective, one means the external researcher's interpretation of the same customs and beliefs; what things mean from an *analytical, anthropological perspective*. The model is evidently applicable also to the study of non-human species.

One of the theses that zoosemiotic research defends is that one can actually approach other animals from an emic perspective. Evidently, this is not an easy task. In ethnological studies, there is still an ongoing, quite animated discussion on the etic-emic issue. One of the main points is that a totally emic perspective is impossible to take, especially when there is no way to establish linguistic interaction with the culture observed, so that its members could describe their own views. The problem becomes even more serious with non-human animals: how can we claim to view matters from any non-human perspective? How can we achieve even the vaguest idea of non-human conceptions of given mental events? In short, the problem is to investigate the possibility of adopting an almost objective perspective towards phenomena that are by nature subjective. The question, as posed by Thomas Nagel's article, is indeed How does it feel to be a bat (a mosquito, a squirrel, or a stork)? :

> It is impossible to exclude the phenomenological features of experience from a reduction in the same way that one excludes the phenomenal features of an ordinary substance from a physical or chemical reduction of it – namely, by explaining them as effects on the minds of human observers. If physicalism is to be defended, the phenomenological features must themselves be given a physical account. But when we examine their subjective character it seems that such a result is impossible. The reason is that every subjective phenomenon is essentially connected with a single point of view, and it seems inevitable that an objective, physical theory will abandon that point of view. (Nagel 1974: 436)

When posed in these terms, the issue goes straight to the heart of the mind/body dualism. Although very reasonable, Nagel's position seems to be only partially acceptable. First, because, from a common-sense perspective (common-sense being used almost regularly when issues of subjectivity are addressed) it may sound a bit simplistic, for it represents a comfortable and socially shared (stereotypical, at times) way of facing a problem that is in fact quite complex. It may be easy to speak of subjectivity as an apparently unavoidable form of interpretation of reality that affects scientific research; however, these kinds of remarks risk being little too defeatist. It is true that there is no way to fully comprehend subjectivity, but is this an absolutely unbridgeable gap between scientific research and a correct interpretation of reality?[3] It is certainly true that a complete interpretation of another's perspective can hardly be achieved, but such a problem is not limited to the study of non-human animals. It applies somewhat to every scientific field. For instance, pharmacology is possibly far more inexact than is zoosemiotics. When a doctor prescribes a drug for a patient affected by a given disease, s/he knows perfectly well that *that* particular drug, made for *that* particular disease, does not necessarily work on *that* particular patient. In other words, an objective interpretation of a given disease does not allow one to grasp all of its possible individual implications. Despite this limitation, though, long live pharmacology, exactly because the practical impossibility of achieving an objective and exact interpretation of a problem should not be considered an impasse for scientific research.

Secondly, as Cimatti emphasises, a distinction between objective and subjective perspective may be simply wrong, for it begs the question: "I believe that

[3] "Among other things, if Nagel was right, not only we could not understand a bat, but other humans either, since we could not interpret their personal perspective anyway. However, it seems that we do understand each other – at least sometimes, so how can Nagel be right?" (Cimatti 2001: personal communication, in Martinelli 2006a).

such a distinction is incorrect. Basically because it presupposes a dualism which is exactly what should be demonstrated, i.e., the dualism between mind and body, and between personal and impersonal perspective. I perceive my mind, from the inside, only through external ways" (Cimatti 2001: personal communication, in Martinelli 2006a). Everything, or almost everything, that can be grasped about phenomena occurring within any organism, is basically the result of observations made from the outside. When, centuries ago, geocentric theories were replaced by heliocentrism, nobody thought it necessary to go to the sun in order to take the perspective of the object observed. Rather, what was considered were causes, effects, and side-effects. All of these elements were considered as a whole and then elaborated theoretically. After that, time after time, ever-more sophisticated technologies have supported those hypotheses. Now, in many cases, a given non-human animal pattern recalls a human one. Modes of production and correlated behaviours of that pattern are also impressively similar, and the same applies to emotional experiences. In addition, the observed subjects, like humans, have minds, [4] feelings, and so on; they interact socially, culturally and so on. In principle, not being human does not constitute a limitation to ascribing the existence of the given pattern to other species as well. Everything must be demonstrated, of course, but, if this is true, using outside clues (i.e., the *body*) to interpret inside phenomena (i.e., the *mind*) is a possibility that cannot be excluded a priori

Thirdly and most importantly: we *do have* clues as to how we might emically study non human species. To start with, we can scientifically study the sensorial organs of animals, the Merkwelt, not only the Wirkwelt. In other words, the way an organism interacts with the environment is largely due to the way s/he perceives it. A fairly recent example (very unethical and invasive, in fairness): in 1999, a team of researchers from the University of California at Berkeley wired a computer to a cat's brain and created videos of what the animal was seeing. By recording the electrical activity of nerve cells in the thalamus (that region of the brain which receives signals from the eyes), the researchers were able to view these shapes. The team used a so-called "linear decoding technique" to convert the signals from the stimulated cells into visual images. Altogether, the scientists recorded outputs from 177 brain cells that responded to light and dark in the cat's field of view. As the brain cells were stimulated, an image of what the cat saw was reconstructed. The image resulting was clearly recognisable as a version of the original scene, i.e., what humans see.

In the cat's brain, as well as in the human one, the signals from the thalamus cells undergo consistent signal processing in the higher regions of the brain that improve the quality of the image perceived. By tapping directly into the brain and extracting a visual image, the team has produced a "brain interface" that may 1 day make it possible to control artificial organs and indeed machines by thought alone. Following the same principle, it should be possible, in the future, to record what one person sees and "play it back" to someone else either in real time or at a later date.

[4]Semiotically speaking, to have a mind implies at least the capacity of (1) guiding one's own behaviour from the "inside", on the basis of projects not directly connected with what happens outside; and (2) elaborating and transforming such representations.

Now, we have two big tools to interpret the cat's behaviour in the light of that visual perception: one is *philosophical*, for we can speculate on how a cat's world can be (and this is not so abstract as one might think); and the other one is *technological*, for we can literally "reproduce" that world, to some extents, starting from the simple projection on a screen.

Further considerations are worthy of mention. It seems that if the taking of an etic perspective is somewhat disrespectful of the subject observed, it is also true that a totally emic approach is by definition unachievable. A good example is zoomusicology, the branch of zoosemiotics that studies the aesthetic use of sound communication among animals. One of the central points of this field is an investigation of musical universals, that is, on those musical structures, processes and experiences that emerge independently in different species (as happens with such traits as scales, rhythmic patterns, dancing movements, coordination between different subjects, etc.). Now, as pointed out by musical semiotician Jean-Jacques Nattiez in relation to human music only, the question itself of defining music and searching for universals is etic in principle:

> The question concerning the universals of music is not and cannot be by itself a universal problem, chiefly because it is related, in the Occidental realm of knowledge, to the existence of a field of research called 'musicology', which considers sound phenomena as musical, independently of the place and conceptualisation which the autochthonous people give them in their own culture. [. . .] the concept of music is culturally determined as much in our own civilisation as it seems to be in others. (Nattiez 1977: 95)

This is even more true for zoomusicology, where not only do we not know *what* other species may call their own music, but we do not even know *if* they call it anything at all. On the other hand, however, one should not be so sure that such a problem must be solved at all costs in order to determine the existence of an idea of music in other animals. In other words, we are not sure that a conception of music, if it ever exists, is separable from the actual, material act of doing music. It may be useful to know how the *aesthetic use of sounds* is defined among the diverse species, but possibly that is not even necessary. We are not necessarily searching for the existence of the signifier "music", i.e., m-u-s-i-c. We are rather searching for the signified "music", the signified of an *aesthetic use of sounds*, which by convention we call here *music*. We could rather call it *Roarrr*, or *Woof Woof*: what really matters is the actual phenomenon. By this it is implied that (1) to create and develop the concept of "music" does not make humans the only musical animal; and (2) it is once again on the basis of a material behaviour that we detect the existence of a given mental entity.

The question posed by Nattiez should not be trivialized, however, for it actually carries within it a far more complex implication. Staying with this example, the musicological problem exists because there is a discipline called musicology, which investigates certain issues on the basis of certain founding paradigms. Not only was musicology born within the human species, but even in a very limited portion of this species (the so-called Occident), which has its own ways of perceiving and decoding the aesthetic use of sounds. These ways are not necessarily shared by (and often are in opposition to) other members of this species.

This is the reason why a good definition of music, without Euro-, ethno-, or anthropocentrisms, becomes a crucial task for zoomusicology, exactly as it was for ethnomusicology. This is also the reason why any zoosemiotic approach should take, as much as possible, an emic perspective, notwithstanding the practical impossibility of totally avoiding the etic point of view. Hence, the compromise is *what we cautiously consider to be an emic approach*, on the basis of similarities in traits, behavioral patterns, and emotional experience, and with the above-mentioned support of technology and philosophical speculation.

In strictly semiotic terms, such a methodological framework demands nothing less than the adoption of *abduction*, the research method so enthusiastically advanced by Peirce. A surprising phenomenon, X, is observed. Among hypotheses A, B, and C, A is capable of explaining X. Hence, there is a reason to pursue A. Although, logically speaking, that does not seem totally trustworthy, abduction, Peirce maintains, is precisely the reasoning procedure that scientists use most, for it is the only method that *explains* phenomena rather than simply classifying them (as induction and deduction do). Abduction is a "logic of discovery", a kind of critical thinking which opens doors of opportunity for scientific research. Abductive strategy tends towards an emic perspective, because it is the only form of investigation which allows for the explanation of phenomena "internal" to the subject observed, according to objective rules decoded from the outside.

So, is the suggestion here that emic is good and etic is bad? Of course not. Let us reconsider, in order to keep a certain coherence in the argumentation, the example of music. One might of course consider etic strategies as the point of departure, and emic strategies as the target to reach. However, within the field of zoomusicology, there are scholars who are contributing to the field as musicians, more than as musicologists (particularly David Rothenberg and Jim Nollman). These people successfully essayed to perform music *with* other animals, and both Nollman (1999) and Rothenberg (2005) wrote extensively about it. A pillar in their arguments is that they always considered their artistic sensibility as the main requirement for defining a certain phenomenon as music. In other words, their basic strategy is based on empathy, which is a very etic approach. If a musician – that is, an individual who is not *alien* to the musical phenomenon, but in fact is an *active part of it* – perceives a blackbird song as music, then there is an acceptable reason to think in general of that song as music. At least if music is considered a universal phenomenon.

It seems too subjective an argument to be taken seriously in the scientific sense. However, the truth is that many are the phenomena for which empathy and subjective recognition are the main (if not the only) criteria for decodification. Zoosemiosis is full of those instances. One may think about emotions like excitement, depression, anger, or about several bodily symptoms of inner statuses. To make another example, most of our interaction with pets is based on this etic-empathic approach.

Of course, it is still to be maintained that etic strategies of this sort are not a sufficient reason for confirming any zoosemiotic hypothesis, but one has to admit their necessity as *one important* argument.

2.3.2 Zoosemiotic Universals

Certain behavioural patterns seem to suggest a clear contiguity between humans and other animals. To deduce them, within most human contexts (except the usual pre- or non-linguistic groups, like infants, primitive communities, mentally-disabled, etc.) is a rather easy task, for they are expressible linguistically. The issue is definitely more problematic within non-human contexts. Theoreticians of different disciplines have tried to interpret these patterns, starting once again from the relation between external manifestations and possible correspondent mental activities. If, in their case, the problem was exclusively methodological, in this case, with these purposes, these interpretations become a precious suggestion in the perspective of the claimed unity mind/body. In ethnological studies, characteristics that are common in every or most human communities are called universals: an application of this concept in music was already illustrated; therefore this is how it is realistic to name them within this wider zoological context. To start with, thus, what we are searching for is semiotic features in non-human cognition that are *homologous* to those in humans. They should not be only *analogous*.

> These parallels immediately raise several problems, the most obvious being whether the animal's behaviour is 'merely' analogous to man's, whether, that is, shifting to a more familiar parlance, [for instance] the label 'dance' is 'just' a colourful and suggestive metaphor – as it must surely be in Frisch's designation of the kinetic component of the communication system of the honeybee as a 'dance' – or whether something deeper is implied, perhaps indeed a remote phyletic homology (Sebeok 1981: 218).

There are two types of homologies: (a) *phyletic* or innate (i.e., transmitted via genes); and (b) *by tradition* or acquired (i.e., transmitted via culture). The traditional example used for distinguishing, and avoiding confusion between, the two concepts, is that of the wings in a bat (i.e., a mammal), as compared to a bird's wings and human hands. In this case, the bat's and the bird's wings are simply analogous, because they look like each other, they serve the same purpose, but they evolved independently from each other, therefore they do not qualify for being considered (phyletic) homologies. On the contrary, homologous are the human hands and the bat's wings, both body parts deriving from the upper arts of the tetrapods, and both showing a continuity in their components (bones' structure, particularly). In other cases, more thorough research in this area has contributed to solve.

When it comes to actual behavior, that is, turning from zoology to ethology, the issue becomes more interesting for zoosemiotics. Research work on analogies and homologies contributed also to dismiss some false myths concerning the specificness of certain patterns. A famous case is that of laughter, which only recently turned out to be homologous, and not just analogous, across primates (for a very recent study, see Davila Ross et al. 2009), while previously it was a common idea that only humans *laugh* in the proper sense, while that facial expression exhibited by other apes is in fact just analogous (a mistake that Umberto Eco also makes when his *Name of the Rose* character William of Baskerville says, in defence of Aristotle, that laughing does not make humans look like monkeys because laughter is a species-specific human expression).

According to Tembrock (1963: 777), who studied the problem in relation to acoustic communication, there are three criteria by which to define homologies:

1. The criterion of position, understood as "the situation that exists at the time when the sound is made. This concept of 'position' would have to include all available external and internal factors" (Tembrock 1963: 777);
2. The criterion of the special quality of the structures;
3. The criterion of interconnection by intermediate forms.

These criteria refer directly to sound forms, but, in Tembrock's view, can be taken as more general principles. Tembrock indicates further, auxiliary criteria that can be used for simple structures: "(a) simple structures can be regarded as homologous if they occur in a large number of nearly-similar pieces; (b) the probability of a homology in simple structures increases with the existence of further similarities of equal distribution in nearly-similar pieces; and (c) the probability of homology of a characteristic decreases with the frequency of occurrence of this characteristic in species which are definitely not related" (Tembrock 1963: 777).

Unfortunately, the fact that certain non-human behavioral patterns may be considered either analogous or homologous to human behaviours does not depend only on the factors listed above. In some cases, the exclusive human-ness or non-human-ness of certain behavioural traits is simply taken for granted – the case of symbolic semiosis was for instance mentioned – so that attempts to detect homologous characteristics in such patterns is a priori accused of anthropomorphism or zoomorphism. This is the reason why a separate treatment for these apparent biases is required (see Chapter 3).

Once established these terminological parameters, one may propose now some possible zoosemiotic universals (i.e., transpecific homologous semiotic traits), as borrowed from Felice Cimatti's theories (1998: 179–190). Particular care should be taken to how such universals imply a continuity between mental processes and actual behaviour:

1. *Sociality*: in order to exist, a semiotic system requires interrelation within a group, even if only for basic purposes (like primary needs' fulfilment). This sociality may increase semiotically, producing more complex forms of behavior such as the exchange of very sophisticated information. Such exchange is possible when sender and receiver are able to represent abstract entities in their mind, i.e., to categorize the environment and recognize the entities within it. Of course, animals must share the same form of categorization (i.e., a *code*). The dance of the honey bee is a sign of something that the other bees cannot see, but do know as an entity, and this entity is abstractly, yet clearly, represented in their minds.
2. *Arbitrariness of categorizations*: the object represented by the sign changes from species to species. What an animal decides to represent through a sign is something that makes sense only in its Umwelt, in its *universe of discourse.* To other

species, the same entity could be represented much differently, or it simply might not be an entity at all. What is an eagle to a human is flying danger to a vervet monkey.

3. *Mind*: In order to be understood, a sign must be interpreted. This requires the taking into account of several variables. Signs may also have no connections with the object represented. Here Umberto Eco's famous definition of semiotics may be recalled – "semiotics studies everything that can be used to lie" – and the various forms of deception in animals (some of which shall be discussed in one of the next paragraphs). The ability of "lying" requires the existence of a mind capable of managing this variability. Humans lie, other animals lie, armchairs do not lie; and we would never lie to an armchair, although we can easily lie to a dog, maybe for teasing him/her, or for some other purpose. This aspect is much less banal than it may seem.
4. *Principle of complexity*: usually, the more intense or complex is the sender's inner state, the more sophisticated is the consequent expression. The dance of the honey bee is much more frenetic when the source of nectar is more abundant.
5. *Ritualization*: Ritualization is defined as a process through which some behavioral patterns are modified, sometimes even in terms of function, for semiotic purposes. Forms of ritualization are the stereotyping of certain patterns, incompleteness and repetition in putting them into action, and – mainly – the metalinguistic use of most signs. Closely related to aesthetic processes such as dance, ritualisation has been detected in practically all species, especially in the so-called superior ones (birds and mammals). For the purposes of this classification, it is important to stress that, through forms of ritualisation, a non-semiotic behaviour may become communicative. Generally, a dog grinding its teeth does not want to fight at all, but this sign is a precise and ritualized message, and the other dog reacts consequently: this is a case in which a non-semiotic, functional act (grinding and then biting) becomes semiotic, bringing out the *intention* of biting.
6. *Self- and Hetero-perception*: This is a general presupposition for the development of a communication system: the use of signs – as abstract entities – requires a sort of "trust" in what they represent (myself, other beings, things, events, etc). Yet, such elements must first of all be perceived. As Cimatti notes, we can talk about the world because we perceive it, and not the other way round (unless we are talking about language, in which case this relationship may also be reversed);
7. *Syntax of perception*: All animals perceive the environment as gestalts, i.e., through a *top-bottom* or, better, *figure-background* framework. As a consequence, the communication of this experience is also gestaltic. Among all data available in the environment, an animal will select only those it is most interested in (e.g., a bee searching for nectar will perceive flowers as a figure and, say, a group of ants as part of the background). This is the very first level of syntax in natural languages, which can be detected both in the simplest interactions of some insects and in the most complex conversations of human intellectuals.

Further important types of universals have also been listed within the field of phenomenology. In this case, even more than in the previous one, the unity mind/matter clearly emerges. Although generally eager to consider non-human animals as "non-subjects", concentrating its attention on life and the so-called human "lived experience", phenomenology, especially Edmund Husserl, has shown special interest on other animals as subjects with which humans share a similar bodily experience. In other words a primary status of "their" and "our" life is the perception of the animated body. San Martin and Pintos (2001), following the indications emerging from Husserl's work, have listed a very interesting *animal ontology* in eleven points:

1. Every animal (EA, hereafter. Of course, the term "animal" includes the human) is a *Körper-Leib*, i.e. a somatic body, animated from within, a living bodily entity, that "feels" and experiences life in an intentional way;
2. EA is a *Ichilich* that rules its own body. The animal is the entity that moves its own body, that "decides" to move;
3. EA experiences a mental life articulated in time units;
4. EA experiences its own body in a direct way. Whether aware or not, the animal "feels" its own body (heart beating, illness etc.);
5. EA – to the letter – is alive. In phenomenological terms, this means that the animal is situated in its body first, then in the surrounding environment. The animal lives in the environment because it lives in itself; by consequence
6. EA is in material relation with the world. Its body is made of matter, like the surrounding environment. It is on a material basis that we are open to something which is not only "ourselves";
7. EA perceptually feels itself as the 0 point of the world. EA is egocentric in principle, and perceives the world starting from itself. Things "are", starting from what "I" am and am not. The relation with the world is based on affirmations and negations of our identity, thus starting anyway from ourselves;
8. EA perceives the world from one point of view. To mention a trivial example, changing physical position, sensorial perception changes;
9. EA experiences a common animal world, and a social horizon of its own community/group/species. The animal knows it is an animal, i.e. it knows to be part of a given category and understands who is a member of the same category, and who is not. The way a dog relates with a cat is clearly different from the way s/he relates with a tree, but has several points in common with the way s/he relates with a human being. Naturally, we cannot be totally persuaded by the idea that a "species" horizon exists. The classification of animals by species is a human invention, therefore one cannot really be sure that such a drastic separation between one species and another that is genetically close to the former, is actually perceived by the two specimens in question. Possibly, the "boundary" between two subjects is perceived on the basis of affective, relational, somatic aspects, and that means that – in principle – taxonomic differences may not be involved at all;
10. EA is semiotic. EA – when interacting – has a set of semiotic problems to solve: it communicates, understands, misunderstands, etc;
11. EA interacts with each other primarily on an affective and emotional basis.

2.3.3 Multimodal Semiosis

Animal semiosis shows further or signs of mind/body coherence when different patterns co-operate to display one message. Zoosemioticians and communication researchers in general tend to categorize signals by the primary sensory channel involved, but – in actual fact – it is rarely so straightforward, for multiple channels are regularly engaged simultaneously. This is especially true in highly social, group-living animals (which does not necessarily corresponds to supposedly smartest species: many insects are definitely more social than the majority of mammals). One of the many things Darwin had anticipated in the fundamental *The expression of emotions in man and animals* (1872) is that multiple, concurrent stimuli are very important in order for the signal to be fully efficient. However, the influence of such stimuli on the signal and its meaning has only recently been closely analyzed (see, for instance, Horn 1983). Behavioral neuroscientists have found that primates (humans included), birds, and insects are especially concerned with the production of information from multiple sensory channels, in the fields of both attention and perception. In particular, posture and movement are the stimuli most frequently associated with the production of the signal. Partan and Marler (1999) have provided a synthetic but exhaustive account of the most important points in multimodality.

Their classification may be summarised in the following scheme (see Table 2.1):

Table 2.1 Multimodal semiosis according to Partan and Marler (1999)

	Separate components		Multimodal composite signal		
	Signal	Response	Signal	Response	Category
Redundancy	a	x	a+b	x	Equivalence (intensity unchanged)
	b	x	a+b	X	Enhancement (intensity increased)
Non-redundancy	a	x	a+b	x and y	Independence
	b	y	a+b	x	Dominance
			a+b	X or x	Modulation
			a+b	z	Emergence

The left side of the scheme depicts redundant (above) and nonredundant signals (below) as separate components (a and b) with consequent responses (x and y – the same letter indicates the same qualitative response; different letters indicate different responses). The right side depicts the responses to combined multimodal signals (a + b). The meaning of a single signal may be either redundant or nonredundant, in that different signals may produce the same message, or two different ones. The advantage of redundancy is the reinforcement of the message and the reduction of the risks of interference (this is why these signs are also called backup signals). Nonredundant signs have the advantage of providing more information per time unit. Empirically, the two typologies of sign can be distinguished by the reaction of the receiver. When emitted separately, redundant signs should provoke the same or a very similar response from the receiver, whereas nonredundant signs should provoke different reactions. The picture may change quite drastically when the signs are combined simultaneously into a multimodal signal (see the right side of the scheme). Redundant signs might provoke the following reactions:

1) *Equivalence* – the multimodal signal provokes the same exact reaction as the signs emitted separately (shown in the scheme as small x letters). Partan and Marler mention the case of courting male moths *Cycnia tenera*, that "elicit equivalent responses from females regardless of whether their pheromones and ultrasonic sounds are presented separately or together" (1999: 1273).
2) *Enhancement* – the multimodal signal produces a reaction that is increased in intensity (shown in the scheme as a capital X). "*Aphaenogaster* ants recruit help for carrying prey by emitting pheromones, but with large prey they also stridulate, producing a substrate-borne vibrational signal. The stridulation has a small effect alone, but both components together recruit more workers. Similar multiplicative effects occur during neural processing of simultaneous visual and auditory stimuli in the superior colliculus of cats" (Partan and Marler 1999: 1273).

A multimodal signal based on the combination of nonredundant signs produce a wider range of possibilities:

1) *Independence* – The two signs are independent and produce distinct reactions, that are not in relation with each other, although they are combined. "Pheromones from female *Cupiennius salei* spiders alert males to the presence of a potential mate. Concomitant vibrational signals from the female then direct males to her location. These two components function independently whether they are perceived simultaneously or not" (Partan and Marler 1999: 1273).
2) *Dominance* – One of the two signs prevails on the other. It is a typical case in play behaviour: "Dogs signal play behavior visually with a bow, and sometimes also growl, normally a threat. Separately, these signals are contradictory, but their combination elicits play, the visual component taking precedence" (Partan and Marler 1999: 1273).
3) *Modulation* – One nonredundant sign affects the other, by modulating its effect: "Male *Alpheus heterochaeli* shrimp respond aggressively to visual cues alone, such as an open claw, but do not respond to chemical cues alone. When the two are combined and the pheromone is from a female, male aggressive responses are suppressed. Neural analogs for modulation include cells in the cat superior colliculus that respond to visual stimuli alone but not to auditory stimuli alone. Auditory and visual stimuli together elicit enhanced responses in some neurons, leave some unchanged, and leave others depressed" (Partan and Marler 1999: 1273).

4) *Emergence* – the multimodal signal provokes an entirely new reaction, that has nothing to do with the two separate signs: "Aromatic pyrazines and red and yellow coloration are commonly associated with noxious insects. Presented alone, neither cue produces aversion in chicks; aversion appears only when the odor and color occur simultaneously. Here, multimodal stimuli evoke a response not elicited by the unimodal components. Similarly, some cat superior colliculus cells respond to multimodal but not unimodal stimuli" (Partan and Marler 1999: 1273).

2.4 A Semiotic Investigation of Deception, Play and Aesthetics

In this research-sample a theoretical model is proposed for analyzing the question of aesthetic behaviour in animals (including humans). The hypothesis is that aesthetics can be explained by starting from the development of lying and playing abilities. In other words, it is suggested that these three aspects lie along some kind of continuum.

In order to defend this thesis, one shall discuss both the perceptive dimension (i.e., the role of the receiver of the deceptive/playful/aesthetic message) and the articulation of the message (i.e., the role of the sender of the fictional/playful/aesthetic message). The latter in particular demands explanation in terms of the messages' interaction (are there fictional/playful components in aesthetics? Are there fictional/aesthetic components in playing? Etc.), and their reciprocal necessity (can aesthetics transcend playful/fictional components? Can lying transcend playful/aesthetic components? Etc.).

The analysis departs from the concepts of lying, playing and aesthetics, and finally illustrates the suggested connection according to a Peircean framework (with a special emphasis on the concepts of firstness, secondness and thirdness).

2.4.1 On Deception

In his *Trattato di semiotica generale* Umberto Eco brilliantly defines semiotics as the study of everything that can be used to lie. In a way, nothing illustrates semiotic phenomena better than a theory of lying. The presence of a lie in a communication system implies the following:

1) There is a code, which is socially shared, that associates a sign to a meaning. Since lies break this association, the latter had to exist previously.
2) The sender of the sign must be able to break this link.
3) In order to lie, the sender must be able to mentally represent not only the receiver's behaviour, but also his/her mind. That is why we can only cheat entities provided with minds.

4) The sender must be able to take advantage of the broken link, making the receiver believe something that is not happening in the way she thinks.

This means that lying, when structured as a semiotic process, is a cognitive act, and cannot be seen as the result of instinct or of a simple behaviourist stimulus-reaction process. More to the point, the prerequisites for a cognitive semiotic (then potentially deceptive) behavior display the existence of the zoosemiotic universals discussed previously (see also Cimatti 1998: 179–190):

1. *Sociality*;
2. *Arbitrariness of categorizations*;
3. *Mind*;
4. *Principle of complexity*;
5. *Ritualization*;
6. *Self- and Hetero-perception*;
7. *Syntax of perception.*

For such reasons, the semiotic analysis of deceptive behavior is a great litmus test for zoosemiotics. In other words, if there is a semiotic phenomenon *there can be* a lie; conversely, if there is a lie, *there is surely* a semiotic phenomenon. To recall the discussion on the "semiotic animal" in Chapter 1 of the present book, one can also say that an animal that is able to deceive in this semiotic sense is, for sure, a *semiotic* animal.

Now, in order not to fall in the pansemiotic trap mentioned in Chapter 1, it must be pointed out that this kind of semiotic lie will not correspond to absolutely *everything* untrue, and a certain semiotic threshold should be established. Maybe not so high as the notion of "semiotic animal" seems to suggest, but definitely not too low either. Above or below this threshold, untrue acts can be considered lies or not. The assumption here is that the *encoding* of the information is the key-point: lies are such when the sender of the message *uses* one or more codes for its purposes. When this does not happen, the subject doing something untrue is not properly lying (make no mistake, however: being below this semiotic threshold does not mean that the patterns in question should not be subject to semiotic research).

Animals adopt several tricks to escape being preyed upon, by *pretending* they are not preys: their untrue action may be semiotic when it metaphorically corresponds to a sentence like "Look at me – I'm not your prey!", but it is hardly semiotic when it metaphorically corresponds to "Let's hope the predator doesn't see me". The latter seems the case with so-called "cryptic mimicry": the animal physically imitates the characteristics of its environment, e.g., the butterflies of the species *Kallima*, which look exactly like leaves. In this case, the hunted animal does not want to establish any interaction with its pursuer, and the information received by the latter (again, metaphorically: "This is a leaf, not a butterfly") is not deceptive in the sense that has here been established.

The Same goes with "Batesian mimicry". The *Clytus arietis* and the *Aegeria apiformis* are two perfect imitations of a wasp. Their *choice* is not to deceive the pursuer, just as being born in Ajaccio cannot be understood as Napoleon's choice.

Other kinds of mimicry ("Müllerian", "Mertensian" and "agressive") present similar features: in all these cases, the trick is a "natural gift" and not a proper choice made by the specimen in question. Having said that, it is another matter to establish whether these insects are aware of their aspects and their impact. In other words, if it can be proved that the *Aegeria apiformis* flies around in a more "light-hearted" way because it *knows* that most likely it will be mistaken for a wasp, then everything said so far about mimicry does not apply, and we are indeed in front of phenomena placed *above* the threshold. But, at the present state of research, this does not seem a credible hypothesis.

A more proper (and somewhat funny) example of deception is the food-call uttered by a rooster in order to signal the presence of food to hens. Usually, the longer or the faster the vocalization, the better and more plentiful is the food. Very often, however, the rooster emits the call even when there is no food available; in that case, the male cheats the females for mating purposes. This very basic theme may present a few variants:

a) There is a new hen in the fowl-yard, but there is no food: the rooster utters a long and intense food-call (the male aims to mate with the new female);
b) There is a new hen and some food: no call is uttered (the rooster does not want to share the food, neither he is interested in the new female);
c) New hen and food, but also another rooster: no call is uttered (the rooster is "chicken-hearted" and does not want to face the new male);
d) New hen, another rooster, no food: a food-call ensues and then a fight with the rival occurs (the rooster wants to establish hierarchies);
e) New hen (or an old-but-still-attractive one), and no matter how much food: no call, but the rooster moves to the hen (not a lie, in fact, but admittedly a nice *coup de théatre*).

Put simply, the rooster seems to do exactly what he pleases, with a rather flexible use of the information signal. And the same applies to the hens, who very often guess the mating purposes of the food-call and simply ignore it – unless, of course, they like the idea.

The motivations that push an animal to lie should be considered once again universal. Lies occur more or less in the same contexts as for human life; or, to put it more fairly, humans lie within contexts that are probably zoologically universal. To start with, animals may lie for the express purpose of survival. If a gazelle is sick or young or weak in general, it may wait for the lion to approach, and then, instead of running, it may start to jump as high as possible. The lion may be impressed by such a display and decide that to chase that prey would be a wasted effort, for the latter looks very healthy.

Keeping something safe and/or hidden (food, for instance) is also another typical motivation for lying. This kind of deception is found also among animals kept in captivity. Belle and Rock, two chimpanzees trained by Dr. Menzel, provide a good example of evolution in the semiotic lie. Since Rock, the Alpha (i.e., dominant) figure, would steal the sweets given to Belle, the latter began hiding her food in a secret place. Everyday, after hiding the sweets, she would go to Rock, let him search her (the strictly semiotic phase: this is where Belle is actually lying to Rock), and then, as he left, she would run towards the hideout. After some time, Rock understood the trick. He searched Belle, as usual, then pretended to move, but then followed Belle, catching her with the sweets. Belle thus changed her strategy, changing her hideout everyday, but the trick did not work for long: Rock had learnt that the signal for the "hidden sweets" was Belle moving after being searched, no matter in what direction. Finally, Belle found the perfect strategy: she changed the hideout everyday, but now, after the inspection, would go to a wrong place, waiting for Rock to come very close, then running towards the sweets, letting him search in vain.

Material objects are not the only things one can hide and lie about. Actions can be hidden as well, and there is no need to say that sexual "cheating" falls into this category. In baboons communities, dominant males are in charge of what can be defined as real harems, and females are expected to "report" every now and then, in order to prove that they are not mixing with inferior males (thus, the show up is a signal standing for "I'm here, doing nothing I'm not supposed to do"). Females, of course, do not necessarily like this condition, but, as occurs with humans, flirtations with other males must be kept secret. What a couple of young "lovers" may do in this case is to couple in a discreet place (say, behind a big bush), take a break, let the female report to the Alpha, and then restart the process.

Another simple knack to avoid the consequences of one's (forbidden) actions is to accuse someone else. The best example is the American Sign Language-trained gorilla Koko stealing food from the refrigerator, then making the sign "Mike did it" (Mike was another ASL-trained gorilla during the same experiment), when she was asked for an explanation by her trainer Francine Petterson.

One can lie for altruism, as well. The well-known *sham behaviour*, displayed by many birds (Fig. 2.4), is a good instance. As a nest is attacked by a predator, the mother may act as if she had a broken wing or similar, trying (often with success) to focus the plunderer's attention to herself rather than to the young chicks (who, being weak and slow, would be a better prey *if* the mother was in good health).

The existence of altruistic behaviour proves, among many other things, that the advantages of lying may not be immediate. The perspective for a, personal or not,

Fig. 2.4 An example of sham bahavior. A mother-bird walks as if her wing was broken

later advantage provides enough motivations for deceiving actions. Such is the case of the ASL-training experience of the chimpanzee Washoe. After months of intensive training, Washoe acquired the ability to cheat with ASL. If any of her trainers would punish her for any violation of the rules, she might use sign-language for getting her revenge. Here is Roger Fouts' full account of one of the most amusing episodes

> When I did report Washoe's obnoxious behaviour to Trixie [Gardner, leader of the Washoe Project, together with her husband Allen] I could always count on Washoe trying to get even with me for making trouble. Once I told Trixie that Washoe had tried to bite me, and trixie gave Washoe a rare scolding. That very night at dinner, Trixie was cooking at the stove. Washoe was in her high chair, at the head of the table, acting like a little angel. I didn't believe this act for a minute, so I was sitting as far from her as I could get.
>
> COME ROGER, Washoe signed to me. PLEASE COME. I shook my head no. There was no way I was getting near her.
>
> PLEASE PLEASE COME ROGER, she tried again. I signed an emphatic NO. At that moment Trixie turned around and saw Washoe making these very sweet and perfect signs.
>
> ROGER, Trixie implored me, WASHOE WANTS YOU! I was trapped.
>
> I began edging around the table, ever so slowly. Trixie went back to her cooking. I kept sliding over one inch at a time. Finally, Washoe couldn't contain herself any longer. She lunged out of her high chair and grabbed me around the neck with both hands. I yanked backward with all my might and broke loose (Fouts 1999: 62–3)

One may also lie out of pure ambition. The vervet monkeys *Cercopithecus aethiops* have a wide range of sounds to warn the rest of the group about the presence of plunderers (sounds are different according to the kind of predator – eagles, snakes, felines; each different predator corresponds to a different strategy of escape). Sometimes when the only specimen in danger is the leader, the vervet monkey who has seen the plunderer does not emit the right sound, or makes no sound at all.

Finally, some room should be left for "lies for lies' sake" cases. Lying can be funny, when structured as a game. The dialogue between the already-mentioned Koko and her trainer Francine Patterson (Fig. 2.5) indicates that lying can be used for amusing purposes as well.

The mechanism of lying occurs through the (sometime conventional) giving of an (actual or ideal) advantage to the receiver. A worthwhile concept to be illustrated is the so-called *handicap principle*, elaborated in the early 1970s by the Israelian biologist Amotz Zahavi (Zahavi and Zahavi 1997). His main question was: why animals produce so often costly, apparently useless and extravagant displays (the typical example being the peacock's plumage) or behavioural patterns (like the above-mentioned gazelles' jumps)? Zahavi maintained that such patterns are signals to other individuals. For instance, the peacock's tail could be considered a signal sent by prospective mates in order to evaluate the individual's overall condition (not to mention the genetic quality). A specimen with efficient sexually-selected characters is a specimen who has *passed a test*. In discriminating between a male possessing a sexually selected character, from one without it, the female in fact discriminates between a male which has survived the test and one not tested. The males with

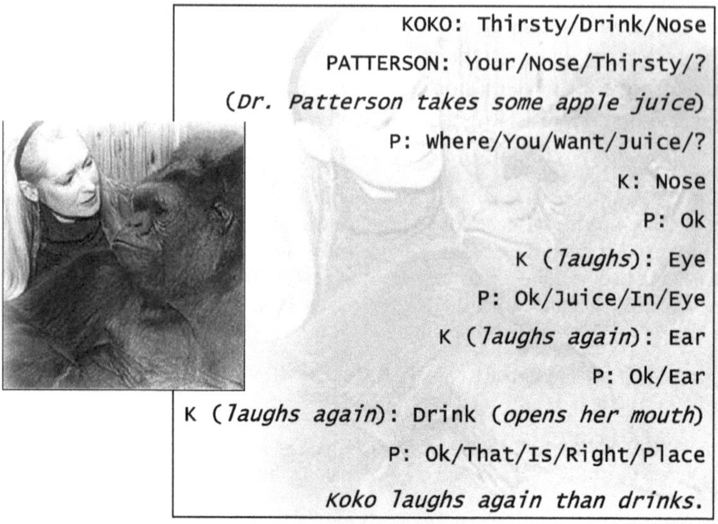

Fig. 2.5 An ASL humorous conversation between Koko and Dr. Patterson

the most developed characters are surely among the best genotypes of the male population.

The implication of such theory, called "handicap principle", is that costly displays or patterns are in fact very useful signals. Grafen (1990) classifies handicaps in the following way:

1) *Strategic Choice Handicap* – This is a form typically considered within current animal-related scientific literature (and often referred to as "costly signalling" instead of "handicap principle"). Here, the sender chooses what type (and *size*) of handicaps to display, taking the receiver's expected reactions to the signal into account. Since the cost of the signal and/or the benefit of the receiver's reactions may vary from one sender to another, different senders normally choose differently-sized handicaps. A good example is the puppy teasing and bothering its parents in order to *measure* their degree of affection and tolerance;
2) *Survival Handicap* – In this case, the possession of the handicap puts seriously into danger the sender's life, so the risk of this lie is very high. Indeed, it is normally "high quality senders" that survive. The receiver is the one who somehow has to establish the quality of the handicap: if the sender survives, then she has in fact proven the efficiency of her handicap. The sham behaviour, or the gazelle's "show-off" fall under this category;
3) *Revealing Handicap* – Here, the sender makes an effort in order to allow the receiver to judge her overall quality. This is of course the case with the peacock's tail, and – in fact – of most courting-related singing;

4) *Condition-dependent Handicap* – In this case, only certain senders (e.g., the high-quality ones) are capable of producing a given handicap. For instance, shiny feathers in birds are a handicap of this type, because their presence imply good health (nutritious diet, low parasite load, etc.).[5]

Finally, it is worth underlining that the word and the concept of "lying" cover in fact a wide range of behavioural patterns. Here, one might mention the typological classification provided by Jean-François Bouvet in his famous *La stratégie du caméléon* (2000). According to Bouvet, a lie may occur through:

1) *performing* – like, typically, in the case of courting, which – if not lying in the common sense of the term – implies at least exhibition (and sometime a true show-off);
2) *hiding* – like in the above-mentioned cases of mimicry;
3) *simulating* – like, basically, all types of playing behaviour;
4) *dissimulating* – like the above mentioned-case of "baboons" harems'
5) *deceiving* – like the above-mentioned cock's food-calls
6) *disguizing* – like in those cases where animals decorate themselves or part of the surrounding environment
7) *pure lying* – like the psuedo-linguistic type of lies displayed by Koko.

2.4.2 On Play

"Playing" is another widely-discussed issue within animal-related studies. It can be defined as one or more behavioural patterns that non-seriously imitate otherwise serious patterns and situations. Animals, including humans, usually play when primary needs do not have to be met. In particular, as Robert Fagen has shown in primates' behaviour (1981, 1993), certain attitudes towards playing reveal the animals' capacity to expend energy not just for utilitarian purposes (including such contexts when the specimen gets literally bored).

Although it is not a fixed rule, young specimens tend to be the most interested in playing (though not the only ones, adult canidae being a popular example). This is one reason why scholars think that the two basic functions of playing are amusement and the learning of rules of social behaviour: both of them are in strict causal connection (i.e., to have fun makes it easier to learn). Normally, playing uses and represents already existing behavioural patterns, such as hunting or fighting. Such representation must be previously established between the two or more participants. One important difference between play and deception, apart from very few exceptions,

[5]One may wonder why shiny feathers are a handicap. The answer is that a "shiny" bird is more likely to be located by predators, but – at the same time – it is the predator herself that soon recognises the prey as a fit one, thus very often renouncing to hunt her.

is exactly that all the subjects involved in the game *share the same lie*, i.e., they agree on the type and the characteristics of the representation. From an evolutionary point of view, to cheat on this aspect (i.e., to lie in the proper sense, manipulating the code without any agreement with the receiver) is definitely not convenient, for the liar would be immediately excluded both in the game and in real life.

In order to establish the basic *plot* of a given game, animals need at least three types of communicative patterns, more or less corresponding to the following sentences: "I want to play", "What I am about to do is playing", "What I just did was playing". Such communicative patterns can be either innate or learned. The sign repertoire is exclusive of play (i.e., not used for any other activity): canidae's play-bow is an exclusively playful sign (Fig. 2.6); pigs announce fight-simulations with a circular dance; great apes display the so-called "game face", which can be seen as homologous to a human smile; and so forth. All are cases of metasemiosis, since they have the double goal to point out the unserious intention of the sender, and refer to the serious pattern they are imitating. Most of these signs, moreover, are of symbolic type, since they have no physical resemblance to the object represented, and they were shaped in consequence of a code agreement.

Game signals are not only visual, but also chemical, acoustic and tactile. The mongoose *Helogale parvula*, for instance, points out its playful intention through a vocalization; the field-mouse *Microtus agrestis* through a pheromone, and so on. Although typical, the use of playful signs only *before* playing, is not exclusive. Species like the squirrel-monkey *Saimiri sciureus* display their playful intentions all over the game-session, the end of those signs actually meaning the end of the game itself.

Normally, the players voluntarily handicap themselves (an action also called *play inhibition*) and establish a situation of balance, so that every participant has

Fig. 2.6 A dog's play-bow

equal opportunities to *win* the game (see the very interesting Aldis 1975). Such silent agreement is usually interpreted as the result of a typical trial-and-error process. Young specimens learn very soon that the game will not last long if the strongest/oldest ones are too energetic towards the weaker/younger ones. The latter will soon escape or – successfully – invoke help from adults. Thus, in order for the game to succeed, it is convenient to create a situation of equality: the stronger participant will bite more softly, show with apparent indifference weaker parts of its body, and so forth. In this sense, one can speak about playing as important in creating a certain morality in animals, mostly through the development or of intersubjective empathy.

As shown in such studies as Gallese (1998), Gallese and Goldman (1998), and Frith and Frith (1999), there is a neurological basis underlying the development of empathy in playing, and its application in other (serious) contexts. The so-called mirror-neurons – located in the cerebral cortex – are maintained to react the same way both when the subject itself or another subject acts in a certain way. If such a hypothesis is acceptable, for all animals one should talk about cognitive empathy, i.e., a capacity of the individual to consciously identify itself with the (physical, emotional or else) state of the other subject.

In any case, the interpretation of playing is anything but easy, and is subject to several classifications. First, *individual play* is distinguished from *social play*. Second, a distinction is made between *kinetic play* (when the players do not use any object) and *instrumental play* (when the players do). Social play can also be divided into *intraspecific* (the most obvious case) and *interspecific* (like the human-dog case).

Most of all, there is an important difference in the interpretation of play, between the *evolutionist theories*, which basically investigate on the evolutionary causes of play, and the *cognitive theories*, which see playing as more integrated with the development of the individual (stressing on communication, adaptation, cognitive processes and so on). Among the evolutionist theories, one might mention

- *Surplus theory* (see Spencer 1873). Playing is considered an outcome of surplus energy no longer required for basic survival.
- *Future use theory* (see Groos 1898). An umbrella expression for what has been also called Practice Theory, Exercise Theory, and Social Functions Play is here seen as an impulse to practice "real life". The animal learns to master its organism (senses, movement etc.) in the way that will be needed later in life.
- *Pleasure theory* (see Grier and Bunk 1992, Patrick 1916). Another umbrella term for different theories (including Patrick's relaxation theory) whose main point is the interpretation of playing as an enjoyable behavioural pattern, or as releasing from tiring mental or physical activities of everyday life.
- *Ectotherm-endotherm Theory*. (see Burghardt 1984). Play developed through the organisms' transition from ectotherms to endotherms. To Bughardt, endotherms required more accurate parental care than ectotherms. A selection pressure towards parental care dependency then occurred. This could have led to a change

of behaviour patterns (response thresholds, effective stimuli etc.), and a strong re-organization of behavioural patterns in the young specimens.

A common point in all the evolutionist theories is that the origins of play are considered purely biological (rather than social, cognitive or emotional). Consciousness-related aspects are not believed to have an important role.

Among cognitive theories, one might mention.

- *Psychoanalytic theory* (see Freud 1959) Play is driven by forces, such as memories, wishes and needs, and fulfils the wish to be *grown up* and to acquire an active role within a given context. This gives the animal a certain feeling of "mastery" of its reality.
- *Constructive theory* (see Piaget 1962) Play is integrated with the development of intelligence, as accomplished through interaction with the environment. Such a process occurs through assimilation (play) and accommodation. Gradually, as the animal grows, its cognitive abilities, such as problem solving and communication-systems learning, develop and its play becomes more realistic. Such a theory can be assimilated with another one, explicitly produced for non-human animal behaviour, i.e., the Theory of Self-Assessment (Thompson 1998). Play is used for self-assessment purposes, in order to practice and improve one's own mental and physical abilities.
- *Socialization theory* (see Vygotsky 1976). Play is a basic activity in the socialization process: the animal achieves meanings and basic social rules. Through play, the animal gradually transforms its motives and needs into socially acceptable modes of relation with reality.
- *Interactionist theory* (see Mead 1934). Play is a relevant aspect in the animal's individual and social development. Such a process is fundamental in the developing of the sense of self. While playing, animals establish an interactional relationship with other specimens, objects and the whole umwelt. By playing, animals make sense out of the world.
- *Communication theory* (see Bateson 1976). Play is a relevant part in the learning process. It "teaches an animal to learn". Such theory emphasises the importance of meta-communication in play (i.e, to communicate that playing is just playing and not the serious behaviour they resemble). Animals learn the ability to frame and reframe roles through metacommunication. In this sense, play can be seen as a form of "semiotic negotiation".

As already mentioned, there is a General consensus that animals mostly (if not only) play if their basic needs are fulfilled. This means that (i) one should exclude that playing is the organism's response to any strictly utilitarian pressure, and (ii) there is specific room for playful activities when the animal is free from unavoidable needs. Actions and situations that normally fulfil real needs, are, during play, put into action without the pressure of the need in question. During the so-called aggressive play ("Catch me if you can"), the chasee tends quite often to slow down, running back-and-forth, or alternating its role with the chaser. Moreover, the behavioural pattern related to the real need is not faithfully imitated, but rather deconstructed, reconstructed, reduced, added with more elements, until the final outcome is something only superficially similar to the original *serious* pattern.

Play therefore appears as not provided with any goal: the factor establishing its end is usually physical tiredness, rather than a functional objective. Playing appears as a self-rewarding activity.

A step forward in the theorization of play, departs from the distinction between *autotelic* (Suits 1977: 125–8, Mitchell 1990: 202–5) and *simulative* play (Mitchell 1990: 202, 205–7). In the first case, the rewarding element of the activity is the activity itself. Exuberant and redundant playful displays, such as leaping, gambolling and twisting in the air fall under this category. Animals do not necessarily need a serious behavioural pattern to de- and re-construct. The key-word is here "repetition": the pleasure of redundancy seems to be the main motivation to obsessively repeat the same behavioural sequence over and over. According to Mitchell (1990: 204–5), autotelic play can be summarized in the following way: an organism O pursues the end E through the activity A, where A can be either E itself, or E is pure fun F (and thus A is performed in frivolous fashion): therefore A is performed for itself and/or to amuse O (Table 2.2):

In simulative play, imitative patterns appear: the playful action simulates a real action, altering of course the goals. Imitated actions, in their serious form, have a strictly functional role. Simulative play implies both interoceptive and exteroceptive (either intra- or interspecific) communication. Once again, such an activity implies a high level of cognitive articulation: the animal is fully aware of the virtual separation between signifier and signified. In other words, "to make a fight" is something different from "fighting". More to the point, the animal must realise that (a) the simulation S of the activity A has properties in common with the latter; at the same time (b) S has properties that distinguishes it from A; (c) S must resemble A in the mind of the receiver, but (d) the difference between S and A must be very clear. A quite complex articulation that requires the use of at least two codes.

What are the differences between S and A? Mitchell (1990: 206–7) indicates four of them:

1) the object pursued by organism O during S is inappropriate, if compared with what O pursues during A (e.g., in simulative fight, O jokingly attack someone who did not do any harm to it and who it has nothing to fear of, at least in that particular context);
2) O consciously introduces unnecessary obstacles (handicaps) in S. For instance, in simulative fight, O shows parts of its body that could be very dangerous to show during A; by consequence

Table 2.2 A schematic summary of autotelic play

$$O \xrightarrow{A} E \quad \textit{where} \quad E = A \quad \textit{or} \quad E = F$$

3) S is "exaggerated", as compared to A (movements are emphasized, energy-waste is higher and excessive);
4) S, like in autotelic play, is redundant, and such redundancy makes S crucially different from A, in terms of finality (e.g., simulative fight does not finish with the victory of one player, and roles are often inverted). Particularly, the scope of S is "benign" as regards to A, i.e., neither the sender or the receiver are affected by S the way they could be by A.

Summarizing: (i) O performs S, (ii) S resembles – and is intentionally designed to resemble – A, (iii) A has a biologically useful function (here called B), and (iv) S is benign s regards to A (Table 2.3):

Table 2.3 A schematic summary of simulative play

$S \sim A$	$A \supset B$	$E(A) \neq E(S)$

Mitchell excludes the possibility that play can exist at a mere morphogenetical level (although this is possible with lying, as in the case of mimicry): at least, a perceptive dimension and a physical-motor coordination are required. With such elements, autotelic (swinging from a tree branch is more the response to a stimulus, rather than the consequence of a learning process) and schematic-simulative types of play (like simulative fight or the "catch me if you can" game) are possible.

After perception and physical coordination, there is of course learning. Many playful behavioural patterns depend not only from the genetic heritage but from external solicitations as well. In the case of autotelic play, gambolling can easily be considered a learned activity (normally happened by chance) that O repeats for fun. As far as simulative play is concerned, one shall mention teasing play (which has the explicit goal of jokingly irritate the receiver, so that the latter reacts and interacts), and mimetic play (like in the well known case of parrots vocalizations).

The next level includes intentional imitation: from now on, according to Mitchell, there is no room for autotelic play. This is the level of fictional play: i.e., a visual-kinetic S of A, like in the case of rhesus-monkeys that play "the mother" by using coconuts in place of infants.

The final, most complex, level is that of the meta-simulation, i.e., a simulation of the simulation. This implies a multiple articulation of the code (something the reader should keep in mind when reading next paragraph). Such is the case, for instance, of those primates (including humans, of course) that pretend that their simulative fight is in fact serious, thus manipulating S twice, and A thrice.

2.4.3 On Aesthetics

The ancient Greek word *aisthetikos* relates to sensibility and perception, but the concept has also (and by now mostly) a connotative relation to the production and the enrichment of art. Arguments against any use of the word "aesthetic" as applied to non-human animals are rather common. The main problem lies in the tension between two apparently opposite ideas:

> Over and over, we keep encountering the same pivotal aesthetic paradox: this emerges from a profound confusion about purpose; it drives us to ferret out compulsively any semblance of utility, usually defined as adaptive value. We find it difficult to conceive of art as a coherent part of animal life and can scarcely imagine it as an adornment of the creatures' leisure. All researchers in this field are stamped by a tension between a deeply felt conviction on the part of many distinguished and sensitive biologists that artistic activity indeed exists in the animal world, and the inability to face its presumed lack of importance, even uselessness. [...] the position assigned to the aesthetic life in Western culture, from Plato onwards, is imbued by an uneasy fluctuation between these two attitudes, that art is at once useless and fraught with significance, purposeless and yet important. (Sebeok 1981: 232)

Even more problematic is the paucity of knowledge about such topics (very few biologists are experts on aesthetics, and very few philosophers are experts on animals), and quite often, the history of animal studies has provided examples of revealing accidents or *gaffes*, like in the case of the paintings drawn by the chimpanzees Congo and Betsy. Exhibited anonymously at a London gallery, they were received by enthusiastic comments like "a brilliant performance of abstract expressionism", except that the same critics were offended when the true identity of the painters was eventually revealed. Any zoosemiotic discussion on aesthetics therefore requires an at least partial demolition of anthropocentric barriers, and a theoretical interface that simply *works* in non-human contexts.

A discussion of this type may start with the following points:

1. According to Charles Hartshorne (1973), at a very biological level, *aesthetic* stands for not merely or too directly utilitarian.
2. The word "aesthetic" – also etymologically – has a clear connection with senses and perception, and particularly with concepts like "pleasure", "perception", "taste", a certain sense of "beauty", etc.
3. According to Roman Jakobson (1963), language has an aesthetic function when it is more concerned with signifiers rather than with signifieds.
4. As Aristotle pointed out in speaking about tragedy, aesthetic messages have to be at the same time *parà tèn dóxan* (i.e., unexpected, contrary to common beliefs) and *katà tò eikòs* (i.e., likely, believable).
5. According to Eco (1968), aesthetic information may be decoded in consequence of three factors: (a) signifiers acquire signifieds only from contextual interaction; (b) the matter of signifiers is not totally arbitrary as regards the signifieds (as in the case of onomatopoeic words); (c) the message involves many levels of reality (e.g., physical matter of the signifiers, denotation, connotation, and so on).

It is a common opinion that human beings are the only animals able to do things just for the sake of doing them, and that art, according to the well-known *l'art pour l'art* idea, is in substance the main form of self-rewarding activity, disengaged from biological and physiological needs:

> To say "aesthetic" is to say "not merely or too directly utilitarian". But we must be careful to balance this consideration against the seemingly connection with utility or it will be unlikely to survive evolutionary change. (Hartshorne 1973: 53)

Aesthetic acts may thus be interpreted as *not strictly* utilitarian (which seems to be Hartshorne's main idea). However, one could also conceptualise the *total* uselessness of aesthetics, envisioning it as something like a cute souvenir. Within such a dichotomy, two large schools – actually, two life-styles – are in competition. Does the "useless" exist? Does it make any sense in life? Can one act in a way totally unrelated to his/her biology? In attempting to answer such questions, those who maintain an absolute link between a living being and its existence – thus, providing every act with a goal – clash with those who think that part of such behaviours are, so to say, "transcendental", i.e., they exist in themselves.

There is actually a third way, i.e., the hypothesis that aesthetics might be a sort of gratuitous complication provided by Mother Nature. Aesthetic sense would have been born as a biologically motivated form of preference, until a form of sublimation turned it into a self-sufficient attitude.

The position defended here is possibly closer to the first, less transcendental, hypothesis, namely, *everything on Earth is useful, including the useless.* Nothing in Nature is without sense, and a biological function can be traced in each act or event. What marks the difference between "eating" and "listening to von Karajan's orchestration of *Le sacre du printemps*" is mainly a matter of levels. The act of "eating" certainly satisfies a more direct and urgent biological compulsion (hunger), whereas the urge to listen to the von Karajan performance is less obvious in its biological function. Nevertheless, it still has one, which can be discovered quite easily if only each degree of an ideal scale was considered.[6]

Any natural (in general) or behavioural (in particular) phenomenon always carry an intrinsic or extrinsic biological value: there is really nothing dangerous in admitting this, and most of all there is nothing reductive in doing so.

This does not mean, however, that one should overlook the presence of some fundamental (and "useless") characteristics of aesthetic activities, such as the evident fact that they produce an intrinsic pleasure. More simply, the thesis is that such pleasure is related to biology, and in fact plays a primary role, since undoubtedly very few things promote the preservation of an individual more than its emotional and intellectual welfare. This is a problem for many scholars, who, in struggling to demonstrate biological necessity in living beings, end up neglecting some fundamental behavioural patterns, just because it could be difficult to interpret them within their theoretical framework.

Desmond Morris, in *The Biology of Art* (1963), indicates self-rewardingness as the very first principle related to pictorial activity.

---

[6]Half seriously, it could be said that the people listening to Von Karajan's version of *Le sacre du printemps* are satisfying an aesthetic wish; the latter makes them feel good; the fact of feeling good helps them to work well; if they work well then they are not fired; if they keep their job, they still make money; finally, if they have money, they can satisfy the primary need of hunger.

> All pictures, whether by young apes or adult humans, must have a self-rewarding element involved as all or part of the motivation of the picture-maker. Other sociological or materialistic motives may or may not be operating at the same time, but if the production of the picture is not also a reward in itself, then its aesthetic value will be impaired. This particular point has been discussed frequently, but judging by the fact that it is so clearly illustrated by the apes, it would appear basic enough. In order to test this point, a chimpanzee was once subjected to bribery with a food reward to encourage it to draw more intensely. The outcome of this experiment was most revealing. The ape quickly learnt to associate drawing with getting the reward but as soon as this condition had been established the animal took less and less interest in the lines it was drawing. Any old scribble would do and then it would immediately hold out its hand for the reward. The careful attention the animal had paid previously to design, rhythm, balance and composition was gone and the worst kind of commercial art was born (Morris 1963: 158–159).

Such a consideration, as applied to music, is enhanced by François Bernard Mâche (1992: 158), as he underlines that non-human animals, like humans, occasionally *play* with sounds, and there is no need to find more "serious" explanations for this. Playing is a justification in itself, and music is a game under several respects. Playing and having fun *are* biological functions.

This conception of music as a game leads to a wider area of aesthetics, comprized of several concepts and concerned mostly with emotional states, but most of all with the idea of "beauty". It seems that the term "aesthetics" is no less vaguely-defined than is the notion of "beauty", and similarly, ideas of "perception", "taste", "pleasure", "sensation", "emotion", "play", and so on. In his philosophical dictionary, Voltaire defines the term "beauty" (*to kalon*) as a totally subjective matter ("Ask a toad what beauty is, the *to kalon?* He will answer you that it is his toad wife with two great round eyes issuing from her little head, a wide, flat mouth, a yellow belly, a brown back").

As aesthetics developed as an autonomous science during the eighteenth century, scholars produced a modern conception of art, as distinguished from the generic Latin *ars*, or ancient Greek *téchne*, founded on the connection of certain activities with beauty, the real common denominator of these activities. Arts became thus *fine arts*, no longer to be confused with science and/or trades, which aim to fulfil human material needs rather than pleasure and beauty.

Beauty, of course, does not designate a unified and objective area of sense. In his definition, Voltaire had already pointed out that the boundaries between the beautiful and the ugly are uncertain, and strongly depend on times, places, conventions and subjects. Beauty is mainly meant as a very general and flexible philosophical notion. Such a notion could be specified in different ways: e.g., as "luminosity" and "splendour" in Homer's poems, in the Platonic tradition, and – partially – in medieval thought; or even as "symmetry" and "proportion", "order" and "limitation", mostly within the long and complex Aristotelian tradition.

Thanks to this flexibility and except for very few cases, the notion of beauty tends no longer to be identified as an objective and invariable quality of being, but rather as an "aesthetic" quality, i.e., as pertinent to sensation, perception and thus subjectivity. One may attribute different specifications to this concept (e.g., involving "pleasure" or “taste”), but all of them always have something to do with the sphere of *aesthesis*, of sensibility. In Baumgarten’s terms, aesthetics is thus an *ars pulchre cogitandi.*

According to Roger Fouts, one of Washoe’s ASL trainers, the sense of beauty is a very basic biological function, related to adaptiveness.

> If we didn’t find things appealing, or beautiful, we would not be attracted to them. What would happen to courtship if perceived beauty did not play a role? The fact that chimps will request specific clothes and dress up displays not only a sense of beauty but also a sense of self, in that they are changing themselves when they dress up or put on lipstick. Adriann Kortlandt reported an observation of a chimpanzee in the wild returning into the jungle with an arm load of grapefruits he had just stolen from a nearby farm. Kortlandt noted that the sunset was particularly dramatic and colourful, and the chimp sat down, put his grapefruit on the ground and watched the entire sunset. When it had passed he got up and moved off into the jungle, forgetting his grapefruit booty. (Fouts 2001: personal communication[7])

The third theoretical dimension in this zoosemiotic sketch of aesthetics makes reference to Roman Jakobson’s scheme on the functions of language. According to Jakobson (1963), there is a type of language usage that is precisely definable as “aesthetic”. It is realized as a message articulated in a rather ambiguous form that directs the receiver’s attention more to the signifiers than to the signifieds, that is, more to the form than to the contents of the message. This does not mean that the message loses its meaning, but rather that such meaning is articulated in definitely more complex and, so to speak, confusing ways.

Within the aesthetic framework, the meaning of a message does not have an explicit function, and the organization of signifiers is led by intentionally “blind” mechanisms and rules, that is, they aim to coordinate the various parts of the message (Cimatti 1998: 99). Imagine being on a ship at the North Pole, surrounded by countless icebergs that do not allow the ship to proceed any further. A very basic consideration of such a situation would be something like, “We’re surrounded by icebergs”, i.e., a message that expresses its contents in a very clear and easy-to-decode way. However, if the one describing such a situation was Samuel Taylor Coleridge, in his poem *The rhyme of the ancient mariner*, the message would be formulated in a very different form: “The ice was here, the ice was there, the ice was all around”. In this case, the message – in terms of contents – is unchanged. What clearly have changed are the words: they acquire much more importance than in the previous statement (“We’re surrounded by icebergs”), and now are in a gestaltic kind of prominent position, leading the reader to pay more attention to them. Now there are contrasts, rhythms, redundancy – what formerly was a simple statement is now poetry.

Felice Cimatti explains the signifier-signified relationship in aesthetic messages in the following way:

> From an evolutionary point of view, the aesthetic use of a language is subsequent to other more adaptive functions, such as referential or conative. A signal arises with a specific function, e.g., to obtain a certain reaction from the receiver. Later in time, such a function becomes less and less concrete and present. Some room is thus left for an (at least partial)

[7] Fouts was interviewed by Dario Martinelli during the preparation of Martinelli (2002).

> aesthetic specialization of the signal, in which the signifier is more and more free from restrictions imposed by its contents, hence gaining autonomy. In this way any further development of the signal can only – or at least mostly – follow the 'traces' left by its physical, formal constitution. The signal is indeed mostly guided by its signifier, and no longer by its contents. (translated from Cimatti 1998: 99–100)

A key concept implied in Jakobson's definition of the aesthetic function is that of ambiguity. Ambiguity has much to do with the process of decoding the message. As Umberto Eco (1968: 63) remarks, the more ambiguous the message the more the information provided. An ambiguous message offers several interpretive options, but may also be interpreted as pure noise, i.e., pure disorder. A productive kind of ambiguity occurs when the attention is led to an interpretive effort, until one finds the right way for decoding, and establishes a more balanced order in that apparent disorder. In that sense, a kind of *pact* is established between the sender and the receiver of the aesthetic message.

This leades to nothing less than Aristotelian theories on tragedy. According to Aristotle, a tragedy must be at the same time *parà ten dóxan* (unexpected, unpredictable, contrary to common beliefs) and *katà to eikòs* (believable, having verisimilitude). In other words, the aesthetic message should be able to surprise, to amaze its receiver, but at the same time it cannot prevent the latter from "accepting" it as true-to-life and from identifying him/herself with it (and, consequently, empathizing, an aspect which recalls the ideas of pleasure, emotion and beauty). In brief, the aesthetic message must be credible in its incredibility. This is mostly because no surprise can occur if there is no layer of reality with which to oppose it.

Dalì's famous melting clocks in the *Persistence of Memory* (Fig. 2.7) are "amazing" precisely because they can be recognized as clocks. It is their similarity to real clocks, together with the manipulation and the re-elaboration of those signifiers, that triggers the mechanism of surprise. Otherwise (a) nobody would have been very surprised, for a melting object is not surprising in itself; (b) everybody would be still wondering what these soft objects in the picture might be. Instead, there are clock-hands, numbers, a roundish shape, all sadly "drooping" on bare trees. Thus, Dalì's clocks are exactly *parà ten dóxan* and *katà to eikòs* at the same time:

The aesthetic message, as mentioned, pays particular attention to its form. This also means that it attracts the receiver mostly (although not only) *because* of its form. A non-aesthetic message – one that is encoded "normally" – does not cause any problem of comprehension, i.e., its meaning flows automatically and points the receiver attention straight to its contents, the latter being conative ("Close the window!"), referential ("What a nice day!"), and so on. On the contrary, the aesthetic message is eccentric and soon grabs the attention, for it is to the letter "peculiar", and its meaning is not immediately clear. In order to understand the message, the receiver must explore it with care, relying on certain factors and resources. Umberto Eco indicates three categories of such factors:

1. Signifiers acquire proper meaning only through contextual interaction: in the light of the context, they are progressively clearer to certain extents and more ambiguous in others; they lead to a given meaning, but, as this happens, further

Fig. 2.7 Salvador Dalì's *the persistence of memory*

decoding options are possible. Finally, if one modifies a single element of the context, the others lose importance, too (Eco 1968: 64). To some extent, then, the aesthetic message functions as a hypertext that, once a concept is clarified, brings more doubts, forcing the receiver to perform a further decoding task. Key-words here are thus *interaction* and *context*.

2. Still, the relation between signifier and signified is emphasized (Eco 1968: 64): (a) the *matter* which signifiers are made of seems not to be arbitrary when compared to their meaning and their contextual relation; (b) the familiarity between two signs, in terms of meaning, can be enforced through certain strategies, like, for instance, the sound familiarity provided by rhymes in poetry; (c) the structure of the sign itself seem to re-propose the evoked sense, (like onomatopoeias within the domain of the sign's sound); (d) the whole set-up of the signifiers, organized according to a given proportion, realises a sound and visual rhythm that is not arbitrary in comparison with their meaning. To clarify this aspect, let us reconsider Coleridge's verse: "The ice was here, the ice was there, the ice was all around". In this case, the parallelism and the redundancy of the words and the structure of this verse go hand in hand with the immense expanse of icebergs that surround the mariners of the poem. In other words, not only are the signifiers here, there and all around, but so are the icebergs! The message, which in "normal" (i.e., non-aesthetic, referential) conditions would not be so redundant, is used now in an eccentric and unusual way, and that is why the receiver is somehow led to conceive a relationship between signifiers and signifieds.
3. The message can put *various levels of reality* into discussion (Eco 1968: 64): (a) the level of the technical-physical substance which the signifiers are made of; (b) the level of the differential nature of signifiers; (c) the level of denoted signifieds; d) the level of the several connoted signifieds; (e) the level of psychological, logical and scientific expectations the signs guide one towards: "Within all these levels a sort of system of homologous structural relations is established, *as if all levels could be defined according to a unique general code which structures*

them" (translated from Eco 1968: 64). Put simply, the aesthetic message is characterized by an extraordinary eclecticism, involving several textual dimensions, not only those predictably related to its meaning. It should not be forgotten that the message is ambiguous. Thus, for instance, it is not only the denotative level that should be taken into consideration, but also all the possible connotations.

2.4.4 A Unified Theory of Deception, Play and Aesthetics

Deceiving, playful and aesthetic behavioural patterns, in humans as in other animals, share a very semiotic characteristic: they are all strongly concerned with the idea of representation. All of them are semiotically representative *of*, *from*, and *through* reality. They represent it in several respects: they simulate it, modify it, manipulate it, and – to a certain subjective extent – improve it. The butterflies of the *Kallima* species are representative of a reality which at the same time is and is not their own reality. This is because their impressive resemblance with a leaf is inevitably their physical and relational condition (maybe cognitive as well, but not much is known about it), yet for the same reasons it is not. To make it clear, let us consider the very probable evolutionary reason why these butterflies look like leaves, i.e., the fact that they are prey for some birds. To resemble a leaf reasonably decreases the chance of being noticed by a bird. Thus, a fundamental aspect of Kallima's reality, to be a birds' prey, is "filtered" – and thus re-articulated – by the representation process that its physical aspect constitutes.

This concerns not only deception, but play and aesthetics as well. When two dogs of very different size and strength simulate a fight where the strongest pretends to be the weakest, they are in fact offering a representation of reality. And such relation S-A (simulation-activity) is not simply an inversion (the strongest is the weakest and the weakest is the strongest): it is sheer interaction. The strongest can play the weakest exactly because it is the strongest, and precisely because the players are representing reality (let alone that they can invert their role whenever they please).

In the case of aesthetics, let us take the example of those birds who practice vocal imitation for aesthetic purposes, like for instance the superb lyrebird *Menura superba*. When this beautiful bird collects, reproduces, and re-elaborates elements of the surrounding soundscape, it is once again semiotically representing reality, through a decontextualization-recontextualization process. This is how, for instance, the sound of a burglar alarm may happen to become the fascinating theme of a lyrebird song.

In fact, the idea of representation is not the sole characteristic that deception, play and aesthetics share. To mention one, all of the three traits, in their evolution, achieved a certain peculiar *independence* from their biological utility, the latter being from time to time so vague that one can be really persuaded that a given action was performed for its own sake. Such is clearly the case of aesthetics, as Morris (1963), Mâche (1992) and others have already proved in the field of non-human behaviour, but – surprisingly enough – that applies to deception and play as well.

Indeed, it is exactly this characteristic that blurs the boundaries between the two traits, as the above-mentioned ASL joking conversation between Koko and Francine Patterson demonstrates (Fig. 2.5).

The main hypothesis of this research is thus the following: *aesthetic behaviour in animals (including humans) can be explained starting from the development of deceiving and playing*. In other words, the hypothesis is that there is a semiotic continuum relating the three activities, according to articulations that will now be illustrated.

First of all, it is already clear that the triadic articulation of such relations can generally be analysed within a Peircean framework. Let us thus put lying, playing and aesthetics as triangle vertices, and see what type of relationship can be established among the three elements.

The first possible relation concerns the *perceptive dimension*, i.e., the role of the receiver of the aesthetic/playful/deceiving message. In this case, the relation is of a hierarchical nature: when dealing with an aesthetic message (let us say a work of art) it is the fictional element, the lie, that catches our attention first. The fact that the message in question is something different from reality, i.e., it "lies" about reality, is possibly the primary perceptive dimension. The playful component then follows, and finally the true aesthetic dimension is perceived. In other words, in dealing with aesthetic messages, one can properly speak of firstness, secondness and thirdness, in the Peircean sense, where lying is firstness, playing is secondness, and aesthetics is thirdness. In addition, as no hierarchical connection is suggested between aesthetics and lying, so that the latter appears as occupying the highest position in the hierarchy, one can deduce that the process of decoding the aesthetic message goes always and only through the lying-playing-aesthetics sequence, and not vice versa.

In order to clarify the whole point, let us reconsider Dalì's *Persistence of memory* in Fig. 2.7. The claim, here, is that the primary form of perception in this message is its "diversity" from, or alteration of, reality. Dalì's painting *deceives* to the extent that it presents situations and objects most of which are clearly non-existent in reality (like the weird face in the middle of the painting), or existent in quite different form (like the melting clocks), or finally non-existent according to our expectations (it cannot be excluded that there might be, somewhere, clocks or faces like those in the painting, but certainly they cannot be expected to be conceived and shaped like that). The next step is the perception of the picture as a "game": Dalì shows his ability to play with reality and its rules by "melting" an object that would not normally be subject to such a process. The observer, i.e., the receiver of the message, after being surprised by the "lie", is now surprised by the amusing and amused characteristics of the drawing. This is the secondness. It is after this, and after this only, that the trully aesthetic component (form, taste, sense of beauty etc.) can finally emerge.

Ça va sans dire, it was Peirce himself who underlined that the first level of perception, and often the second one also, can be easily be "taken for granted" in the receiver's mind. In other words, as far as this issue is concerned, the observer might not be aware s/he is going through a process like the one described, i.e., lying-firstness, playing-secondness and aesthetics-thirdness. Such sequence, it seems, is present cognitively, but not necessarily rationally. There is too much of our socio-cultural structures and super-structures in the decoding of aesthetic messages to even think we can be fully aware of this process: in approaching a painting like Dalì's, we already expect a work of art, we are probably in a place where works of art are expected to be displayed, we see a picture which is no more challenging to our conception of art than a post-modern work, for instance. Very little room is left for a rational awareness of the perceptive process we are going through.

One might thus wonder why this explanation started with an example like *The persistence of memory*, that evidently does not illustrate the point in the clearest of ways. The reason is that Dalì's picture is possibly in the middle of a series of cases where this triadic relation can be extremely clear, or apparently quite neglected. For instance, in the Italian pictorial movement named "Verismo", in the first half of 1800s, it appears that the surprising element, the Aristotelian *parà ten dóxan* dimension, associated with the idea of fictional representation, is quite differently articulated, if not seemingly absent. Close similarity with reality was indeed a programmatic intention of the veristi. The limited space of this section does not permit us to go at the heart of the matter of this example, which is still considered pertinent, although certainly not prototypical.

On the contrary, the focus will move on other art genres, whose articulation shall make the point a bit clearer. One of these genres is surely the caricature. Figure 2.8 shows a caricature of Umberto Eco: here is another type of representation, that soon catches our attention because of its "fakeness". There is a substantial difference between the cartoon-like character of the picture in the figure, and the photograph at its left: both are certainly representations of Umberto Eco, but the caricature tends

Fig. 2.8 A picture and a caricature of Umberto Eco

to alter reality much more than the photograph does (due to technical reason, and not to the intention of the photographer). On the contrary, we understand that the intention of the cartoonist was exactly that of altering the image of Umberto Eco. The cartoonist is possibly "lying" about Umberto Eco's physical appearance.

Perceiving this, our mind is then attracted by the playful elements of the representation (somatic traits are emphasized in an amusing fashion; a pencil in Eco's mouth jokingly reminds to his intellectual work, but also to his addiction to cigarettes; the triple crown, full of archaic alphabets and signs, recalls Eco's semiotic works, etc.). At this point, the properly aesthetic dimension is perceived. Being caricature a more functional type of art, explicitly aimed to amusement and/or satire, the playful-secondness dimension is in this example quite clear, and most probably rational in our decoding process.

Independently from the various examples discussed above, it is quite useful to stress that the context referred to here still aims to be the zoosemiotic one, therefore, to provide too species-specific human examples makes surely the issue more comprehensible, but also, on the other hand, risks being methodologically misleading. The concepts of art and aesthetics are thus to be kept within a precise threshold, which can be transpecifically recognizable, and thus definable according to the theoretical principles defined in the previous paragraph. Such principles support the thesis defended here with enough clarity, starting from the idea of biological utility of aesthetic behaviour and its semiotic relation with lying and playing activities. If we ventured into the post-modern problematics of the most recent definitions of art and aesthetics, we would make two very basic mistakes:

1) We would refer to an extremely culturally-shaped idea of art, where its relation with biology is made too foggy and confused;
2) We would risk, through a simplistic "all or nothing" policy, invalidating a whole discourse on the basis of instances which are of course interesting and deeply fascinating, but still controversial and statistically of little relevance.
   In other words, the fact that the conception of art is always (and thankfully) "pushed to the limits" by modern avant-gardes, does not deny, in fact reinforces, its theoretical status, and even more – for the sake of this companion's arguments – its biological foundations.

Having said that, the focus shall now be on a second analytical level of the triadic model proposed here. This time, the attention converges on the *articulation* of the aesthetic message, i.e., on the role of the sender. In this case one needs to investigate (i) how signs interact with each other, and (ii) how they depend on each other. The former shall be called the *level of interaction*, and the latter the *level of necessity*.

In the case of interaction, the situation seems the most egalitarian, i.e., all of the three elements are in reciprocal relation. Practically speaking, this means that there can be aesthetic and/or playful components in a lie, deceiving and/or playful components in aesthetics, and so on. Some examples shall be provided below.

In terms of necessity, the picture changes: the more articulated the message, the less the autonomy, and vice versa. What appears to be the simplest formulation, i.e., deception, is also the most solid, in the sense that it does not require the support of a playful and/or aesthetic component in order to exist. Lying can *also* be playful and/or aesthetic, but there can still be a lie without aesthetic or playful components. The food-call of the rooster, when deceiving, is neither aesthetic or playful: it is just deceiving. Conversely, play needs a deceiving component, but not necessarily an aesthetic one (in all cases, including the autotelic one, play is somehow fictional, and requires fictional elements, starting from fantasy). Aesthetics, finally, is the most complex of these elements, and yet the "weakest", because there is no way it can exist without playful and lying components to contribute to its formation. Indeed, aesthetics – the way it was defined in the last paragraph – implies characteristics such as (a) a manipulation of the code, a manipulation of the expectations of the receiver, some sort of biological advantage for the sender, that are typical and necessary for lying; (b) an active interaction sender-receiver, the presence of rules, biological advantages for the receiver also (all typical and necessary for playing); and finally (c) all those features which appear to be typical and necessary for the aesthetic activity itself, such as the multiple articulation of the code (or codes), a certain ambiguity – in the sense meant by Eco, a relation with the animal's biology that appears at least indirect, and so forth.

A very good anthroposemiotic example is in Marcel Duchamp's *LHOOQ*, universally known as the "Gioconda with moustache" (Fig. 2.9). This drawing is:

Fig. 2.9 Duchamp's *LHOOQ*

1) *A lie* – Duchamp is *lying* about Leonardo's picture and most probably about the real Mona Lisa – unless one proves she had a moustache. Our system of expectations is manipulated and so is the code. The "advantage", for the artist, consists in his drawing catching our attention and surprise, thus satisfying the

typical egocentrism of most works of art (if not all of them). The lie perpetrated upon Leonardo's original (more than the lie about the real Mona Lisa, who we have not encountered either in person or in any other portraiture) is what we perceive first in approaching *LHOOQ*: "Hey, this Gioconda has a moustache!" is possibly the first comment of those who have never seen Duchamp's work before.

2) *A game* – the sender (Duchamp) and receiver (ourselves) of the message (*LHOOQ*) establish a deal: to play a simulative/social/instrumental game with Leonardo's Gioconda. Amusing elements are added (the moustache of course, but also the title, which – read in French – sounds like "*Elle à chaud au cul*"), the "rules" of Leonardo's picture are decomposed and recomposed, thus new rules are established: e.g., the pole of attraction is not anymore Monna Lisa's enigmatic smile, and not even Leonardo's marvellous painting technique, but rather the appearance of the moustache, and the artistic outrage implied by it. Moreover, there are advantages for the receiver as well: s/he is interested and amused.
3) *An aesthetic message* – the code is multiply-articulated, and several levels of interpretation are involved (the idea of the outrage itself is a whole issue: not only Duchamp is jokingly offending the masterpiece par excellence, he is also doing it through a typical childish vandalism, i.e., adding unexpected somatic features to a portrait. This makes the anti-artistic gesture by definition a part of the artwork: so, is this art? Is this *all* art? And what is art? What is the boundary between art and joke, or between art and provocation? Is there any boundary? And so on). The ambiguity is almost total, and once again displayed on several levels: for instance, moustaches are conventionally associated with men: in the past – and still nowadays – there had been some discussion on the possibility that Mona Lisa was not a woman, or – on the contrary – that she was the portrait of an ideal woman; also, the title – in its evident sarcasm – creates even more ambiguity: what does it *really* mean? What is the relation with the picture, or with the original painting?

Finally Turning the attention back onto zoosemiotics, few arguments are more convincing than the case of the passeriform Satin Bowerbird *Ptilonorhynchus violaceus* (Fig. 2.10). As Karl von Frisch put it: "[. . .] those who consider life on earth to be the result of a long evolutionary process will always search for the beginning of thought processes and aesthetic feelings in animals, and I believe that significant traces can be found in the bowerbirds" (Frisch 1974: 244).

During the mating period of the year, the male specimens (different from females because of their more uniform and bluish plumage) build a sort of nest, called a pergolate or bower, that functions as a "garçonnière" for an imminent romantic meeting with the female. Apart from the building technique, that results in a sort of tower,[8] what is really amazing in pergolates is the precision and meticulousness

[8]Some other species of the same genus erect cone-shaped maypoles around a tree, that may end up being higher than a human being. Pergolates are so skilfully built that their discoverer – the Italian naturalist Ottaviano Beccari, around the end of nineteenth century – at first thought he was dealing with dollhouses built by creative young girls.

Fig. 2.10 A male bowerbird decorating his pergolate

with which the bird decorates the inside and the outside of his construction. All types of objects are utilized, from those already available in Nature (twigs, shells, flowers, etc.) to small human artefacts (fragments of bottles, pens, pieces of clothing, etc.). The objects are usually brightly-coloured, preferably dark-bluish, which seems to be a favourite colour for this species. Indeed, the entrance, and often other parts of the pergolate, is blue, for the bird paints it with a special paint obtained by chewing some berries, and distributed by means of a pseudo-brush (a chip of wood) held in his beak.

The choice of where to place each object seems not to be random: very often, the bird can be seen putting a given element in a given place, then taking a couple of steps backwards, carefully surveying the whole thing, then perhaps not content with the general effect, deciding to change the position of that element. Such a description may sound a bit anthropomorphic, but it is well-backed by the following, definitely more authoritative, one, provided by Von Frisch:

> Every time the bird returns from one of his collecting forays, he studies the over-all colour effect. He seems to wonder how he could improve on it and at once sets out to do so. He picks up a flower in his beak, places it into the mosaic, and retreats to an optimum viewing distance. He behaves exactly like a painter critically reviewing his own canvas. He paints with flowers; that is the only way I can put it. A yellow orchid does not seem to him to be in the right place. He moves it slightly to the left and puts it between some blue flowers. With his head on one side he then contemplates the general effect once more, and seems satisfied. (Von Frisch 1974: 243–244)

Even more impressive is that withered flowers seem not to be tolerated, for as soon as one dries up, the bird replaces it with a new one. It should be underlined that none of the decorative objects have structural importance to the whole construction; i.e., it stands either with or without them, the principle of construction being similar to that of ordinary nests, whose stability is basically due to interlaced twigs. The presence of such decorations is thus unnecessary in practical terms: withered flowers pose no danger, the collocation of each object does not affect the balance of the pergola, and the blueberry paint does not serve as a glue.

After finishing his job, the male stands awaiting the female, who, if attracted by the pergolate, heads towards him. As a grand finale, before starting his typical courting dance (another aspect deserving further analysis), the male welcomes the female by holding another fresh flower in his beak. The choice, the use and the qualities of the decorations and the actions performed by the bowerbird emphasize thus all the characteristics of lying and playing that were mentioned: from the manipulations of the code to the re-establishment of rules, and so forth. As far as the aesthetic component is concerned, let us now reconsider – with particular care – the theoretical model proposed in the last paragraph, and apply it to the case of these amazing birds:

1) It is hard to maintain that the whole process of construction and decoration of the bower, plus the courting ritual of the male, are "strictly" or "too directly" utilitarian behavioural patterns. Among birds, an almost endless series of different and more "economical" and "direct" courtship procedures are available. One can always say that the whole activity is typical of the instinctive-behavioural background of the species *Ptilonorhynchus*. Such a statement is however a bit problematic, since: (a) This observation clashes with the wide variability of pergolas, which change according to the individuals involved (there are much more and much less accurate bowers, with different features and decorations even when the same resources are available); (b) The same argument could be applied to humans, too, since courtship rituals are always more or less varied, but they are still courtship rituals; and simply (c) a semiotician should not be very eager to work with anti-semiotic concepts like instinct (see Chapter 3 of this companion). To accept and employ the notion of instinct relegates to non-semiotic a phenomenon which throughout this whole section has been defined and characterized in a semiotic sense.
2) Even considering the somewhat murky definition of concepts like "beauty" and "pleasure" (see previous paragraph), it must be underlined that the data collected strongly suggests a relationship between the bird's choices and an idea of beauty and pleasure. Not only does he decorate the pergolate with extreme care, but he makes adjustment to it if the final result is not satisfying. It may again be anthropomorphic to attribute to this bird forms of reasoning like "choice", "indecision", "correction", and "control". Yet, ethologists and even anthropologists are prone to think that way (e.g., Dobzhansky 1962), and any alternative explanations are in fact much less convincing;
3) That the signifier-signified relation is more oriented towards the former seems to be beyond discussion. It has been underlined, for example, that decorations have nothing to do with the stability of the building, thus, in terms of meaning, a withered flower has the same value as does a new one. If the bird decides to replace the former with the latter, it is definitely because he makes it a matter of form.

4) The female selects a pergola among the many offered by the males around her. It is important, ethologists say, for her to be impressed by the beauty and novelty of the selected pergola. At the same time, a pergola must be recognized as such, otherwise the female would not notice it at all. Here, then, is a typical, Aristotelian-framed aesthetic example: unexpected but believable, pergolas are *parà ten dóxan* and *katà to eikòs* at the same time;
5) All the factors for decoding the aesthetic message, as indicated by Eco, seem to apply to this case. For instance, one can point out (a) the various levels of reality (that of the bowerbirds, the way they "see the world", their Umwelt) involved in the construction of the pergola; (b) the levels of denotation (i.e., a place to mate and couple) and connotation (e.g., health conditions and resources of the male); (c) the physical resemblance between signifiers and signifieds (dark blue is the favoured colour, like the males' plumage; surely, it is evolutionarily convenient that the female appreciates this colour, finding it an aphrodisiac, so to speak, for it to be all around the pergola).

A couple of further considerations: firstly, Theodosius Dobzhansky observes that "It is undeniable that a well-decorated pergola gives the bird a pleasure that can only be defined as aesthetic" (1962: 215). Secondly, Finnish architect Juhani Pallasmaa suggests that since male bowerbirds "have a fairly inconspicuous plumage, the display function of the feathers of the male has been transferred to an elaborate external contruction" (Pallasmaa 1995: 53). It would not be a surprise if, sooner or later, a similar theory was applied to human aesthetics.

Concluding this section of the book, there is another reflection that this model – particularly the relation that we called "necessity" – suggests (a suggestion that should be handled with care, since it is just a supposition not supported by any evidence): there could be a phylogenetic relation among the three elements lying-playing-aesthetics, i.e., one feature could have been the evolutionary cause of the following one, with lying as the oldest element, and aesthetics as the youngest. This is a possibility that deserves a more rigorous investigation.

Chapter 3
Anthropological Zoosemiotics

3.1 General Problems Addressed by A.Z

Anthropological Zoosemiotics represents the zoosemioticians' contribution to the dicussion of the most critical topic in all studies concerning nature: the problem of the approach. Before starting any kind of discussion on other living forms, we should ask ourselves *how* to approach them. Is it possible, as humans, to study them in an adequate way? Are we able to observe other animals without being conditioned by a human interpretation of reality? Questions concerning the approach arise whenever zoosemiotic issues are discussed, and – as seen in Chapter 2, with the paragraph "Emic vs. Etic" – are unavoidable also at the level of Ethological Zoosemiotics. However, if in that case the question mostly requires biological (scientific in the strict sense) investigations, in AZ the problem appears in two clearly distinguished guises.

As already anticipated in Chapter 1, there is one form of AZ called *communicational* (cAZ), which regards areas of inquiry like applied zoosemiotics and interspecific communication. In this case, the question of the approach takes mostly the same shape it has in EZ: it is the biological sciences that take charge of the most important issues. This type of AZ therefore aims again to be of the scientific type.

The second sub-category, which is named *significational/representational* (srAZ), concerns the vast areas of myths, tales, allegories, ethics, and systematic classsifications, in which non-human animals are for humans mostly a philosophical problem. And philosophical is thus also the main attitude required to a zoosemiotician to deal with these issues. SrAz, therefore, represents the purely humanistic side of zoosemiotic research, the area wherein the philosophical value of doubts and speculation acquires a crucial, and unavoidable, importance.

Although, in the form presented here, the development of AZ, particularly srAZ, is very recent and took a recognizable shape only in the twenty first century (pioneered in Martinelli 2004b, and developed in Martinelli 2006a) it is once again Sebeok who must be thanked for raising the issue:

> How man communicates with animals, and vice versa, has, so far, been of marginal concern to both semiotics and ethology [. . .]. Yet in all arenas of life the relation between man and animal – from protozoan to primate – is now decisive. Wherever they meet, whether

D. Martinelli, *A Critical Companion to Zoosemiotics*, Biosemiotics 5,
DOI 10.1007/978-90-481-9249-6_3,

> man is an animal's scourge or its prey [. . .]; whether one is a parasite of the other [. . .]; or whether one species accepts the other as a conspecific [. . .], or, to the contrary, as an inanimate object [. . .], the liaison implies that each must, perforce, learn, if not totally master, the essential elements of the reciprocal's code. After establishing loose contacts with an animal, man, if he so wills it, can follow up by taming it, a process that can be defined as a systematic reduction of flight distance achieved by conscious manipulation of the animal's code.
>
> A tame animal can then be subjected to the purely synchronic process of training, and that in one of two complementary ways: apprentissage or dressage, i.e. for scientific testing or for performing in exhibitions, the two being polar opposites [. . .] The two procedures are distinguishable in at least two respects: the semiotic character of the sign which initiates the requisite action (unmarked vs. marked), and ,the degree of emotional intensity coupling the interactants (minimal vs. maximal). Next, if economic circumstances dictate the planned development of form with certain properties man deems desirable, he superposes a diachronic dimension: by selective breeding, or alteration of the genetic code, domestication (with flight distance approaching constant zero) will ensue. The final, alas irreversible, step may be the cutting off of all further communication with members of the species —the persecution, then eventual extermination, of its feral ancestors. (Sebeok 1990: 45)

As Sebeok himself always admitted, the main forerunner of this kind of cAZ, was the Swiss zoologist Heini Hediger, in such seminal studies as *The psychology and behavior of animals in zoos and circuses* (1968), where he warns that

> [. . .] man should learn far more than anywhere else about the most important ways in which [animals] express themselves, and so be able to form a picture of their internal condition from these external signs. [. . . T]he greatest attention must be paid above all to the animals' expressions in the zoo, as these provide an important factor in determining the animals' mental and physical well-being. Characteristic changes in facial expression, and in the body, are of frequent occurrence in most animals. (Hediger 1968: 140)

In Chapter 3 of this book, the focus shall be on the two main topics of AZ that semioticians have already devoted their attention to (the concept of "animal" in the semiotic perspective, analyzed by Sebeok, and the *super*-critical topic of "Language"), while room will be left in the "Chapter 5" for a new topic in the field (that is, the ethical agenda and implications of AZ).

3.2 The Concept of "Animal" from the Semiotic Perspective

It is appropriate to start the discussion on AZ precisely from the reflections anticipated by Sebeok, in several of his works. That is, a semiotic definition for the term *animal*, which is also a most prominent topic in the subcategory of srAZ. To say that this word is quite hard to define is probably the understatement of the year. Much research has been done on the matter (see, for instance, the highly interesting Ingold 1988), but the discussion is still very lively.

However trivial, the first step suggested here is to look up the word "animal" in a dictionary. According to the Webster's Revized Unabridged Dictionary an animal is *n.*

1. *An organized living being endowed with sensation and the power of voluntary motion, and also characterized by taking its food into an internal cavity or stomach for digestion; by giving carbonic acid to the air and taking oxygen in the process of respiration; and by increasing in motive power or active aggressive force with progress to maturity.*
2. *One of the lower animals; a brute or beast, as distinguished from man; as, men and animals.*

adj.

1. *Of or relating to animals; as, animal functions.*
2. *Pertaining to the merely sentient part of a creature, as distinguished from the intellectual, rational, or spiritual part; as, the animal passions or appetites.*
3. *Consisting of the flesh of animals; as, animal food.*

The American Heritage Dictionary of the English Language provides the following entry:

n.

1. *A multicellular organism of the kingdom Animalia, differing from plants in certain typical characteristics such as capacity for locomotion, non-photosynthetic metabolism, pronounced response to stimuli, restricted growth, and fixed bodily structure.*
2. *An animal organism other than a human, especially a mammal.*
3. *A person who behaves in a bestial or brutish manner.*
4. *A human considered with respect to his or her physical, as opposed to spiritual, nature.*
5. *A person having a specified aptitude or set of interests: "that rarest of musical animals, an instrumentalist who is as comfortable on a podium with a stick as he is playing his instrument" (Lon Tuck).*

adj.

1. *Relating to, characteristic of, or derived from an animal or animals:* animal fat.
2. *Relating to the physical as distinct from the spiritual nature of people:* animal instincts and desires.

And so on (other dictionaries do not present significant variations). In each entry a reference to the Latin origin of the term is indicated as well, i.e. the word "animale", which stands for "living", "provided with soul" (soul is *anima*, in Latin).

Besides several theoretical implications, which will be discussed later on, these authoritative definitions emphasize the existence of two different semantic typologies associated with the term: a denotative and a connotative one. In other words, in the first case the scientific meaning of the word is stressed, i.e. the animal as a multicellular organism belonging to the animal kingdom, thus different from plants and other organisms. In the second case, a whole series of more or less metaphorical semantic nuances are presented, with associations to the most diverse contexts.

Starting from the denotative dimension, it may be useful to call to mind the taxonomic classification of the so-called "kingdoms" of the biosphere, which Sebeok himself mentions in some of his works (rather extensively, for instance, in 1998: 61–67). In his opinion, a good classification of living systems is not only a matter of (mostly evolutionary) levels, but also of *types*. Types are essentially determined by the modes of nutrition. Such a criterion leads to the articulation of three complementary macroscopic entities, named super-kingdoms:

1) *Plants*, i.e. producers. Through photosynthesis, they obtain nutrition from non-organic sources;
2) *Animals*, i.e. ingestors (so called for they ingest food in their body, and there it is digested). They eat other organisms, and can be divided into *herbivores* (ingestors of plants), *carnivores* (ingestors of animals) and *omnivores* (ingestors of both plants and animals);
3) *Fungi*, i.e. decomposers. They do not ingest, but rather decompose externally their food through apposite enzymes, and eventually absorb the molecules that result from this process.

Since this classification, as probably every other classification, in the end excludes some (actually, many) life forms, Sebeok suggests applying the principle of the evolutionary levels and defining the remaining forms (the inferior ones) as non-plants, non-animals and non-fungi. In detail:

1) *Protista*, i.e. micro-organisms lacking embryogenesis, are heterogeneous from the point of view of nutrition. An example of protista are algae;
2) *Prokaryotes*, i.e. monocellular organisms (like bacteria). Nutritionally heterogeneous like protista, prokaryotes are however incapable of ingestion.

It shall be useful to underline the fact that there is no scientifically-acknowledged taxonomic classification that refers to an autonomous category named "human beings" (not that this should surprise anyone), and moreover all classifications include human beings in the animal kingdom. A suggestion we are given is thus that, scientifically and biologically speaking, the human being *is* an animal. Ever since Darwin, with the exception of a few fundamentalist religious and/or ultra-conservative groups, no member of the scientific community has dared deny that human beings are animals. However, a totally different issue is to see whether this scientific assertion is acknowledged within the cultural dynamics of one or more human communities. In fact, the issue is highly complex.

This is why a semiotic analysis of the various connotative dimensions of the term animal is useful. It is at the connotative level where most of the semiotic (significational/representational) human-other animal relations occurs. Sebeok (1998: 67) is certainly right in saying that the animal easily becomes a cultural object, a kind of information that – as a result of a given system of socio-cultural values – determines a gap, if not a straight reconfiguration, between signifier and signified, a concept that MacCannell (1976: 110) properly names *countersign.* In this sense, synonyms of "animal", like "beast" or "brute", are also of full significance. Although it must be pointed out that the following list is certainly incomplete, at least eight connotations of the word "animal", as used in western cultures, can be identified:

1) *Animal as any other animal except humans* – This is of course the most common use of the term (in fact, it is probably more common than the denotative level itself). Whatever the context, including specific fields like the biological or the zoological one, people often refer to the term "animal" as a semantic whole that clearly excludes humans. In fact, most of the time, the term is explicitly chosen to establish comparisons: "the human-animal relationship...", or "we are humans, not animals ... ".
2) *Animal with reference to exclusively human characteristics* – However weird it may sound, the term "animal" is often used in reference to supposedly species-specific aspects of the human being ("the human being is a political animal", "human beings are thinking animals" etc.).
3) *Animal as a description of a particularly uncivilized human being* – A typical epithet addressed to bad-mannered and/or uncultivated people is exactly "you are an animal" or "he is such a brute".
4) *Animal as an indicator of aggressive and/or violent attitudes* – Quite normally, especially in the mass-media, serial killers, sexual maniacs and other similar characters are described as "wild animals", "cruel beasts" and so on.
5) *Animal with reference to natural and/or instinctive attitudes* – Expressions like "animal instinct" or "animal strength" are often used to describe certain types of behavior that appear ancestral or anyway pre-cultural. The song *The Animal Instinct*, performed by the Irish rock band Cranberries, refers, for instance, to a wish for escape and freedom. "Animal instinct" is also an expression often associated with mother-child interactions.
6) *Animal with reference to remarkable physical (mostly sexual) performances* – As a consequence of an equal-but-morally-opposite principle of point 4 of the present list, a human being (mostly, a man) who appears as particularly gifted in physical-erotic terms can easily be nicknamed "animal" (particularly erotically-gifted women, for some reason, tend to be mostly identified with felines like tigers or panthers).
7) *Animal with reference to the ability of a human being to adapt to a certain context* – A particularly gifted actor or singer is often defined a "stage animal"; someone who is professionally keen to work during nighttimes is often referred to as a "night animal"; and so on.

8) *Animal in the sense of a zoomorphic non existent creature* – Such is the case of unicorns, centaurs and all those mythical creatures that result from the narrative elaboration of the myth. In certain times and places, they are or have been considered *existent,* thus, in semiotic terms, they do in fact exist as objects.

If we consider the ethical nuances of these eight connotations, it seems quite clear that the first four points imply a negative consideration of non-human animals, while the following three appear as mostly positive (the last case, that of mythical animals, is probably to be considered of a neutral type, for the association can be either negative or positive, or ambiguous, depending on the case and the creature described). In other words, in the first four cases (a) distances (mostly cultural) are established between the human being and the non-human animal, in quite a strong fashion; and/or (b) conventionally negative and unacceptable behavior is associated with non-human animals, rather than human beings, almost as if the latter could act that way only when lacking or losing their properly human (and humane) characters. In the following three cases, on the other hand, (a) conventionally positive aspects of the extra-cultural dimension of human behavior are underlined; and/or (b) certain scientific (thus, in part, denotative) aspects of the concept "animal" are underlined, through the use of metaphors, like in the case of adaptation in point 7 of the list (e.g., "night animal").

Another issue that arises from this list but is apparently paradoxical, is that *animal* is a very general concept in cultural contexts (as a single community of people can be, for instance), thus corresponding to more or less socially-shared meaningful unities; while it is an extremely detailed and strict concept when referred to wide intercultural contexts. Any person or any group of people perceive the animal according to a set of criteria and particulars; at the level of inter-cultural comparison, any or several of the particulars accepted by one group may be rejected by another (some, for instance, make a clear distinction between animals, fish and birds, animals being the terrestrial ones). Both groups, however, perceive something that is considered an "animal". Within each community (and, by this word, we also mean groups *within* the same culture), *animal* may be a general word, but, cross-culturally, it acquires a specific meaning. This aspect, although not exclusive of the term "animal",[1] in any case carries a bunch of interesting implications (starting from the difficulty, if not impossibility, of defining what an "animal" is in a culturally universal sense.[2] Moreover, "animal" is also – and mostly – a concept emerging from a textual and contextual interaction: trivially, mussels are animals for the zoologist and "frutti di mare", i.e. sea food, for the gourmet[3]). However, most notably, it seems clear that "animal" is not a term easy to pin down in a dictionary definition,[4]

[1]To mention one, a similar discourse was made about the term "music" (Martinelli 2002: 147–63).

[2]Even in strictly taxonomic terms, consensus is not global. Some cultures, for instance, refer to cetaceans as fish, rather than mammals.

[3]Note that "frutti" is the Italian word for "fruit".

[4]In Aristotelian terms, such a definition would be constituted by the so-called *necessary and sufficient features*, and it would be characterized by a closed category of the term, exactly like in a

except for its strictly denotative dimension (and, anyway, we have seen that there are exceptions in that case as well).

"Animal" is then to be included in a series of concepts, like "music", "game", "freedom", that demand a definition by prototypes. Starting in the mid-1970s, a theoretical school nicknamed the *Roschian Revolution* began to enjoy greater popularity (see Rosch et al. 1978). Eleonor Rosch is quite a famous psycholinguist, and her revolution derives from the categorization principles she postulated, which indeed provided quite a crucial change to classical Aristotelian-framed analyzes of meaning.

Before discussing Rosch, one needs to mention a better-known figure, the philosopher Ludwig Wittgenstein (particularly 1978). Wittgenstein was one of the first to question Aristotelian categorizations, which at that time were absolutely dominant in philosophical and logical speculation. Such categorizations were based on the idea that concepts can be explained according to necessary and sufficient features. Together with these, a similar type of categorization, called componential analysis, gained much credibility. Componential analysts postulated that the meaning of a word could be established according to a series of meaning components, which are not exclusive to a single word, but shared with others.

The meaning of the lexeme *mother*, for instance, can be divided into the following semantic components: HUMAN, FEMALE, ADULT, ANIMATED. The feature HUMAN distinguishes the meaning of the word *mother* from the meaning of the word *elephant*, whereas FEMALE distinguishes the meaning of the lexeme *mother* from the meaning of the lexeme *man*, etc. By means of the componential analysis, words are divided into binary, oppositional components, stressing the complementary features of the world.

Wittgenstein considered componential analysis interesting but in many respects inadequate. In order to prove his point, he constructed the famous example of the concept of "game". According to Wittgenstein, a concept like "game" cannot be defined by necessary and sufficient features, since it consists of a complex web of cross-referencing and overlapping similarities, both in general and in particular. Indeed, if game A shares some characteristics with game B, and B with C, but not A with C, and all of them are still "games", then it is easily demonstrated that there are no traits without which, or thanks only to which, one can or cannot speak of a "game". In short, the boundary between game and non-game is not clearly marked: "What does still count as a game and what does no longer? Can you tell the boundary? No, you can draw one; for none has so far been drawn. (but that never troubled you before you used the word game.)" (Wittgenstein 1978: 67)

In addition, a concept like this describes not only the structure within a given category of meaning, but also the way this latter should be learnt: "game" is not a concept that can be learnt by distinguishing it from a "non-game". What is possible is to categorize it on the basis of examples.

---

dictionary: I call "animal" *always and only* a concept provided with the features *x, y* and *z*. Just *y* and *z* would not be enough, and *w, x, y* and *z* would be too much.

Wittgenstein did not mean to exclude componential analysis in all cases (e.g. a category such as "triangle" *can* be distinguished from the category of "non-triangle"). Yet, categories such as "game" are definitely more blurred in their definition. The point is to understand what elements actually affect the determination of a word's meaning. As Wierzbicka points out (1989: 7), "Language doesn't reflect the world directly: it reflects human conceptualization, human interpretation of the world". Not only does the meaning of a word reflect the relation between language and reality, it also reflects the way in which reality is structured and categorized in human cognition.

At this point, Eleonor Rosch's theories come into play. The best possible explanation of her ideas is by means of a game. If we ask a friend to name the bird species that first comes to his/her mind, then most probably our friend will recall a sparrow, or a nightingale, a canary, an eagle, a hawk, etc. Very seldom do we think of penguins or ostriches. The game usually works, and, according to Rosch's studies, those who immediately think of ostriches and penguins are very rare indeed.

Consider two birds, an ostrich and a sparrow: the two creatures have some features in common, like ANIMAL, LAYING EGGS, FEATHERS, BEAK, WINGS etc. However, as in Wittgenstein's example of *game*, ostriches and sparrows, although both members of the category "bird", do not share the same amount and type of features: e.g., both ostriches and sparrows have wings, but only sparrows fly. Consequently, the above-mentioned features cannot be called "necessary" ones. In addition, the characteristics featured above are not sufficient for the category "bird", since they do not allow a clear distinction between other categories; snakes, for instance, lay eggs as well. Thus, it seems impossible to find necessary and sufficient features to cover the sum of all potential birds in the world. In short, componential analysis is not applicable for the categorization of natural kinds.

In Rosch's opinion (1978: 27), human categorization "should not be considered the arbitrary product of historical accident or of whimsy but rather the result of psychological principles of categorization". This means that any process of categorization, starting from linguistic categorization, is first of all determined by human cognition. Since human language seems to have a special tendency for categoriality, there must be a good reason why humans categorise the world. Within this framework, Rosch et al. (1978: 28) points out that "the task of category systems is to provide maximum information with the least cognitive effort". In contrast to the traditional theory of categorization, Rosch maintains that we assign an object to a category not by verifying a set of criterial features, but by comparing that object to a prototype of the category. The prototype is considered the best exemplar, which is generally associated with the category concerned.

Rosch made several experiments with the category of birds. In one of these, she investigated how long it took her students to verify category membership. Students were asked to press a button to indicate true or false in response to a list of statements, such as "A penguin is a bird" or "A sparrow is a bird". Here, response times were generally shorter for more representative exemplars, such as sparrow, than for less good exemplars, such as penguin.

Now, similar considerations can be made with *animal*. To set the limits between animal and non-animal is – in most cases – a purely subjective matter, even though there might be an almost absolute consensus in certain ones. The concept is far more structured as "game" than as "triangle". Animal is not a fixed concept, and the category "animal" carries a series of features which different examples of animal do not necessarily share. In addition, every culture and every individual categorises the concept, according to some central prototypes: a mussel is probably far less central than a tiger, and so on.

According to Rosch, the five main features of prototypicality are:

1) A category has a prototypical structure. A prototype is regarded as the most representative member and is the central entity around which all other members are organized. Thus, some members are *more members* than others. Who knows how many animal species can be taken as prototypes, but we can be sure that pets (particularly dogs and cats) have a prominent role in this sense. And we can also be sure that mussels and humans (for different reasons) are much less "animals" in people's minds.
2) Prototypical categories cannot be defined by means of a single set of necessary features, since members do not share the same amount and type of features. For example, all existent living beings provided with wings are animals, but not all animals have wings.
3) Prototypical categories exhibit a family resemblance structure; their semantic structure can take the form of a set of clustered and overlapping meanings. This is particularly true with the concept "animal", if we only consider the amount of information produced in association with the term.
4) Prototype categories exhibit degrees of category membership; not every member is equally representative of the category. In contrast to the classical approach, the members do not have the same status. The question of membership is not a question of either-or, but a matter of more or less. As already suggested, a mussel *is* an animal, but it is probably *less* of an animal than a tiger.
5) Prototypical categories are blurred around the edges: i.e. one category merges gradually into another category. The mammal-fish confusion concerning cetaceans, or the mammal-birds one regarding bats is quite representative in this sense.

3.2.1 Forms of Human- Other Animal Interaction

Once one has explained what an animal is (or, in fact, once one has explained how hard it is to say what it is), the next step is to consider the various forms of relationship that human beings establish with other animals, and how they are semiotically articulated. As with so many other issues, it was still Sebeok who gave a significant input to this type of inquiry. The task shall now be to shape more systematically his reflections, trying – when possible and necessary, and with due humility – to update his analysis, in the light of the most recent technological and/or cultural changes.

First of all, it will be interesting to establish a general set of motivations that push human beings to interact with other animals, whatever form these interactions may assume. Without any particular order, at least eight typologies of motivation may be distinguished:

1) *Adaptation* – The human adaptation in, or – more and more often – the anthropization of a given environmental context, always implied and implies a meeting/collision with other animal species. Some of these species are "useful" to the anthropization, others are driven away, or even eliminated (partially or totally), others are needed to a greater degree.
2) *Progress* – Scientific and/or technological. Human beings believe it is useful to exploit other animals in order to support their own evolution. Both scientific and industrial research make a wide use, in different ways, of non-human species.
3) *Work* –Non-human animals have always (although in constantly decreasing numbers) been part of the human productive cycle, or at least they support it.
4) *Needs* – Primary or not, real or not, many human needs are fulfilled by, or thanks to, non-human animals. We eat animals, wear them, and so on.
5) *Pleasure* – Many animals are exploited (which means killed or imprisoned, as well as just used as graphical models) in order to satisfy purely hedonistic wishes. The list includes pets, zoos, cartoons, toys and many others.
6) *Tradition and culture* – Religion, myth, folklore, literature, art. . . it is practically impossible to find even one human community that does not interact with other animals from a cultural point of view. Such an interaction is either abstract (e.g. tales) or concrete (e.g. sacrifices).
7) *Philosophy and research* – Almost a special case of point 6, the scientific-philosophical relation with other animals is established in order to know more about them, or know more about ourselves as humans (up to the point of establishing our identity on the basis of this relation, as seen with such notions as the "semiotic animal" and others);
8) *Daily life* – Even though this one may appear as a super-category that includes all the other points in the present list, it is still worthwhile pointing out that non-human animals are often full part of our life, independently from our choices and needs. They are part of our surrounding landscape and our actions: "While I am writing in this very moment, I can see (a) a notebook whose cover is a drawing of two dolphins; (b) the advertisement of a company whose logo is a rabbit; (c) a series of books of animal-related studies; (d) a pair of leather shoes; (e) a moose soft-toy; (f) a few CDs of animal sounds; (g) a cartoon-cat as a virtual assistant of my Microsoft Word program; (h) the reproduction of a Magritte's painting representing a dove, whose body is coloured like the sky; plus many other things I am probably not able to notice right now, for they are so much part of my home environment, that my perception tends to 'take them for granted'" (Martinelli 2006a: 224–5).

Having established the motivations, let us see *what* these interactions are in detail. As already said, Sebeok (for instance in 1990: 107, or more extensively in 1998: 67–73) was the first to deal with this issue from a semiotic perspective. To his list, consisting of eight categories, more suggestions can be added, basically as a result of the different times we are living in nowadays (with consequent new forms of relationship). The additions comprise the last four points of the following list:

1) The human being can be a *predator* of other animals, and vice versa – By this, Sebeok refers to all the reasons why a human being exploits, through killing it, the life of another species: food, clothes, research, etc. At the same time, the human being also happens to be a *prey* of certain non-human species: such is the case of sporozoites (those protozoa responsible for the malaria disease, transmitted to humans through the mosquito *Anopheles*), but also of those occasional instances when specimens of different species attack humans in order to eat them or – more often – defend themselves.
2) The human being can also be a *partner* of other animals – Partnership relations between humans and other animals date back to ancient times. Sebeok mentions four particular typologies of partnership (1998: 68–9): (a) host-guest relation, as in the case of pets; (b) mutual dependence, as in the case of bee-breeding or dogs for blind people; (c) sexual relation, as in the case of certain communities or certain pornography; and (d) relations aimed at social facilitation, as in the case of pets used to facilitate the approach with other people (this is for example the case with children who can be more easily approached using a cute dog or cat as an excuse).
3) The human being can exploit other animals in *sports* and *hobbies* – This category includes both very cruel examples like bull-fighting (corridas) and circuses, as well as very innocent ones like feeding pigeons or bird-watching.
4) The human being can be a *parasite* of other animals, and vice versa –This category includes cases like reindeer-breeding, where humans exploit every single resource offered by the species in question; or – in the opposite relationship – cases like the hair-follicle mite *Demodex*, fleas, lice, etc.
5) The non-human animal can establish relations of *conspecificity* with the human being – This happens mostly with pets, who tend to identify their human host as part of their own species (or, more rarely, themselves as part of the human species). This phenomenon is known as zoomorphism, but – as discussed later in this book – one has reasons to be sceptical about the use of this term, as referred to in this and many other contexts. We do not have clear proof of whether humans can see other animals as conspecific, as well: however, such could be the case with children, for instance, or with humans bred in the wild by some non-human community, in the same way that inspired Edgar Rice Borroughs' famous character of Tarzan;
6) The non-human animal can establish relations of *insensibility* with the human being, and vice versa – Some non-human animals may consider human beings as an inanimate part of their Umwelt (Sebeok mentions the case of birds who perch on a human body as if it were a tree). In fact – one shall add to Sebeok's formulation – it occurs even more often that human beings tend to be physically

insensible towards other animals. This is mostly due to ignorance, so that some species (especially marine ones) are often mistaken for plants or others.

7) The human being can *domesticate* other animals – Domestication is defined as the "reduction or possibly total elimination of the reaction of escape in a non-human animal from a human one" (Sebeok 1998: 70). Taking the definition in its general form, it is fair to state the opposite as well, but one should be able to establish how active is the animal's role in reducing or eliminating the reaction of escape in the human being.
8) The human being can *train* other animals, and vice versa – The non-human animal is more or less forced to learn a given behavioral pattern on the basis of quite drastic options (life if s/he learns, death if s/he does not, grant/punishment, etc.). According to Heidiger (1968: 120), there are two types of training: (a) *apprentissage*, i.e. laboratory training; and (b) *dressage*, or circus-like training. The difference consists basically in the typology of relation established between trainer and trainee: almost inexistent in the first case, quite intensified in the second case.[5] Although he did not mention it, it is fair to guess that Sebeok had nothing against the idea that human beings can also be trained by other species. In domesticated animals, for instance, it is quite clear that the relation is conducted by both parties in terms of reciprocal and constant training. People who takes their dogs outside for the daily walk precisely when the dogs approach them holding the leash in their mouth, are rather evidently humans trained by dogs dealing with dogs trained by humans.
9) The human being can also be a *manipulator* of non-human animals. It is quite probable that Sebeok took for granted this option within the first and/or the third category of this list, but – especially in the light of the most recent technological progress – it seems now fair to say that "manipulation" is a concept that deserves a section of its own. Cases like *Dolly the sheep* suggest that human beings now have quite a strong control of the genetic heritage of non-human species (in fact, of the human species as well). However, without only referring to very recent times, the domestication process of many species occurred and still occurs through a constant manipulation/selection. Finally, even single parts of characteristics of the non-human animal (especially physical and somatic ones) are manipulated for essentially aesthetic purposes (ear-pointing, tail-cutting, etc.).
10) The non-human animal can be a *source of knowledge* for the human being, and vice versa – The already-mentioned zoomusicology is a good example of (non-invasive, in this case) exploitation of non-human animals aimed at the enrichment of the knowledge about our Umwelt (in the specific case, origins and development of the musical phenomenon). Invasive practices like vivisection also fall into this category. The opposite is acceptable as well: the presence

[5] Such an affirmation implies Sebeok's conviction that scientific experiments involving a trainer-trainee emotional relation are not serious. In this book, the issue will actually be approached in a different way.

of and the danger represented by cars, for instance, is a source of knowledge for many species who live in anthropized contexts. To cross a street, for a dog from the twentieth century, requires quite a different attitude than the same action, as performed by a dog of the nineteenth century. Obviously, the presence of a source of knowledge within a given Umwelt implies important changes in the construction and the representation of that Umwelt. Similarly,

11) The non-human animal can be a *significational source* for the human being, and vice versa – The transformation of the non-human animal into an allegory, metaphor, myth, etc. is certainly a significational process, whose source (not the sole one, though) is the animal in question. However unexpected this may appear, such a process is probably not humanly species-specific, i.e. other animals, too, can in principle use humans as significational sources. The superb lyrebird *Menura superba*, who – among many others – imitates human sounds in order to compose his personal "serenade" for the female partner, is in fact using the human being as a significational source in every respect. Finally,
12) The human being can be *defender/protector/promoter* of other species, and vice versa – Such is the case of animal rights organizations, wildlife protection, and so on, but it is also the case of different individual attitudes. Non-human animals can also be at least defenders and protectors of the human being: typical is the case of a dog barking at or attacking an individual who is potentially dangerous for his/her human host.

Another form of semiotic investigation to discuss in this paragraph refers to the Greimasian tradition, through the re-adaptation of the semiotic square (Greimas 1987: xiv, 49). The human-other animal relationship, as any other relationship, is articulated in temporal-spatial units, i.e., in any relationship there is a *where,* a *when,* a *where from*, a *how long*, and so forth. Since it seems that long-term temporal units are applicable to the human being only (at least, we do not have proof of *historical* awareness in other species), the following model will be considered in a unilateral sense. In other words, only the human approach to other animals, and not the other way round, will be classified. In speaking of space and time, thus, the reference is to human spaces and human times, while – when speaking of non-space and non-time – the reference is to non-human (i.e., non-anthropic or non-anthropized) spaces, and to the absence of temporal continuity (i.e., human-other animal relationship based on extemporaneity and/or discontinuity). The result is the square, wherein the following relations can be detected:

- *Space/Time*: most forms of domestication, manipulation, training and significational exploitation belong to this category. The relationship between humans and cats, for instance, was articulated continuously in time (producing significant genetic and behavioral changes in the species *Felis catus*), and used – in most cases – typically anthropic spaces, such as houses, farms, etc.
- *Space/Non-Time*: The notion of non-time (or extemporaneity) is to be interpreted in two ways: on the one hand, it refers to relations that are circumscribed in time, like those with extinct species, for example; on the other hand, it emphasizes how

certain relations, although continuous in time, are strongly related to the context in which they occur, so that neither the human nor the non-human individual have an idea of "precedent" in their genetic or behavioral repertoire. In other words, one thing is to meet a dog, another is to meet a shark. Examples for the space/non-time category are the recent trend of having exotic animals as pets, certain forms of amusement like zoos or circuses, etc.
- *Non-Space/Time*: Many forms of relation, e.g., prey-predator interaction (where humans are normally predators), occur in more or less non-anthropic areas, but – at the same time – they have been continuous in time, and have ended up modifying the behavior of certain species (for instance, certain areas that birds would use as migration resting places, are not used anymore by those species because of the presence of hunters).
- *Non-Space/Non-Time*: Such is the case of any extemporary relation occurring in non-anthropic areas between humans and – normally – wild species (including the not-so-pleasant meeting with a shark). This category should include most forms of defence/protection/promotion of non-human species, from the prohibition of hunting for endangered species (which is evidently circumscribed in time), up to the famous rammings performed by Greenpeace activists against whaling ships.

3.3 Interspecific Communication and Language

Turning attention to the field of cAZ, therefore to that field where an actual semiotic interaction between humans and other animals occur, and thus the research tools of natural sciences are again preferrable to those of humanities, there is one of the most problematic questions of zoosemiotics, in fact of semiotics in general: is language a species-specific human feature? And, most of all, *how* species-specific is it? Although this is a very open question in nearly all disciplines that deal with language and communication, there is a certain consensus, among semioticians, that there is a threshold which qualitatively separates human language from non-human forms of communication.

It is not one of the aims of this book to go through the (several) definitions of language and the descriptions of its features, if not in terms of a very general introduction to the subject: after all, this is not specifically a topic of zoosemiotic concern, but rather one of the basics of general semiotic (and linguistic) knowledge. This section will instead focus on the entire, specifically anthrozoosemiotic, discussion on attempts to teach human language to other animals, with the explicit goal of re-opening the question of language and Interspecific Communication Experiments (ICE, from now on) from a zoosemiotic point of view, starting from the realization that crucial aspects of the issue have so far been underrated or even missed. In particular:

- The specifically semiotic achievements of the ICE;
- The connection between these achievements and the notion of language;

- The consideration of ICE, as research carried out on individual non-human subjects, with specific background, attitudes, etc., in relation to the results achieved;
- The several implications of ICE, in terms of methodology, outcomes and, in a general scientific sense, ethics.[6]

Most of the reflections here proposed are – to say the least – controversial, and exactly for this reason they fully qualify for a "critical" approach: the discussion on these matters certainly deserves a more thorough scrutiny, and – perhaps – a different approach altogether.

The semiotic discussion of language, and its supposedly human species-specificity, relies – for the most part – on Sebeok's reflections about language itself, and about the ICE carried out by several psychologists and primatologists in the second half of the twentieth century. Sebeok's arguments, later reinforced by many of his followers, can be summarized in the following points:

1) *Language is a species-specific human device. The ability of language acquisition is a result of the mental capacities of the human being.* "The word 'language' is sometimes used in common parlance in an inappropriate way to designate a certain nonverbal communicative device. Such a usage may be confusing in this context where, if at all, 'language' should be used only in a technical sense, in application to humans. Metaphorical uses such as 'body language', 'the language of flowers', 'the language of bees', 'ape language', or the like, are to be avoided" (Sebeok 2001b: 14);
2) *Before being a communication tool, language must be considered a modelling system, specifically, the, very human, secondary modelling system.* "Language is, by definition, a secondary cohesive modelling system providing humans with the resources for extending primary forms ad infinitum. [. . .] From a biosemiotic perspective, the language code can be defined as the cohesive system providing the modelling resources for converting what von Uexküll called 'concrete living existence' into 'active plans'. (Sebeok and Danesi 2000: 108)". Identifying language with communication was, according to Sebeok's literary words, "a vulgar error" (1991: 71);
3) *The difference between language and non-human forms of communication (and/or modelling systems) is qualitative.* Quite simply, humans have language, other animals do not, rather than saying that language is a more refined form of communication (or modelling system) as compared to other animals' devices, as a Darwinian approach might suggest. It is precisely its nature as secondary modelling system that makes language a uniquely human feature: "All the animals paleontologists classify generically as *Homo*, and only such, embody, in addition to a primary modelling system [. . .] a secondary modelling system, equivalent to

---

[6]As already mentioned, a specific discourse on practical ethics will instead be proposed in Chapter 5 of this companion.

a natural language. The difference amounts to this: while the Umwelten of other animals model solely a (for each) 'existent world', man can, by means of the secondary system, also model a potentially limitless variety of 'possible worlds'". (Sebeok, 1996: 106);

4) *Such a capacity to model possible worlds is expressed via three major (and again, exclusive) characteristics of language:* what one may call *distant space-time semiosis*, *narrativeness*, and *linking signs*. Distant space-time semiosis refers to the ability to keep track, transmit and reconstruct both recent and remote past events and places, and the ability to articulate projects and expectations regarding both immediate and remote places and future events (in a way not dissimilar to what Hockett called "Displacement"). Such an ability is both direct and indirect, the former being related to the personal experience of the subject, and the latter referring to experiences that the subject has not lived or will hardly live personally. In other words, language allows human beings to talk not only about their childhood and about their intention to live in a country-house when they will finally retire from work, but also to discuss the defeat of Napoleon in Waterloo and to wonder about the day when Martians will finally land on Planet Earth.
5) Narrativeness refers not only to storytelling (which is still an important feature itself), but mostly the general capacity of accessing and describing alien umwelten, either imaginary or not. Regardless of whether these descriptions are trustworthy or not, what matters here is the fact that in principle, any dialectic account of a given context, environment or reality is *possible* in the human semiosis. The descriptions of the Umwelten of both the species Peking Duck *Anas platyrhynchos*, and of the fictional character Donald Duck (inspired by the very same species) are possible only thanks to the existence of such a device like language.
6) Finally, linking signs (or linking words) refer to a specific aspect within the broader concept of syntax (which, as seen in Chapter 2, is not an exclusively human feature), namely those signs that in verbal speech are known as conjunctions, transitions, and prepositions. Linking words are a form of para-signs that do not refer to any other existing entity apart from themselves, and whose function is to create meaningful relations among signs that, by contrast, stand for something else than only themselves. In other words, linking words are yet another confirmation of the capacity of language to create possible worlds, even when those "worlds", as in this case, are simply signs that are untied from tangible entities.
7) Most semioticians agree that Any other claims concerning the uniqueness of language can be dismissed, once and for all, as anthropocentrically-biased semiological, rather than semiotic, blunders. Those include the use of symbolic signs, the ability to consciously deceive the receiver of a message, the ability (and the tendency) to put into action semiosis of aesthetic type, the so-called metasemiosis, and so forth (more details, from the zoosemiotic point of view, can be found, apart from several points of this book, in Sebeok 1990: 77–98, 1981: 210–259, Martinelli 2006a, Cimatti 1998: 59–106, 179–189, 205–210).

8) *The ICE*, carried out throughout the last decades, *were either a failure or, when apparently successful, fatally biased by misinterpretations of the Clever Hans Effect* (CHE, from now on) *type*. However controversial such a position may be (the experiments conducted were of the most diverse types, and one might suggest that it is dangerous to draw one single conclusion from), there seem to be an almost absolute consensus on this point, especially among Sebeok's followers who do not hesitate to label those experiments as anti- or pseudo-scientific. Besides this book, Felice Cimatti (1998: 107–165) seems to be the only exception to this rule.

In the present section, thus, it will be argued that these five pillars built upon our (i.e., human) concept of language, present elments of inaccuracy and incompleteness, needing additions in some cases, and revisions in others. The following theses will be defended:

1) As anticipated in Chapter 2, in the end of the paragraph "Symbols, syntax and names", *language should rather be considered as a species-specific human NEED*. ICE have not only demonstrated the cognitive and practical abilities to acquire language on the part of other species, but also that the human being appears to be the most motivated species in using and developing this tool. Moreover, even though not specifically denied by semioticians, *the constitution of the human vocal apparatus in the acquisition of language plays a crucial role*, possibly more crucial than the human brain (that is, it was the possibility of talking that favoured the natural selection of language-inclined brains, not vice versa);
2) Certainly, there are very good arguments for maintaining that language is more a modelling system than a communication device. However, *the phylogenetic and ontogenetic bases for language to be acquired remain those of the communication system*. In other words, language proves to be more efficient as a modelling system, and other sign systems prove to be more efficient than language as communicative forms. *But*, it should not be forgotten that (a) language originated as a communication system, and (b) communication is the primary reason why language is passed on from parents to offspring (that is, parents teach language to their children for communicating, first of all: exploring and categorizing reality are merely consequences of this process).
3) Still, if language is – also or mostly – a modelling system, then, in evaluating ICE, *what shall be analyzed more thoroughly is if and how non-human animals use language to map their Umwelt, and – possibly – to access the human Umwelt*. It is curious that it is precisely semioticians, who are the primary promoters of the notion of language as modelling system, that discuss ICE almost exclusively in terms of "comunication" experiments;
4) *Linking signs seem to be the only specific characteristic of language that is mastered by human beings only. Distant space-time semiosis and narrativeness, on the other hand, are not exclusive of humans, although reinforced and improved by the employment of language.* More specifically, linking signs are

the very upgrade (therefore, once again, the terminology employed here *does not* imply qualitative differences[7]), added to a natural communication system, that allows an impressive enrichment of the narrative and distant-space-time semiosic potential of an individual.

5) *The discussion conducted by semioticians on ICE is at times generic and approximate, and requires a certain revision.* Topics that need to be re-visited are at least the following: evaluation of the methodologies and results of the ICE themselves; distinction within different ICE; the CHE and the entire discussion on anthropomorphism; and the consideration of otherwise ignored aspects and outcomes of ICE.

Most of the following discussion shall focus on the last point of this list, for several conclusions (or so it seems) about language, and its supposed human species-specificity, did actually depart from this very argument.[8]

3.3.1 Semiotic Scepticism

The history of ICE is pretty long and articulate: it involved various animal species (mostly Great Apes, like chimpanzees, gorillas, orangutans and bonobos, but also cetaceans and birds), several different tools and methodologies (first spoken language, then sign language and material symbols); it departed from the most diverse working hypotheses and ended with the most diverse results (some positive and some negative); and, scientifically speaking, the results were quite often influenced by the time in which the experiments were performed, going hand in hand with contemporary theoretical trends in psychology, ethology and other disciplines (for instance the programs were very behavioristic in the beginning, and very cognitive eventually). In the glossary of this book, several entries are devoted to this history and its most relevant cases, which of course should be read, if not already familiar, before continuing with this section.[9]

Given such a complex scenario, it is a bit surprising that the account given by semioticians of these experiments has most of the time been generic, reductive,

[7] Apologies for insisting on this point, but it really is the core of the matter, and there should be no misunderstanding whatsoever, in any of the topics discussed in this book. It seems this is the lesson semiotics is most resistant to learn.

[8] It will probably upset many colleagues that the reflections that will follow from now on are slightly, or sometimes openly, critical towards that part of Sebeok's work that discussed ICE and CHE. Although within the rules of academic discussion there is never a need to justify these action, one should still remind how crucial every single theoretical formulation or reflection that Sebeok produced was for the development of zoosemiotics (and this whole book proves it). The fact that (very humbly) one portion of his work is criticized should only be read in the direction of encouraging a discussion that he himself activated.

[9] The hypertext of "people, paths and ideas" related to ICE is best started with the entry "Interspecific communication experiments".

and – from time to time – simply inexact.[10] Most of all, what surprises is how generically ICE are treated. Very seldom are they taken case by case (and subject by subject, a point which is not irrelevant): what we are normally offered is one single pot with all the experiments in it, and a set of conclusions applied to all of them indistinctly. Sebeok almost seems the only one to have acquired his knowledge from direct first-hand sources. For quite a few of his followers, it appears that all they learned they learned it from him.[11] What are these conclusions? Larry Trask (1995: 21–22) summarized them in four basic points:

1) Much of the evidence in ICE is purely anecdotal: "…it consisted of reports that some particular animal on some particular occasion had been observed to do something-or-other pretty damned impressive" (Trask 1995: 21). Anecdotes, Trask points out, are not of scientific relevance;
2) The criteria for evaluating IC tests were too broad to be reliable. "For example, if a signing chimp was shown an apple and asked (in American Sign Language) 'What is this?', the experimenters frequently counted as a correct response any sequence of signs including the sign for 'apple', which is a far cry from the sort of response usually heard from a human child learning a first language" (Trask 1995: 21);
3) In many cases, the claimed ability of certain non-human animals to actually perform the linguistic task assigned to him/her was reported by the experimenters only, so one has "nothing more than the experimenters' own word for it that the apes were making any signs at all" (Trask 1995: 21);
4) The most important reason why ICE should all be considered misleading and anti-scientific is their incapacity to avoid the CHE and – more generally – anthropomorphic interpretations of the results of the experiments: "the critics discovered that the experimental procedures typically used to test the animals

---

[10] *I will never forget a lecture on the subject, given by one of the greatest living semioticians, in which subjects and methodologies of different ICE were quoted at absolute random: Washoe, who was trained with American Sign Language, was reported to have communicated with lexigrams (which, in reality, was the case with another chimpanzee, Lana); Koko, a gorilla, was quoted as a chimpanzee; and still Washoe, who was trained by Allen and Beatrix Gardner and Roger Fouts, was reported to have been trained by Herbert Terrace (who, in reality, was training another chimpanzee, Nim Chimpsky). Nobody is asking for indisputable precision, of course, but how seriously can we take – say – a musicologist who, during a lecture, claimed that Hey Jude was written by the Rolling Stones; Jimi Hendrix was the drummer for Led Zeppelin; and Abba was a German duo?* (Martinelli 2006a). And, it must be feared, it is no coincidence that the only ICE scholar mentioned in that occasion was Herbert Terrace, since he is the one who was most often mentioned by Sebeok in his writings. Unfortunately, one has to infer that that lecture was exclusively based on second-hand material.

[11] A most remarkable exception to this rule is represented by Felice Cimatti, who indeed should not be considered a Sebeokian or a Peircean (his theoretical bases deriving more from the likes of Jakobson, Gozzano and others), and whose book *Mente e linguaggio negli animali* (1998) is a rather distinct case in the entire semiotic panorama. A short look at his list of references easily proves that Cimatti did his homework on Savage-Rumbaugh, Premack, Pepperberg & co. very well and accurately.

> were so slipshod that an animal under test could often see its human handler unconsciously forming the required response with her or his own hands, so that it could see what to do" (Trask 1995: 21–22)

To Sebeok, the "pervasive, insidious penetration of Clever Hans" (1990: 68) in ICE is the main problem, and he reminded us of it in practically every essay he wrote on the subject. Perhaps, thus, the issue of anthropomorphism deserves a deeper discussion, which is what will be provided later on. First, however, the first three points of Trask's list should be noted.

The first point reveals a certain lack of attention in reading the actual texts reporting ICE. A more careful scrutiny would have suggested that most of the anecdotes are not episodic at all, both within the same experiments and, more importantly, across different experiments. For instance, the use of signs in a creative manner to name unknown objects starting from similar known ones is an "anecdote" reported in several cases: Koko used "white tiger" for a zebra, "finger bracelet" for a ring, "elephant baby" for a Pinocchio doll, or "eye hat" for a mask (see Patterson and Linden 1981); Lana asked for an "oranged-coloured apple", not knowing the lexigram for "orange" (von Glasersfeld 1978: 732); Washoe used "drink fruit" for a watermelon, or "water bird" for a swan (Hill 1980: 336). Then again, it is true that Koko called a zebra "white tiger" only once, or Lana an orange "orange-coloured apple". But that is because they were soon taught the signs for "zebra", and "orange", so why bothering to still call those objects in their own temporary way? What counts is not the redundancy of a specific example, but rather the redundancy of a rule (or, more precisely, a modelling strategy): when Koko sees a zebra, but does not know the ASL sign for it, she starts reflecting upon it (metaphorically: "Well, it's clearly an animal, it runs... it has stripes like a tiger, but it's white..."). Quite exactly, Koko models her experience through the language. It must also be mentioned that other ICE subjects, like Sarah, were trained to use signs (plastic symbols, in her case) for "Name of", through which they could ask how to call a certain object. This way, Sarah did not need to create white tigers or water birds, but she would simply ask directly for the right plastic symbol representing the unknown object.

As for the second point of Trask's list, here is a typical example of the situation that will be explained more thoroughly in the next sections: in this case the fear of making anthropomorphic mistakes becomes an even greater mistake. Everyone who is learning a new language (or any sign system) goes through a transitional period when confusion is more regular than precision. Terms, syntax, pronounciation, not to mention the infamous "false friends", are all problems whose solution may take a significant amount of time. The situation described by Trask of the ICE subjects randomly performing signs until they catch the right one is quite normal, as normal is the criterion applied by the researcher in evaluating these attempts: it is exactly what occurs among humans, with children or with foreign language students. We encourage the right answer, as it arrives, with a "yes" or a smile, and we are keen to accept a few wrong attempts before the right word is pronounced when the trainee is just at the beginning of his/her learning process.

Trask's third point can hardly be taken seriously. So, we cannot really trust Premack, Gardner, Patterson and company, since they were the only ones observing their trainees performing the experiments, and they may be lying. Never mind that a lot of footage documenting the experiments is available (a lot is on-line, too); and never mind that the projects were never carried out by one single researcher, but by entire teams, whose members would partly or totally change throughout the years (to mention one, Washoe's project has been going on since 1966, and still continues after her recent death, as in the meantime more chimpanzees have been added to the community), and young assistants, whose names would otherwise remain obscure, could easily gain newspaper headlines with scoop interviews of the type "My boss is bluffing about the apes". What is suggested here is that, since we cannot check, there is a reason to think that ICE are not reliable. If that is the principle, however, nearly any empirical finding, in any scientific field, is liable of suspicion, with the sole exception of those performed in public arenas or reality TV shows.

3.3.2 The Case for Anthropomorphism

Every discipline has its original sins. An average student in social psychology learns very quickly that a research that is not based on empirical data (possibly collected in quantitative-statistical fashion, and measured through a scale that someone else has invented) is not going to be given any consideration at the academic level.

The original sin for anybody involved in studies of animal behavior is anthropomorphizing the interpretations of observed behaviors. If this happens, the scholar will, slowly but inevitably, slip into the realm of researchers who are seen as näive and not scientifically credible. If this can occur to a zoologist who deals with the functioning of the digestive apparatus of amphibians, just imagine how delicate a zoosemiotician's position is.

The ancient Greek *morphé* means "form", "shape", "appearance". According to Webster's Dictionary, anthropomorphism can be defined in two ways: (1) the representation of the Deity, or of a polytheistic deity, under a human form, or with human attributes and affections; and (2) the ascription of human characteristics to things not human. The dictionary of ethology, edited by one of the most authoritative Italian ethologists, Prof. Danilo Mainardi, defines anthropomorphism in the following way:

> [Anthropomorphism] is the tendency to interpret animal behaviors by ascribing to them typically human motivations and goals. Occasionally, this led to the evaluation of animals' behavior in moral terms—and to persecuting them.[12] This mistaken habit of seeing animals in human terms has long affected the interpretation of their behavior, and the most evident examples are medieval bestiaries. Research in child behavior has demonstrated that the tendency to anthropomorphize animals arises spontaneously, and that it disappears only in adulthood, thanks to appropriate education. The same tendency is detectable in primitive civilizations and plays an often relevant role in the relation between humans and pets.

---

[12] See the extremely interesting Evans (1906).

Dangerous anthropomorphic interpretations have been, and still are, present in many cases of zoological exposition. (translated from Mainardi 1992, 48)

At least three points in this definition should be highlighted and eventually explored in more depth:

1. Anthropomorphism is considered, almost without exception, a scientific mistake, which, historically, seems to have damaged or delayed a fair interpretation of animal behavior.
2. According to the terminology used in Chapter 4 (see also Martinelli 2002: 61–62), it seems rather clear that the attitude emerging from this definition is of a binary and qualitatively anthropocentric type. Such a framework should be further taken into account.
3. Although Mainardi considers anthropomorphism a mistake to correct with "appropriate education", he recognizes the spontaneous emergence of anthropomorphic attitudes, both psychologically (during human childhood) and anthropologically (in primitive populations).
   On a very general level, it is evident that anthropomorphism is perceived by the scientific community as a serious mistake, unable to make any useful contribution to ethological knowledge. As emphasized by Tom Regan (1983: 6–7), if other animals are described as having characteristics that are also human, but not only human, such as "being alive", such attribution does not constitute an anthropomorphic mistake. The problem concerns more critical characteristics. Regan mentions the example of consciousness:

Now, if consciousness is a characteristic of humans only, then we are guilty of anthropomorphism if we regard animals as conscious; we make more of animals than what they are; we erroneously picture them as humanlike. Suppose the charge of anthropomorphism is made against those who view animals as conscious. How can it be met? Surely not by repeating the attribution, no matter how many times it is repeated, and no matter how many different people repeat it. All repeating the attribution could show is how many people view animals as conscious, and this fact, while of interest in some contexts, is impotent in the present one; however many people regard animals as conscious, it is quite possible that the view held by these people is anthropomorphic. Given the charge of anthropomorphism, and given the deficiency of trying to meet it by insisting oneself, or getting others to insist, that animals are conscious, it should be clear that another way must be found if this charge is to be met. (Regan 1983: 7)

In any case, to anthropomorphize apparently means to commit a sheer scientific crime. Not without a touch of irony, Jeffrey Masson and McCarthy (1995: 32–33) entitles a paragraph of his bestseller *When Elephants Weep*, "Anthropomorphism as contagion", and quotes several authoritative scientific essays claiming that anthropomorphism is a "disease", a "trap", an "illusion", a "womanly bias" (*sic*!), or simply a "lie".

The real milestone for the enemies of anthropomorphism is the Clever Hans Phenomenon. Hans was a horse, owned by a certain Wilhelm von Osten during the first half of the twentieth century (Fig. 3.1). Hans came to be quite well-known for his purported mathematical computing skills. When asked to solve any kind

Fig. 3.1 Clever Hans with Mr. von Osten

of arithmetic task, from simple addition up to calculating the cube root of numbers like 103.823 (Cimatti 1998, 146),[13] Hans would amaze everyone by striking his hoof against the ground as many times as the exact result was. His fame grew greater as ever more difficult tasks were asked of him and different verification tests were set (e.g., the tasks were written on a blackboard instead of being vocally pronounced). This went on until a scholar named Oskar Pfungst realized that when Hans was interrogated by someone who did not know the right answers, his hoof strikes were wrong. Working from this observation (repeated and confirmed several times), Pfungst realized that Hans's arithmetic competence was in fact quite low, but that his psychological skills were superior to those of most humans.

Clever Hans, indeed, did not have the slightest idea of how to solve arithmetic problems, but was remarkably skillful at detecting the exact moment when the interrogator was pleased with his response (i.e., the right number of hoof strikes). Unintentionally, the interrogator would send some kind of sign (facial or postural, mainly) which displayed surprise in seeing that Hans had actually reached the right number of hoof strikes.[14] At that very moment, Hans would stop striking the ground with his hoof.

Unfortunately, instead of *also* pointing out how exceptional such psychological skills were, scholars only use this episode to recall how dangerous and misleading anthropomorphism can be (in this case, attributing mathematical skills to Hans). From then on, in experimental contexts, one speaks of the Clever Hans Effect every time the animal ends up being the manipulator, rather than the subject, of the experiment, and – more generally (and more specifically, as far as zoosemiotics is concerned) – when the interpretation of a given animal behavior is biased by the illusion that the animal in question has done something a bit too extraordinary.

---

[13] Which is 47, for the record.

[14] Possibly, a typical and evident display would be the opening of the eyes and mouth.

John Andrew Fisher's "The Myth of Anthropomorphism" (1990) represents a very efficient attempt to classify the several nuances of the concept of anthropomorphism. Fisher is aware of two crucial aspects: (1) there is extreme confusion in the use and definition of the term; and (2) however one may define it, every scholar takes great care to avoid being affected by it, as one would normally do when dealing with diseases:

> Anthropomorphism is usually regarded as an embarrassment to be avoided. Philosophers and scientists often approach anthropomorphism as an obstacle to be overcome by those who wish to attribute cognitive or emotional states to nonhuman animals. Thus Donald Davidson suggests that "Attributions of intentions and beliefs to animals smack of anthropomorphism". Even those who favor animal rights try to avoid being accused of it. Annelle Baler, for example, feels obliged to say, "I see nothing at all anthropomorphic or in any other way absurd in saying that one may break faith with an animal, exploit its trust, disappoint expectations one has encouraged it to have". And Mary Midgley asserts: "There is nothing anthropomorphic in speaking of the motivation of animals". Contrary to this loose consensus, I will argue that there is a considerable amount of confusion about anthropomorphism. I will argue that the mistake or fallacy of anthropomorphism is neither well-defined nor clearly fallacious. There are many different conceptions of anthropomorphism and the common ones do not support their common rhetorical use. (Fisher 1990: 96)

A first separation is made between *interpretive anthropomorphism* and *imaginative anthropomorphism*. Interpretive anthropomorphism refers to "all of the usual cases of ascribing mentalistic predicates to animals on the basis of their behavior" (Fisher 1990: 100). Typical in this case is the description of certain behavioral patterns as intentional actions. Imaginative anthropomorphism is in turn defined as "the productive activity of representing imaginary or fictional animals as similar to us" (Fisher 1990: 100). Such is the case with the whole tradition of fables, myths, cartoons, and fiction in general.

Interpretive anthropomorphism is in turn divided into *categorical* or *situational*. The first case (Fisher 1990: 101) consists of "ascribing mentalistic predicates to creatures to which the predicates don't ever in fact apply". The problem is precisely categorical, for given behavioral patterns cannot be attributed to given species without falling into the anthropomorphic mistake. In the case of situational anthropomorphism (Fisher 1990: 101), an animal's behavior is interpreted "in ways that could possibly apply to that animal in other circumstances, but which do not in the situation in question". It is thus anthropomorphic to ascribe a given pattern to species X in situation Y_1, but would not be in the situation Y_2.

Lastly, Fisher divides categorical anthropomorphism into *anthropomorphism by species* and *anthropomorphism by predicate*. The former (Fisher 1990: 101) consists of situations when "application of mentalistic predicates could be counted as anthropomorphism depending on the species. What wouldn't be anthropomorphism concerning a chimp might be concerning a worm". The latter (Fisher 1990: 101) describes situations when "application of mentalistic predicates could be counted as anthropomorphism depending on the predicate. I have in mind applying the wrong types of predicate".

Then, how dangerous is anthropomorphism for scholars? Quite dangerous, it seems, if some scholars even went as far as to formulate theorems to prevent anthropomorphic (or similar) attitudes. The so-called *Morgan's canon* (named after the English psychologist Conwy Lloyd Morgan, whose life bridged the nineteenth and twentieth centuries) is one of the most classical strategies. It says that "in no case should actions or behaviors be interpreted as the result of a superior psychic faculty, when it is possible to interpret them as a result of an inferior faculty". This statement will be discussed further on in the present section. Along with Morgan's canon, another method, quite typical of ICE programs, is the creation of the so-called *Emotion-Free* experimental context. Every possible unwanted input from the experimenters that might result in a CHE (facial expressions that might unconsciously suggest the subject of the ICE the solution for a specific task, affective – therefore, not anymore scientific – relation established with the subject of the ICE, etc.) is made sure to be avoided. Researchers, then, wear masks, observe the experiment unseen from another room, leave the subject of the ICE alone in the laboratory, avoiding any kind of interaction, and so forth. Not unpredictably, the completely Emotion-Free ICE built up throughout the last decades ended up in a failure, which, according to many semioticians, is the ultimate prove that: (a) nonhuman animals are not able to learn language; and (b) The CHE is the real core of the problem. When avoided, the ICE reveal their true nature, i.e., misleading non-scientific practices.

3.3.3 Critical Remarks on the Interspecific Communication Experiments

Sebeok's criticism of ICE was not limited to CHE, and – by all means – was articulated in a much deeper way than Frisk's summary (which was intended as a summary of a number of positions on the topic). In 1980, a collection of essays entitled *Speaking of Apes*, edited by Thomas and Jean Sebeok, was published and soon became the major point of reference for semioticians on the topic of ICE. It is in the extensive sixty-page long introduction (Sebeok and Sebeok-Umiker 1980: 1–60) to this book that one finds the most detailed illustration of Sebeoks' views on ICE. In the text, the Sebeoks go through nearly the entire panorama of ICE, presenting their doubts or occasionally their resolute rejection. Their arguments are in all cases well-formulated and convincing, and it is probably this characteristic (together with the undeniable sacredness of Sebeoks' work in general[15]) that made the approach of Sebeoks' followers on the topic as confident as previously described. Semioticians' next observations on ICE proved to be rely entirely upon this very text, entitled "Questioning Apes", showing little, if any, advancement from there, in

[15]This being said it with no negative connotation whatsoever. Only the greatest scholars become academic saints.

a time (the 1980s and then the 1990s, up to nowadays) when ICE programs were growing rapidly and – in most cases – successfully.

Thirty years after, it thus seems reasonable to take a slightly more proactive attitude towards that text, in the light of what has happened in the meantime (let us here recall the importance of the birth – in 1976 – and the definitive establishment – in the 1990s – of the cognitive branch of ethology, within animal studies), but also as a consequence of a more critical reading. Maybe some things, although still well-formulated, are not so convincing as they seems.

To start with general remarks, "Questioning Apes" may appear not to respect in full the non-written etiquette of scientific compilations. *Speaking of Apes* is a collection of essays that intends to gather the main positions against *and* in favour of ICE. The editors of such a work, in writing the introduction to the book, should certainly summarize and contextualise all these positions, but, at the same time, they should not to take too strong a stand in one direction or another. Or, if they do, that should be made by referring to external sources, in order not to discredit the book itself, as a collection made of good essays for one half, and rubbish for another. In other words, the editors should not exploit the chance of writing their introduction *after* all the other texts have been delivered to them.

Regrettably, "Questioning Apes" takes all possible advantages to make it clear how ICE programs are here regarded as anti-scientific and fallacious, and how naïve the pro-ICE essays within the same book are. Thus, by, say, page 20 there has already been attempt to prejudice the reader against an essay that appears on page 200.

The only ICE scholar spared Sebeoks' criticisms, and in fact highly praized, is Herbert Terrace, who conducted an ASL-based ICE program on the chimpanzee Nim Chimpsky (a witty pun on Noam Chomsky, of course). The program ended up being a complete CHE-affected failure, and Terrace – with rare professional honesty – admitted it with no excuses, thus becoming a hero of anti-ICE supporters:

> The honesty of Terrace and his colleagues in reporting this and other procedural steps which others, we suspect, omit from their accounts, is to be applauded. It makes it possible for the reader to assess more accurately the reliability and import of the data presented. (Sebeok and Umiker-Sebeok 1980: 13)

The procedures that the Sebeoks are here referring to do indeed reflect the serious bias of the ASL trainers on Terrace's program: they tended to revise their reports after discussing among themselves, thus, in the end, they agreed on many more points than in their previous observations of Nim's behavior. Indeed, comparing each other's observations may be useful for correcting (rather than provoking) individual mistakes, but clearly the chance of messing up the data collected is higher than the chance of improving them. However, the additional step taken by the Sebeoks in commenting upon Terrace's case, is to use it as a synecdoche for all the other ICE programs. The history of ICE reports both unsuccessful and successful programs, the latter being more recent and numerous. Terrace was simply one who made some mistakes and admitted them. That makes him a highly-respectable and honest scholar, but not the proof that ICE programs are all biased and deceiving.

A last general remark, before getting to more specific considerations, is the fact that the Sebeoks seem not to have been extremely interested in visiting the sites of those research programs. The only instance they mention is a 1969 visit to the Gardners' house in Reno, Nevada, where the Washoe-project was taking place at the time. The description of that visit does not seem too far from how Soviet authorities would selectively guide delegations of western communist parties through the (few) decorous areas of USSR:

> During our stay in Reno, in 1969 [. . .] we were, as guests, necessarily at the mercy of our gracious hosts, the Gardners, in terms of where and when we could view Washoe, and what additional experiences of the project – in our case, heavily edited films of the chimpanzee, casual testing of her signs in our presence using a box of well-worn objects, and discussions with some of the animal's trainers – we were allowed to have. (Sebeok and Umiker-Sebeok 1980: 29)

This version of the events may be counterargued in at least six ways:

1) In 1969 the Washoe-project was at the very beginning (it started in 1966 and it is still going on, with many other chimps – including Washoe's offspring – being introduced into the project), and it was only during that year that the first report of the project was published (Gardner and Gardner 1969);
2) A research laboratory, even if located in a private house (as in Gardners' case), has its rules and limitations, and people cannot simply do and check what they want just because they do not trust their hosts;
3) The "heavy editing" of the films was simply meant to spare the spectators hours and hours of totally uninteresting material;
4) A chimp is after all still a chimp, i.e., a wild animal, and it is not always predictable how s/he would react to the presence of a stranger.[16]
5) This is a scientific experiment, not a circus show, and the chimpanzees are not trained to *perform* before any audience, but they rather need to trust their interlocutors before actually interacting with them; finally
6) What makes a reader uncomfortable is how energically the Sebeoks express their reservations not only about the experiments themselves, but also about the honesty of the IC researchers.

More importantly, there is no information about any other visit to any other IC project, although invitations were not lacking. In a 1998 interview for the New York Times, Sue Savage-Rumbaugh, responsible for the Kanzi project, when posed the

---

[16] As a general rule, not only in Gardners' case, no visitor is allowed inside the enclosures where chimpanzees are kept. Chimpanzees are some seven times stronger than humans, and they also have denser bones and thicker skin. In most instances of interactions, chimpanzees must restrain themselves to avoid hurting humans. When playing, chimpanzees normally throw, slap and playbite each other. However, if these actions elicit laughter among them, a human would be seriously hurt. The fact that they may master a bit of human language, unfortunately does not mean that they will use rethorics and diplomacy when the situation becomes critical.

question "Many in the scientific community accuse you of over-interpreting what your apes do" answered as follows:

> There are SOME who say that. But none of them have been willing to come spend some time here. I've tried to invite critics down here. None have taken me up on it. I've invited Tom Sebeok (of Indiana University) personally and he never responded. I think his attitude was something to the effect that, ' It's so clear that what is happening is either cued, or in some way over-interpreted, that a visit is not necessary.' I would assume that many of the people associated with the Chomskyian perspective including Noam Chomsky himself have the same approach: that there's no point in observing something that certain doesn't exist. (Dreifus 1998: newspaper article)

Let us now enter a more specific realm of the matter and go through some of the assumptions that seem not entirely accurate in "Questioning apes". On page 10, a severe position is taken against the ICE program carried out by Allen and Beatrice Gardner (whose essay appears on page 287 of the book). The methodological foundations of the project are called into question, as likely to produce "distortions", altering the results of the experiment, or – even more often – focusing on a wrong pattern or an irrelevant priority:

> One observer may prime another to inadvertently create situations in which a newly reported sign might be likely to recur by, for example, using a certain tool when the sign for that object has been reported by another observer. This outcome is especially likely owing to the fact that trainers were instructed by the Gardners to record the context in which a new sign was observed as well as the sign itself. Should the ape actually produce a sign in this sort of context, it would not be considered the result of outright prompting, molding, or the CHE, all of which the Gardners deny played a role in the reporting of new signs, and yet such innocent provision of opportunities for corroboration of other observers' records would certainly influence the overall course of the study (Sebeok and Umiker-Sebeok 1980: 10).

Basically, the negative aspects are two: encouraging – in whatever form – the production of a sign instead of another; and the presence, during the ICE, of contexts or situations that *facilitate* the emission of signs. The counterargument here is that this is exactly what human parents do: (a) they encourage the emission of certain words instead of others, even more insistently than the Gardners (as in the words "Mum" and "Dad"); and (b) they continuously create situations and contexts that facilitate the emission of signs and words (a recurrent instance being those language-related toys, such as picture books whose images are named in big fonts). It is to be inferred that this *is* a succesful strategy for teaching language.

It also remains to be explained what is so wrong in *facilitating* the emission of signs. If – say – a dolphin D is trying to teach a human being HB the language of dolphins, HB would not really mind if D somehow encouraged him/her to emit "dolphinesque" signs. In establishing how scientific an ICE program is, one should always take into account two main factors:

1) all the difficulties related to learning a communication system that is not only different: it is alien, it belongs to another species. It is much more difficult than learning a foreign, but still intraspecific, language; and

2) simply enough, the trainee subject might just happen not to be a genius. Not only may Washoe be supported in the emission of human signs: she *must* be supported, as any person who is learning an alien language. After all, are we after teaching language to Washoe, or are we after teasing her?

The argument here is that the scientific cleanliness of any experiment should not be exercized purely for the sake of cleanliness. The target of the experiment is much more important than its aesthetics. If what we want to check is (a) whether non-human animals are able to communicate with human language; (b) whether they can ask what they need/want through the use of human language; and (c) whether they are able to model their perception with human language (as well); then our ICE program shall achieve a degree of methodological cleanliness that does *not* interfere with, or negatively affect, these targets. On the contrary, we shall create the most encouraging and inspiring environment for these target to be reached. If a chimpanzee, or a specimen of another species, learns the human language despite all the obstacles that a semiotician claims are absolutely necessary for the scientific validity of the test, than what we prove is simply that the chimp is a genius, not that chimps are able to learn human language.

On page 14, again commenting upon the Washoe project, the Sebeoks agree with Eric Lenneberg (1980: 80) in negatively evaluating Washoe's spontaneous ASL-signs productions. If the Gardners look favourably at the creation of new signs in Washoe, as symptoms of enthusiasm towards, and cognitive appropriation of, the new communication tool, Lenneberg does not hesitate to apply Ockham's Razor and Morgan's Canon to such instances, and finds that "we are simply testing our own ingenuity to assign interpretations to productions that might, for all we know, have been emitted randomly".

The nature of these tools will be discussed at length in one of the following paragraphs: what however strikes one immediately is the fact that such a paradigmatically creative entity as language is here contested in its very nature. Language is by definition something flexible and imaginative: its codes – so useful for associating a given sign to a given meaning – are constantly challenged by neologisms, metaphors, loans from other codes, and other factors. When people speak a foreign language with native-speakers of that language and they have problems in finding the right word, it is precisely their creativity that rescues them from misunderstanding: they will use the word they have in mind in their own language hoping that it is somewhat similar to the other idiom, they will make gestures, they will point their finger towards something similar, they will describe that one word with an entire sentence, they will make up a new word if necessary... certainly they will come up with something, and in the end – rest assured – their interlocutors, although maybe a bit amused, will understand what they have in mind.

It is this very quality that makes language an always-in-progress entity. In Italy, there was no such thing like a straight transition from Latin to Italian. Uncountable hybrid idioms were created in the meantime, and even today what we call "Italian" is not entirely the same collection of words that it was only a few years.

The spontaneous creation of new ASL-signs is, therefore, rather to be welcomed as a very positive reaction from Washoe to the human language. She might not have learned the right sign for the right word, but there she goes trying to make her point anyway. And this is especially remarkable when the so-called "innovations" take place. These cases were already mentioned in the paragraph entitled "Semiotic scepticism": an innovation takes place when the trainee does not know the sign for a given object, therefore simply makes up a new sign, based on a cognitive association (*Finger bracelet* for a ring, *White tiger* for a zebra, etc.). This is a rather convincing proof that language is used precisely as a modelling system, other than a mere communicative device. Of course, for someone who is at the same time opponent of ICE and promoter of the conception of language as a modelling system, this is hardly acceptable:

> Unfortunately, these instances are generally reported in such a way that not enough is learned of the context of occurrence to enable us to rule out the possibility of either trainer suggestion or overinterpretation. [. . .] there is available a more parsimonious explLanation for Fouts' report that Washoe was creating a new compound lexical item when she produced "water" plus "bird" in the presence of a swan and was asked, "what that?". Since Fouts provides no evidence that Washoe characterized the swan as a bird that inhabits water, we may just as well assume that Washoe, who was very familiar with the question form what that?, was merely responding by first identifying a body of water, then a bird. (Sebeok and Umiker-Sebeok 1980: 15)

Again, counterarguments are many, and, it must be stressed, *particularly* when it comes to this point:

1) It is simply untrue that the instances of linguistic inventions are badly or insufficiently reported. A few have already been reported here, and there are more. One has to simply check the relevant texts for confirmation;
2) Of all the possible examples, the Sebeoks chose to discuss one of the very few instances of linguistic inventions in which the interpretation is subject to ambiguity: "water" and "bird" are indeed rather general concepts, so it is in principle possible that Washoe could have referred to the body of water first, and to the bird afterwards. But then, what about "drink fruit" for a water-melon? "Drink" is an action, not a tangible object. What could Washoe possibly have meant, if not that kind of fruit that is actually so juicy that its consumption is more similar to drinking than to eating (at least in comparison with other fruit)? What about Koko's *white tiger*, then? Did she see a "body of white"? And where did she see the tiger?;
3) Even considering the case of "water bird" alone, can we really accept the (Morgan's canon-friendly) explanation that Washoe, once asked "What that?", referred to two different objects, one of which is at least as big as a pool (if not as a lake, or even as the sea)? How did it happen that, all of a sudden, Washoe became so generic and scarcely-selective? Imagine: somebody takes Washoe out in the garden and points at a football, asking "What that?". Should we really expect that she comes up with something like "Grass ball", or "Garden ball"? And why are the birds that fly not called "Sky birds"?; in addition,

4) Was there nothing else than just a body of water and a bird? Not a tree, not the sky itself, not some clouds, not another animal, not a single element that might have caused Washoe to call the poor swan something like "water tree cloud bird"?; finally
5) The incidental comment "Washoe, who was very familiar with the question form *What that?*" should not go unnoticed. That is quite a statement, if one considers that Sebeok does not believe that Washoe (or any other language-trainee) does actually understand the questions that she is asked. His idea, as firm supporter of the CHE, is that Washoe reacts to that question by randomly making gestures, until she notices that her trainer is satisfied. How come she is now "very familiar" with the question? Could it just be that she understands it?

On pages 15–16, the Sebeoks' review focuses on the question of language manipulation for the purposes of amusement, what has been nicknamed "Monkey humour". In a famous dialogue (here reported in the paragraph "On deception" in Chapter 2), the ASL-trained gorilla Koko teases her trainer Dr. Francine Patterson, by pretending she wants her apple juice in the most unexpected places (nose, eye, ear, and finally mouth). The manipulation of a code for personal amusement, as applied to non-human animals, is something that does not convince the Sebeoks, who again seem more at ease by calling into question the CHE:

> If a sign or other response produced by an ape appears to be inappropriate [. . .], human trainers appear all too willing to stretch their imagination in order to make the animal's performance "fit" conversationally. [. . .] Thus, anomalous chimpanzee or gorilla signs may be read as jokes, insults, metaphors, or the like, much as the not infrequent offenses against the very elements of counting and the fundamental arithmetical processes made by Clever Hans were regarded in part as intentioonal jokes and by an authority in pedagogy as a sign of independence and stubborness which might also be called humour. (Sebeok and Umiker-Sebeok 1980: 15–16)

Further on, that very example of Koko's humour is reported as a typical CHE-affected misinterpretation: the truth, it is maintained, is that Koko was simply unable to perform the correct sign for "mouth". As before, a number of remarks can be made:

1) As the request for food or drink is by far the most recurrent ASL interaction between Koko and Patterson, it is extremely unlikely that Koko has not mastered it properly, and that this episode is something other than just an amusing diversion to a routine action. In addition, Patterson in any case reports other cases of her gorilla's humour, like when she was asked "to place a toy animal under a bag, and she responded by taking the toy and stretching to hold it up to the ceiling" (Patterson 1978: 456);
2) These examples show a typical and basic feature of humour: turning things upside down. A asks something to B, and B does the exact opposite. B has fun by teasing A, and teases A by having fun. Nothing is more elementary than this, in humour;
3) If Koko was just making mistakes, why on earth would she laugh?

Staying with Koko, and signs of "independence and stubborness", there is another critical remark on page 37. In commenting on one of the ASL tasks proposed to Koko, Francine Patterson (1977: 10) noted that the gorilla was not particularly attracted by the exercise, and after a while (five trials per day, maximum, and two sessions per week) she would stop being co-operative and start performing the same sign over and over, or not performing any sign at all, or generally displaying boredom. Excluded these instances – said Patterson – Koko was able to perform the right signs in 60% of the cases. The Sebeoks claim that the this method of counting was incorrect:

> Koko is said to have scored correctly 60% of the time, or above chance, on the series of double-blind tests which were administered to her. It is reasonable to ask, however, how Koko's instances of avoiding the test, e.g., by responding to all objects with the same sign, were scored. We are not told whether or not such inappropriate responses were discounted as avoidance measures or counted as errors. [. . .] We can only guess whether or not Koko's performance would have been below chance were a less biased accounting to have been made (Sebeok and Umiker-Sebeok 1980: 37)

From a general understanding of Patterson's research, the Sebeoks are certainly right in suspecting that she did not count Koko's reluctance as a mistake. But the question is, why should she have? Why should the lack of enthusiasm to perform a given task be equalled to a wrong performance of it? Let us compare the situation to an ordinary human instance of two people (X and Y) playing a game, for example chess. As the game proceeds, X makes both good and bad moves: in one instance X successfully captures one of Y's castles, on the next move X makes a mistake and loses a knight, and so on. At some point, 60% of X's moves have proved to be correct, and 40% definitely have favoured Y. So, overall X is leading in the game. Except that, well, X gets a bit bored and wants to quit. Y, however, insists that they continue, despite X's attempts to persuade Y to the contrary. X therefore refuses to make any further moves, but again Y insists. The two go on like this for a while, until X gets a bit irritated and decides simply to move the queen back and forth until the game is lost and an end can finally be put to it. At last, Y understands and allows X to leave.

Now, X's attitude can certainly be criticized from several points of view – most of all for lack of politeness – but no one could really say that X's reluctance to play corresponded to a series of *wrong* moves on the chessboard. Not even when the queen was moved redundantly back and forth: it was very evident that X was not interested anymore, and was just trying to persuade Y to finish the game. An ideal observer recording the successfulness of the two people's game will definitely agree that X's percentage of good moves is still 60%, and that the final sabotage does not really count.

On page 16, the Sebeoks' target is mistakes and imprecisions:

> The determination of trainers [the Premack's team, this time] to show their animal [the chimpanzee Sarah] in the best light has even led to the denial that errors are mistakes, again bringing to mind one of the assumptions which sets research on psychic phenomena apart from normal scientific procedures. The so-called psychic's mistakes on tests of her powers are frequently used to prove that those powers are real, based on the assumption

> that if the performer was using mere tricks she would be correct every time. (Sebeok and Umiker-Sebeok 1980: 16)

Once again, there can be a line of reasoning that make Premack's attitude perfectly logical: if the trainee of a given experiment performed all his/her tasks without a single error, one really would think that s/he had found a way to cheat. If however the performance displayed a minimal percentage of mistakes, one would be more reassured that the ape, who – it must be assumed – is more likely to be normally-gifted, intellectually speaking, than a genius, has done his/her job in the best possible way.

Besides, the point is that the percentage of tasks performed correctly was significant, (i.e., more than 60–70%): if one answers correctly 6–7 questions out of 10 in any school exam, one will normally get a good mark.

Staying for a moment with recollections of school, it is difficult to agree with a remark made a few pages further on:

> A related issue concerns the personal preferenecs exhibited by the apes for certain project members. All the animals used so far in the ape "language" projects have shown some preference for certain trainers, performing better for these "favorites" than for the rest, but there has been no systematic attempt to account for these differential responses on the part of the animals. (Sebeok and Umiker-Sebeok 1980: 22–23)

What seems to be forgotten here is that the trainees are first and foremost subjects, with their characters and preferences. They cannot be the infallible machines that the Sebeoks are demanding in order to take their learning-efforts seriously. We all have our personal recollections of school subjects that we liked, but we were not able to learn properly *precisely because* certain teachers were not kind to us. The converse is true of subjects that we did not like, and yet we managed to learn well thanks to teachers who were both kind and competent. Not to mention that more often than not, our professional choices (especially academic ones) become clear only when we find a true mentor, who understands our potential and point us in the right direction. In other words, having preferences for one trainer instead of another is not only normal and scientifically-acceptable. It is also healthy.

Incidentally, it is also mentioned (Sebeok and Umiker-Sebeok 1980: 23) that the trainers, too, have their own preferences, among the trainees. Again, do we not remember our schoolmates who were *teacher's pets*?

One final remark in the "Questioning apes" text calls for some further reflection – the observation on page 30 that every now and then, the apes use signs from their own non-human repertoire in order to support their ASL, which results in the observer being "unable to determine which of the actions he sees performed by such an animal are part of its natural repertoire and which are the results of special training" (Sebeok and Umiker-Sebeok 1980: 30). Once more, it is difficult to understand why this should be a problem, when in fact it could be regarded as a bonus. It shows active (possibly enthusiastic) participation of the trainees in the communication process, and – most of all – it shows very clearly that they have fully understood that it *is* a communication process. That is why, when unable to make themselves clear by means of the human code, they do not hesitate to use their own sign system, hoping, perhaps in vain, that their interlocutor will finally get their point.

In concluding this section, it should not surprise anybody that Sebeok, the man who so convincingly argued in favour of the fully-semiotic qualities of non-human communication is here so much against the extension of the notion of language to other animals. We cannot be surprised because his definition of language has always been very clear, as something whose communicative characteristics are not at all important, in comparison with its features as a modelling system. Therefore, talking about communication is an entirely different issue than talking about language.

What perhaps *is* surprising is a certain inaccuracy and a kind of apriorism in his arguments against ICE, as if he has already made up his mind and is not going to change it, whatever happens. To sum up Sebeok's arguments in few words, it turns out that, for him, the true learning of a language entails the following conditions:

1) the avoidance of any encouragement or support during the learning period;
2) the prohibition of creativity and imagination in producing signs;
3) code manipulations must be understood as linguistic mistakes;
4) boredom and insubordination are taken as linguistic mistakes;
5) the acceptance only of instances of communication where not a single error is made;
6) the prohibition of any form of personal preference towards any trainer;
7) the prohibition of the use of one's own original sign repertoire to support the learning of the new idiom.
   In such a learning environment, one must conclude that it would not only be Washoe who would fail to progress in human language studies, but perhaps even Umberto Eco.

One last general note about the CHE. Even if, for the sake of argument, one accepted that ICE are CHE-biased, the fact is, there is a very thin line between imitating and understanding. Even imagining a fully-CHE-biased interaction, it is hard to find a significant difference between performing a sign because "this is how I get the apple", and performing it because "I know this stands for 'give me an apple'". As a matter of fact, both approaches prove that the trainee has learned how to achieve what s/he wants through the code s/he is expected to use. As soon as s/he has more than one sign (say, just to be banal, A for apple, B for banana, and C for candy), and processes the information in such a way that s/he performs B when s/he wants a banana, and not when s/he wants a candy, then the task is successful, whatever the exact configuration of his/her reasoning is, either it goes like "I will perform B, so I get a banana", or like "I will perform B, which stands for banana". The only difference is that the former relation is indexical, the latter is symbolic, but the connection sign-object is there anyway, which is exactly what ICE detractors deny. We might certainly accept that a conative type of communication proceeds by indexes, especially when the trainee proves s/he can use this foreign code in a symbolic way as well (one would in any case hope that by now, enough evidence in that respect has already been provided).

3.3.4 *Subject Matters*

Another aspect of the Sebeoks' review that somehow surprises is that they seem to care a lot about aspects that are probably not that important. The aim of this paragraph is thus to point out which ICE-related issues have been either underrated and/or insufficiently dealt with. In that respect one shall not spare the ICE scholars themselves a certain amount of criticism. The claim here is that several aspects involved in these projects were overlooked or misunderstood. Now, that does not apply only to Sebeok and other semioticians, but to the scientists involved in these projects too.

To start with, even though many of these scholars seem to feel genuine love and compassion towards the non-human animals they are dealing with (which is certainly the case with Jane Goodall, Roger Fouts, Sue Savage-Rumbaugh and Francine Patterson, to mention some of the most familiar cases[17]), it must still be emphasized how ethically questionable it is to propose that a non-human animal may be stolen from his/her own environment and put in a cage for most of the time, forced to establish relations and perform tasks that s/he probably does not want or need, and exposed to a great degree of boredom and depression, if not – unfortunately – violence and deprivation.[18] It is no chance that the most successful ICE (Washoe, Koko, Kanzi, Alex, etc.) were those where the trainees were given the best possible quality of life.

Apart from this, there are other issues that probably deserve a more careful scrutiny. One of those certainly concerns the application of the Umwelt theory (apologies for basically repeating reflections that were already proposed in the paragraph about the "semiotic animal", in Chapter 1). It has been said several times (including by Sebeok himself of course) that belonging to a different Umwelt is a sufficient reason for a non-human animal not to be able to access a sign system that relates exclusively to the human Umwelt. Furthermore, the human Umwelt is, by contrast, able to access any other Umwelt, at least in principle, thanks to the inherent characteristics of language itself, in particular what is in this book called "Narrativeness". And if one considers, somehow, past and future (especially future) as "possible worlds" (after all, possible-to-happen future or possibly-happened past are *things* that are not yet or not anymore *objects*), then also the quality that was here called "distant space-time semiosis" belongs to this discussion.

Moreover, even if this has been emphasized far less often, there is definitely an intrinsic difficulty in configuring an Umwelt that is different from one's own (and that applies both in sensu stricto, i.e., accordingly to Uexküll's formulation of the concept, and in sensu lato, for instance when we talk about cultural or personal Umwelten). Now, even though most of these statements are difficult to deny, they

[17]The mentioned scholars are also the ones who have the greater media exposure.

[18]An entire chapter of Roger Fouts' beautiful best-seller *Next of Kin* is devoted to the cruel conditions in which Washoe had to live during the period she moved from the Gardners' house in Nevada, to William Lemmon's laboratory in Oklahoma.

can still be reformulated in a more accurate way. Firstly, as already mentioned, it is certainly true that the decoding of alien Umwelten is almost a contradiction in principle. Taking Uexküll's theories very literally, one might deduce that if we, members of the Umwelt A, manage to decipher the Umwelt B, then we cannot really talk about different Umwelten anymore, because the occurrence of this very process would prove that the Umwelt B is simply part of the Umwelt A, therefore not "alien" (and in fact not Umwelt). When we talk about *things*, in a way, they are no longer *things*, but already *objects*, or at least in a philosophical sense.

Secondly, the Narrative and Distant Space-Time semiosic qualities of language (the former, in particular) seem to create a certain uniqueness in the human Umwelt, in that they allow a strong dialectical, philosophical and even rhetorical configuration of alien Umwelten. *But* (as was once again discussed in Chapter 1): dialectics, philosophy and rhetorics are not enough to ensure *full* access to another Umwelt, even if we can make some significant observations. As already said, particularly from an ethical point of view, this should not be forgotten, as we like to think that, when we investigate other animals, we are able to understand *all* that is there to understand. How does it feel to live in a world dominated by olfactory stimuli, rather than visual ones? Are we sure language is enough to fully grasp that? And what about channels that are completely alien to our semiosis, like the electric one? *How does it feel to be a torpedo?*

More importantly, for the purposes of the IC topic, the question about Narrativeness and Distant Space-Time semiosis is that one cannot be at all confidently answered, as these features of human language are *qualitatively* different from other animals. And that is because ICE, in most cases, turned out to be successful attempts to teach human language to other animals. So, if language supposedly belongs to the human Umwelt, then it follows automatically that Kanzi and c. *did* access the human Umwelt, from the front door (that is, through the feature that by definition characterizes the human species). Mastering language, in the extensive way humans do, leads to increasing narrativeness and distant space-time semiosis, therefore the more refined and articulated our linguistic knowledge, the deeper our narrative abilities. However, the outcomes of ICE call all this into question, probably preventing us from claiming a qualitative *exclusivity*. If anything, they prove that narrativeness and distant space-time semiosis are language-specific features. But that is an entirely different matter, Because now the difference is between language and other communication and modelling systems, not between humans and other animals. If other animals were able to learn language, at least to some extent, that means that access to more sophisticated semiosis is clearly within their cognitive potential.

Then why, as a matter of fact, did language develop in humans and not in other species (as far as we know)? And why does language not arise spontaneously in other animals (since they were able to acquire it only after human's guidance, in contexts of captivity)? Obviously, those questions deserve a separate and thorough treatment, as they are hardly either anthrozoosemiotic or (in all likelihood) zoosemiotic topics. Nevertheless, one or two reflections are possible. It goes without saying that one can only attempt a purely speculative hypothesis (and it is comforting to know that nearly no hypothesis about the origins of language have any empirical support, either).

Language – it is this book's claim – arose in the species that more than any other *needed* it, not in the only one that *could* invent it. The existence of some 3,000 different idioms in human communities led to the common place that language is a purely cultural phenomenon. In actual fact, no less than forty universal fonems exist, and a selective attention towards them among infants from all over the world has been empirically observed, proving the spontaneous, thus biological, basis of symbolic language (see Mainardi 1992: 452). To accept this point makes it easier to accept that language must have been born as a result of an evolutionary need.

One opinion (see also Malacarne 2005: 243–247) is that the need for a language came as a solution to the many physical limitations of the human being and as a consequence to a few advantages. In most of the problems that any other animal can solve alone (feeding being the most obvious example), human beings need cooperation. In that respect, language could have been subsequent to strictly adaptive functions of communication in general, such as the referential or the conative one, which required more complex specifications, due to the inability of humans to perform a given task on their own. In particular, the referential function (this one, too, along with the conative, being found in many animal species) had a fundamental adaptive value (see also Cimatti 1998: 71–2), in that it allowed a true body extension to times and places that were not directly attended by the interlocutors (a human or other animal can be *informed* of the presence of food in a certain area where s/he has not been, or is not at the very moment). The more efficient and articulated the communication, the more successful the cooperation (as one can also see in species like whales or wolves, both deciding the hunting strategy before actually putting it into action). A signal arose with a specific function, e.g., to obtain from the receiver a given tool that the latter had to go and collect from another place. Later in time, such a function must have developed in more articulated ways (passing through stages of generalization, abstraction, etc.), going hand in hand with natural selection and phylogenetic evolution of the species. The most successful human subjects were those who could communicate in the most precise and articulated way. Operatively speaking, the very *selection* consisted in the development of two brain areas, the so-called Broca (that controls the motor expression of sounds) and Wernike (that is involved in the perception of sounds) , and – above all – the vocal apparatus. When scientists say that we are similar to Chimpanzees for around 98–99% of our genetic heritage, they normally fail to add that within that 1–2% of difference there is precisely the vocal apparatus and these two brain areas. The causal relation between these elements and the creation of language is reciprocal: the former were responsible for the development of the latter, and the latter was responsible for the development of the former.

And it was of extreme importance to have that kind of resources in order to develop a device such as language: the vocal apparatus, in particular, made it possible to exploit the richness, the flexibility and the practicality of the vocal-acoustic channel. In other words, having such a flexible organ for speaking, human beings, literally, *spoke*. A cognitive use of the language was thus made possible, and the process of extension and enrichment began, evolving into the virtually limitless articulation of nowadays: it was during this process of expansion that – one may guess – language became more important as a secondary modelling system. The very same process, as applied to hands, is described by Friedrich Engels to tool-fabrication and work (1896: 545–554): hands became at the same time the cause and the product of work.

Related to this issue is also another, quite crucial, consideration: so far, ICE have been evaluated only at an ontogenetic level, and not at a phylogenetic one. When we think about the transition from the birth of the genus *Homo* (4–5 millions of years ago), when the first – very rudimental – traces of a language presumably appeared, to the *Cro-magnon* (125,000 years ago), with whom we probably have the appearance of a less-embryonic form of language, we understand that it took millions (!) of years for language to take a recognisable shape. The above-mentioned anatomycal and neurological features developed with an extremely slow pace (and randomly, if one takes a strict Darwinian perspective): apes were introduced to language, in a methodologically proper way, only half a century ago, with very few subjects, and in captivity. One cannot reasonably expect perfection already from this first generation of language speakers. Plus, even if at some point (in who knows how many centuries) we will witness in apes linguistic capabilities comparable to an average human adult, we would still have to see whether, reintroduced in the wild, these subjects will benefit from such competence, that is, if their species *needs* language. Finally, if we find out that they do, we will have to wait another shot of who knows how many millenia before natural selection creates a language-inclined non-human species. Now, with the due respect, *this* is a fair, non anthropocentrically-biased, line of reasoning.

Anyway, leaving aside these speculations on the origins of language, there remains the claim that narrativeness and distant-time semiosis are not features that characterize exclusively the human language. *Sic stantis rebus*, a "qualitative" difference between human language and other communication systems would probably appear to be the existence of these linking signs mentioned at the beginning of this discussion, that is, those solely-interoceptive signs whose exclusive function is that of connecting signs that also, and mainly, have exteroceptive meanings. Curiously enough, the big advantage of linking signs, and language by consequence, is not the ability to make meaningful signs, but exactly the opposite, i.e., creating meaningless signs. Signs that do not refer to anything when taken alone (unless a specific metaphoric use is made of them, but that is an entirely different issue), but can make a great deal of difference when combined together with meaningful signs, in certain numbers and sequences. Charles Hockett, in defining the famous (or even infamous) sixteen design features for human language (1960a), had a very similar concept in mind when describing the 12th feature, duality. Non-human semiosis seems to target most of all economy and ergonomicity: signs, when used to communicate, must all make sense. Roughly speaking, a non-human animal emits five signs if s/he has five things to say. Humans may employ a great deal of signs just for expressing one thing. But it is exactly this "meaninglessness" factor that increases exponentially the semiotic potential of human language: it creates specifications, differences, appartenances, separations, distinctions, and so forth. In a word, it creates a virtually unlimited number of relations.

So, must one finally surrender to the anti-Darwinian notion that, after all, there is some qualitative difference between humans and other animals, even if by now it is reduced to articles and prepositions only? Of course not. Also in this case, even if there was a total lack of evidence, it would be scientifically too risky to launch a new "ultimate sign of human distinction" campaign, as it was ironically called in Chapter 1. Enough of that mistake. Moreover, the fact is, there *are* cases, within the non-human animal world, in which an apparent anti-economic use of communication occurs, and where the sounds employed do not necessarily correspond one by one to a specific meaning, while the entire formulation of a phrase *does* say something. It is, predictably, the case with birds and cetaceans, particularly those communication instances that we call songs.

The humpback whales, *Megaptera novaeangliae*, are passionate manipulators of their compositions. They provide their songs with continuous variations, so that, after about 5 years, they sing a totally new song and, as far as scholars could notice, they never come back to the original version. Variations occur during the mating season. A phrase may slightly be shortened or slightly prolonged, or suddenly interrupted and replaced with a new one. A theme may be enriched with new sounds and then inserted into the whole context, or a part of it can be changed in sequence. Every whale updates and includes new melodies in its song. They are performed faster and the variation is kept within the song until the next mating season. Then, old songs are reprised at the very point they had been left last year.[19]

Now, there are certainly no elements to affirm that those extra meaningless sounds emitted by a whale are something qualitatively comparable to some human "of", "the", "with", "to", i.e., to "linking signs". The evidence is however that there are cases (and not only a few) of non-human animals using communication in an anti-economic way, cases, that is, where they say more than what they mean. Not to mention (although it does not count for the type of argument that is being expounded here) that anyway language-trained animals easily managed to learn linking signs like "this/that" (Fouts 1999: 293) or "on", "and". "in" and "of" (Mainardi 1975: 125–9), For what we know, anyway, human language remains a highly specialized form of communication/modellation in this very respect.

Coming back to the issue of Umwelt, in the strict sense, there is another issue that both IC researchers and semioticians seem to have overlooked, and that is the importance of the way the perceptual field of a given specimen (therefore, probably, of an entire species) affects the actual reception of a linguistic sign. In the paragraph entitled "Signals, indexes and icons" in Chapter 2, it was mentioned that Washoe, once learned the ASL sign for *flower*, took to name that way everything with a strong

[19] For an extensive description of the concept of variation in zoomusicology, see Martinelli 2009: 145–53)

smell, therefore using a form of iconic categorization that is not visual, as it would be almost systematic in a human being:

> When FLOWER started to become a common item in Washoe's vocabulary, it was reported in several inappropriate contexts that all seemed to include prominent odors: for example, Washoe signed FLOWER when opening a tobacco pouch and also when entering the Gardner kitchen at a time when there was a chicken boiling on the stove. Taking our cue from these observations, we modeled and molded the SMELL sign in appropriate contexts. For some time, FLOWER [. . .] persisted as an error, but gradually, a highly reliable, well-formed SMELL sign emerged. (Gardner et al. 1989: 82)

Washoe's signing must have certainly looked "inappropriate" to the specimens of a species which basis so much of its perception on the visual channel. We are given a flower, and we are told "this is a 'flower'": in 99% of the cases we will recognize other members of the category "flower" on the basis of a visual resemblance: a picture of a flower, a painting of it, a plastic flower, and maybe even an object that is not designed to resemble a flower, but – as *it kind of looks like one* – we can still label it as such in order to clarify what we mean to an interlocutor. Washoe obviously reflected in a totally different manner. The relevant characteristic of that object, to her, was its smell, i.e., the olfactory feature of the flower. Consequently, the semantic field called "flower" became to her the field of smelling objects, no matter what their visual appearance (as for humans the semantic field of flowers is a field with certain visual features, no matter what their olfactory appearance). A bottle of perfume, in this sense, fits more to Washoe's semantic field than a plastic flower. Nothing, in this process, is less "clever" or logically-legitimate than the type of categorization established by humans.

This instance is not isolated. A similar episode happened when Washoe, who was also one of the chimpanzee-painters (see Morris 1963), was asked to draw a ball. Again, as humans, our first idea for representing a ball is iconic: we trace a circle on the paper, and perhaps – if we are football fans – we add black pentagons and white hexagons inside it. But to Washoe, it seems, the most relevant iconic characteristic of the entity "ball" is kinetic, not visual, as she drew a series of progressively smaller arcs, i.e., the trajectory of the ball (Fouts 2000, personal communication). Once more, such a reasoning process is perfectly acceptable: in fact, it gives us precious information about the chimpanzee's Umwelt (or perhaps only about Washoe's Umwelt, but that would still be very interesting).

In this respect, the fact that many of the language-trained non-human subjects have proved in several different cases a high degree of flexibility in accessing the human Umwelt, and reconfiguring their own accordingly (for instance, accepting the symbolic representation of a series of objects that they would usually represent indexically), shall be saluted as one of the most remarkable achievements of these ICE.

Finally, however banal it may seem, it is important to emphasize that the identity of the ICE trainee as an individual, and not as a member of a given animal species, is by all means crucial, but has definitely been underrated by both IC scholars (at least some of them), and semioticians. The human language has not merely been taught to any chimps, but first of all to Sarah, Washoe, Moja, Lana, etc. Koko and

Mike are Koko and Mike, and only then they are gorillas. And Kanzi is Kanzi first of all, and only then he is a bonobo. This list of tautologies is quite important here, since it reminds us of the following things:

1) The trainee is first of all someone who has emotions and feelings, and is exposed to experiences and events, *then* s/he is someone on whom an ICE program is performed;
2) The individual responses of the trainees may be related to factors that are external to the experiment environment, and may simply be the result of a personal experience, or be part of the cognitive-emotional background of the subject in question; in particular
3) Observations and results (either negative or positive) of ICE have often been inappropriately generalized to an entire species, at least, if not to entire orders, or to the whole Animal Kingdom *tout court*. In simple words, when Moja fails to perform the right ASL gesture, it does not mean that all chimpanzees have failed. Likewise, if Moja will someday compose a rhyming sonnet in ASL, that does not mean that we should expect the same from Lana, Lucy, Sarah and the others;
4) An extensive account of the personal history of the subject should always be integrated with the rest of the ICE, as a fundamental and unavoidable element for the correct interpretation of the trainee's response. The success of Fouts and the Gardners with Washoe and other chimps, and of Patterson with Koko and Mike, is possibly due to the fact that this aspect was fully taken into account. As an example, one may read how carefully Roger Fouts (in 1998) relates Washoe's progresses with ASL to specific events (positive and negative) in her life.

Still, many seem to forget these apparently basic notions. Perhaps, the most evident instance where the individuality of the trainee is totally ignored concerns our expectations towards his/her behavior. It is what we may half-jokingly call "thirdness-only expectations", with obvious reference to Peirce's notions of Firstness, Secondness and Thirdness. We expose a non-human animal to a problem-solving task, by – for instance – introducing a variable in his/her usual environment. We do not only expect the trainee, who *does not* know s/he is supposed to perform a task, to focus quickly and only on that very variable, but, while doing that, we also expect him/her to privilege a line of reasoning which is by no means the most obvious and spontaneous one. This so-far-ambiguous concept can be exemplified through the words of someone who was able to depict it with extraordinary efficacy.

Sometimes, in fact quite often, philosophers and artists are able to see much further than scientists, and this is probably because, or so we believe, they do not see empathy and sensibility as biases. They rather use them as devices for a better comprehension. Thus it should not come as a surprise that no one has succeeded better in picturing the paradoxical situation just described than the writer and Nobel-prize winner John M.Coetzee. Apologies for the long quotation here, but it really makes the point clear. Plus, it is a great piece of literature:

> Let me recount to you some of what the apes on Tenerife learned from their master Wolfgang Köhler, in particular Sultan, the best of his pupils, in a certain sense the prototype of Red Peter. Sultan is alone in his pen. He is hungry: the food that used to arrive regularly has unaccountably ceased coming.
>
> The man who used to feed him and has now stopped feeding him stretches a wire over the pen three metres above ground level, and hangs a bunch of bananas from it. Into the pen he drags three wooden crates. Then he disappears, closing the gate behind him, though he is still somewhere in the vicinity, since one can smell him.
>
> Sultan knows: Now one is supposed to think. That is what the bananas up there are about. The bananas are there to make one think, to spur one to the limits of one's thinking. But what must one think? One thinks: Why is he starving me? One thinks: What have I done? Why has he stopped liking me? One thinks: Why does he not want these crates anymore? But none of these is the right thought. Even a more complicated thought — for instance: What is wrong with him, what misconception does he have of me, that leads him to believe it is easier to reach a banana hanging from a wire than to pick up a banana from the floor? — is wrong. The right thought to think is: How does one use the crates to reach the bananas?
>
> Sultan drags the crates under the bananas, piles them one on top of the other, climbs the tower he has built, and pulls down the bananas. He thinks: Now will he stop punishing me? The answer is: No. The next day the man hangs a fresh bunch of bananas from the wire, but also fills the crates with stones so that they are too heavy to be dragged. One is not supposed to think: Why has he filled the crates with stones? One is supposed to think: How does one use the crates to get the bananas despite the fact that they are filled with stones?
>
> One is beginning to see how the man's mind works.
>
> Sultan empties the stones from the crates, builds a tower with the crates, climbs the tower, pulls down the bananas. As long as Sultan continues to think wrong thoughts, he is starved. He is starved until the pangs of hunger are so intense, so overriding, that he is forced to think the right thought, namely, how to go about getting the bananas. Thus are the mental capabilities of the chimpanzee tested to their uttermost.
>
> The man drops a bunch of bananas a metre outside the wire pen. Into the pen he tosses a stick. The wrong thought is: Why has he stopped hanging the bananas on the wire? The wrong thought (the right wrong thought, however) is: How does one use the three crates to reach the bananas? The right thought is: How does one use the stick to reach the bananas?
>
> At every turn Sultan is driven to think the less interesting thought. From the purity of speculation (Why do men behave like this?) he is relentlessly propelled toward lower, practical, instrumental reason (How does one use this to get that?) and thus toward acceptance of himself as primarily an organism with an appetite that needs to be satisfied. Although his entire history, from the time his mother was shot and he was captured, through his voyage in a cage to imprisonment on this island prison camp and the sadistic games that are played around food here, leads him to ask questions about the justice of the universe and the place of this penal colony in it, a carefully plotted psychological regimen conducts him away from ethics and metaphysics toward the humbler reaches of practical reason. And somehow, as he inches through this labyrinth of constraint, manipulation, and duplicity, he must realize that on no account dare he give up, for on his shoulders rests the responsibility of representing apedom. The fate of his brothers and sisters may be determined by how well he performs. (Coetzee 1999: 126–127)

Leaving aside the specific case of Köhler and Sultan, what matters here is the line of reasoning. Hardly, if ever, the emotional condition of the trainees, as subjects, is considered important. We do not expect Firstness from them: we have a straight expectation for Thirdness. We expect them to react to our stimulus in the way we have planned (and received funding for), while the rest is either uninteresting or wrong. Why? The same principle applies to many of the previous critical remarks on Sebeok: why should a trainee not have his/her own preferences? Why should s/he not be supported in the learning process? Why should we expect him/her to be a language-machine when s/he also has a life and a personality, other than a predisposition or none to learn ASL or lexigrams, or whatever they are?

3.3.5 Morgan's Canon and Other Forms of Scientific Fear

During a biosemiotic seminar in 2004, in Estonia (see Martinelli 2006a: 276–288, for a lengthier discussion), the issue of the Morgan's Canon was raised as a form of anti-semiotic methodology, and a proposition of its exact contrary (which therefore sounded like "in no case should actions or behaviours be interpreted as the result of an inferior psychic faculty, when it is possible to interpret them as a result of a superior faculty") was named by biosemiotician Kalevi Kull *Martinelli's Canon* (a nod to the then contemporary notion named *Bankov's Razor*, already mentioned in Chapter 1). The objective of that notion was to formulate the main semiotic contradictions (and therefore incompatibility) of Morgan's Canon:

1. When Morgan talks about the "possibility" of interpreting animal behavior in a given way, he is referring to those cases where a *speculative* choice is necessary, that is, where a non-human behavioral pattern that is analogous to a human one, does not give empirical and unquestionable evidence that it is also a homologous pattern. In other words, he refers to those cases where the animal *seems* to behave or think in a certain way (similar to the way a human would behave or think in the same situation), so the point is: does the animal *seem*, or is s/he *really* behaving like that?

In these instances, evidently, all we have is this similarity between the human and the non-human behavior. Thus, logically speaking and when counterproof is missing, the starting hypothesis should be: the two patterns *are* homologous, precisely because they are similar, and that is the only thing that we know at this point. Then, we can pursue the necessary kinds of research, and try to understand if the hypothesis is confirmed or not. Still, at the very beginning, and the very beginning is the research phase to which Morgan's Canon refers, we shall think of a homology.

Let us take an example: on one side we have a car, on the opposite side we have an object that has a slightly different shape, that *seems* to move, *seems* to have four wheels, *seems* to have seats, steering wheel, windscreen wipers and a car-stereo. At the moment, that is all we know. What are we going to do? Perhaps, we should start from the idea that this mysterious object is actually a car, maybe from a different brand than the one we have on the other side, but still a car, and our research should be oriented in that direction (are those really wheels? Are those really seats? Is there an engine, a brake pedal, etc.?). That seems to be a more convenient procedure than to think that the object we are observing is actually a skateboard, so those things that look like seats are actually a simple flat board, and the wheels are much smaller than they seem.

Still, this is what Morgan's Canon does. Not only does it suggest that we should not trust appearances, it also tells us that reality is simpler (or more reduced) than it appears to be. Felice Cimatti (1998: 147–151) proposes an interesting and somewhat ironic example to illustrate the applications of this canon to the case of Washoe. According to the supporters of Morgan's canon, at the very moment Washoe is asked to perform the ASL gesture corresponding to a given object, she does not have the slightest idea of how to perform such a task, since she is not able to understand the relation between signifier (the ASL gesture) and signified (the object). Thus, she proceeds by attempts, moving her arms at random, until – unconsciously or not – her trainers display apparent satisfaction, which lets her know what the right gesture is. Here is a typical example of the CHE, which supporters of Morgan's canon find the most suitable interpretation. Unless one is able to invalidate any risk of the CHE, they say, ICE are a waste of time. Most of all, the more or less emotional relation established between researcher and animal should be avoided by creating an "emotion-free" context. Of course, to consider the use of a communication system as unrelated to social and emotional interaction is a self-evident contradiction and a more serious mistake than anthropomorphism. What is the point in learning a communication system if you have no one to use it with, do not receive adequate stimuli, and, in the end, do not have any reason to learn it? In such a case, says Cimatti, the mistake is not only methodological, but also theoretical: "if language is a system based on emotional interaction [. . .], then it is simply impossible to study it without considering this aspect" (translated from Cimatti 1998: 149).

In other words, here we face the type of situation that Paola Cavalieri (1999: 24–5) warns us against: the aprioristic parsimony of Morgan's Canon requires very often more complicated explanations for the phenomena it comments upon, therefore contradicting its seemingly economic philosophy. This way, inevitably, something of that object we were observing in the example, remains unexplained. For instance, what is that thing that looks like a car-stereo? What about those look-alike windscreen wipers? How do they fit in with the idea of a "skateboard"? Scholars in animal studies invented an *ad hoc* solution to these disturbing problems: a big, fat black box where they put everything that cannot be explained. The most common name for this black box is "instinct".

Although seemingly the most easily definable concept, at least in non human-related contexts, instinct is in fact the most complex and tricky one. Decades of lively discussions among scholars have not yet produced an even approximate definition. Instinct has been defined as a "voluntary but never-previously-put-into-action behavior", a "behavioral impulse for the accomplishment of biological functions", a "stereotyped behavioral pattern", and a "pattern put into action without a precise idea of its results". It is already evident that these definitions are more hiding than showing: they just describe an action, but they do not give account of causes and articulation. Moreover, some of them are in contradiction with each other. What seems to be certain is that instinct is often a functional and easy-to-use theoretical tool for the explanation of behaviors that are difficult to account for. As Gregory

Bateson (1969) properly puts it in one of his metalogues, instinct hence ends up being a real explanatory principle:

> Daughter: *Daddy, what is an instinct?*
> Father: *An instinct, my dear, is an explanatory principle.*
> Daughter: *But what does it explain?*
> Father: *Anything – almost anything at all. Anything you want to explain.*
> Daughter: *Don't be silly. It doesn't explain gravity.*
> Father: *No, but this is because nobody wants instinct to explain gravity. If they did, it would explain it. We could simply say that the moon has an instinct whose strength varies inversely as the square of the distance...*
> Daughter: *But that's nonsense, Daddy.*
> Father: *Yes, surely. But it was you who mentioned 'instinct', not I.*
> Daughter: *All right – but then what does explain gravity?*
> Father: *Nothing, my dear, because gravity is an explanatory principle.*
> Daughter: *Oh.*
> (Bateson 1969: 11)

A better explanation of the concept than Bateson's can hardly be given. It is clever, witty, and ironic enough to emphasize how exaggerated a meaning we provide "instinct" with. Considering the number of actions, from the simplest to the most complex ones, that are described as instinctive, it shall be much easier to define the term in one of the following ways: "Supernatural power", "Invincible weapon", or – more humbly – "Ultimate solution to every problem". To which, it should also be added (a) where one can buy some instinct, (b) how much it costs, and (c) if it is legal.

The problem with instinct is its absolute conceptual flexibility. Once we have established that it is an impulse that drives animals (including humans) to perform a (limited) number of actions, and it can definitely be accepted that there are actions that we perform quite *before* having a mental representation of them, it becomes extremely easy to apply it to every case where a mental representation is not detectable (i.e., it is not verbalized by those who experience it).

Here lies the difference: if we produced our hypotheses in accordance with our observation, we would look at a beaver carefully cutting its pieces of wood in such a way that they all weigh the same (which is what actually happens: they *do* weigh the same. See the very interesting Richard 1967 on the topic), and we would say that, indeed, the beaver is carefully cutting his/her pieces of wood in such a way that they all weigh the same. With Morgan's Canon, however, we are faced with a problem: our hypothesis now implies a superior psychic faculty. And this is very bad, if it is *possible* to interpret the same pattern as a result of an inferior faculty. We need an inferior one, that is, a faculty that does not require such an articulated mental activity (provided that estimating the weight of a piece of wood is "an articulated mental activity"). So, we buy a dose of instinct: the beaver does not have the slightest clue of what is going on, but s/he has this supernatural power, which drives him/her to chose this piece of wood instead of that one, which is too heavy and too long.

Now, apart from the general theoretical remarks on instinct itself[20] (which hopefully are sufficiently illustrated, by now), the question is: was it *possible* to apply an inferior psychic faculty in this case? Because that is what Morgan says: we do not have to apply a superior faculty, when it is *possible* to apply an inferior one. Otherwise, one might say, we shall just leave the superior one, right? On the basis of which principle are we allowed to apply the notion of instinct on the beaver's weight estimations? Was it *possible*? Well, of course, everything is possible if we want it to be: but that is exactly how Bateson's explanatory principle works. It explains something simply because we deliberately designate it as explanatory. It does not explain it in a very scientific sense. So, what did Morgan mean by *possible*? Was it a *scientific* possibility or simply a *rhetorical* one?

2. What do we mean by "superior" and "inferior" psychic faculty? Do we mean "more" or "less" similar to the human psychic faculties? If yes, as one may bet, then we are typically facing that anthropocentric bias that anyone who has fully accepted Umwelt theories should refuse a priori. If there is a transpecific paradigm to interpret psychic faculties, that should probably be adaptive skills. The more adaptive the subject, the more "superior" his/her psychic faculties. Gregory Bateson again makes an excellent point in this respect. In experimental contexts, he says (Bateson 1972: 368–370), the animal's intelligence is tested through four main points: (1) the animal "may or may not perceive a difference between the stimulus objects X and Y"; (2) the animal "may or may not perceive that this difference is a cue to behavior"; (3) the animal "may or may not perceive that the behavior in question has a good or bad effect upon reinforcement", that is, that doing the right action will be granted with (usually) food; and (4) the animal "may or may not choose to do 'right', even after he knows which is right". As it can easily be deduced, the four points are set in order of priority, that is, 4 is possible if 3 is accomplished, 3 is possible if 2 is accomplished, etc. Therefore, step 4 reveals a superior psychic

[20]The discussion on instinct should not end with this note, as the abuse of the notion is a real threat to scientific inquiry. Scholars should take a very critical approach towards those colleagues who so easily refer to this black box. In general, every application of the notion of instinct to complex multi-phased actions should be forbidden or accepted *only if* very convincingly and empirically motivated: it is all too easy to open the black box and put in processes and actions that we are not able to explain otherwise. Except that this is not science: it is much closer to religion. What is the difference between explaining a complex phenomenon with the existence of instinct and explaining it with the existence of God? And why not aliens? It is necessary to be very firm on this also because in order to protect the notion of instinct itself from massive trivialization. There *is* a number of human and non-human phenomena that originate at a pre-cognitive level and produce an actual physical action. But if one wants to make sense out of these instances, they first of all must be circumscribed: extemporaneous escape from danger, movements supporting physiological needs, innate competencies on immediate parental care . . . This is what should be called "instinct". Hunting strategies, elaborate parental care, establishment of social relations. . . if one calls those instances "instinct", there is no real need for any ethologist or zoosemiotician. A priest is enough.

faculty than 3, and so forth. Now, are all these steps really tested in experimental contexts? Do trainers really go through the whole procedure? Bateson is pretty doubtful.

> Let me now consider for a moment the art of the animal trainer. From conversations with these highly skilled people – trainers of both dolphins and guide dogs – my impression is that the first requirement of a trainer is that he must be able to prevent the animal from exerting choice at the level of step 4. It must continually be made clear to the animal that, when he knows what is the right thing to do in a given context, that is the only thing he can do, and no nonsense about it. In other words, it is a primary condition of circus success that the animal shall abrogate the use of certain higher levels of his intelligence. The art of the hypnotist is similar. (Bateson 1972: 369)

3. Any action, even the simplest one, never has a unique impact, and by consequence can never be interpreted as monolith. A Canon that promotes simplicity is in practice a canon that prefers to "chain" observation in such a way that it gives the least possible disturbance.

Clever Hans was undoubtedly unable to perform mathematical calculations, so, to Morgan, all we need to know is that this horse was not a good mathematician (what would have been a superior psychic faculty), but simply manipulated his observers (what is an inferior psychic faculty). Period. And one cannot help wondering why are we so superficial with regards to that episode. And why we should evaluate the events only in a negative sense, as something (mathematical calculation) that *did not* happen. What we also have is a horse that, in front of dozens of observers, was always – or almost always – able to (a) detect an even microscopic facial or postural expression, (b) process these expressions mentally, and (c) understand which one, among the many, was to be interpreted as a message of approval.

And maybe there was even more to it than this. It was not only the ability to detect approval, it was also detecting the right degree of approval. Let us picture the scene: somebody asks Hans to indicate the square-root of 441 (21). The spectators gathered around the horse are very sceptical and determined not to applaude or congratulate unless the horse really stops his counting at 21. Thus, Hans starts to hit his hoof on the ground. After two or three hits there is already a quite interesting range of expressions among the observers: somebody is chatting, somebody is laughing, somebody is betting that the horse will not even reach 10, and so on. Everything is accompanied by some non-verbal language that varies according to the related emotional state, but also to the subject involved (one mat just think of how many ways to laugh exist). Hans understands that all these expressions are not relevant, and keeps on hitting his hoof. After some ten hints, somebody already goes like "Well… not bad!", and in general there is an increasing attention towards the scene. In this case, too, Hans could not care less, and proceeds with his counting. Here we are at 18–19 hits: people grow excited and surprised: "hey, he's doing it! He's really making it!". These are all expressions that one might easily mistake for a "definitive" approval, the one that Hans is looking for. But no, the horse does not stop yet. He reaches 21, and then, only then, he understands that the type of expression he was looking for is now printed in almost everybody's face: it is the final approval, the one that suggests he should stop, because people are happy with his performance.

Now, an hypnotist or a psychologist could not do any better: this is for sure. Still, to Morganists, all this seems not to have any importance: what matters is that Hans was not an Einstein. To be a Freud is not enough. Why?

The contradictions of Morgan's Canon certainly provide solid ground for an alternative concept (an alternative Canon) to grow. However, the development of the concept should not be exclusively in opposition to something, but also in favour of something else. The issue of anthropomorphism, which has already been illustrated, presents for example several obscure points. What is really anthropomorphic, given that many characteristics are human but not only human? Is anthropomorphism really to be condemned in toto? And most of all, could the fear of anthropomorphism be more dangerous for scientific research than anthropomorphism itself? In the last 20–30 years, several scholars in animal-related studies developed a less hostile attitude towards anthropomorphism. They for instance realized that promoting a certain empathy between humans and other animals does not necessarily constitute an anthropomorphic mistake, and that neglecting it might be even more misleading. An example that is really worth mentioning, called "critical anthropomorphism", is analyzed by Luisella Battaglia, in her essay *Etica e diritti degli animali*:

> Critical anthropomorphism aims to use human experience in a critical manner, in order to recognise emotional manifestations, putting into relation our most immediate subjective intuitions with comparable notions and data provided by neurophysiology, ethology, zoology etc. This way, empathy, typical of classical anthropomorphism, is integrated with the most recent scientific research on animal life and behavior. In support of such an approach, we could say that if human-animal similarities are accepted for scientific experimentation purposes, they should also be accepted in the field of emotional sensibility. In any case, any doubt about emotional sensibility should be gauged so as to benefit the weakest subject. In particular, the presupposition of similarity, when there is no clear counterproof, should be interpreted in favour of the animals. (translated from Battaglia 1997: 123–124)

In other words, if, as often happens, scientific research involving non-human animals – either aiming to study animals themselves, or aiming to study humans *through* other animals, as in medical research – is based on presuppositions of similarity between humans and other animals, then it shall be fair to apply the same principle on a more consistent basis.

> It is not fair [. . .] to maintain that there is a resemblance between humans and animals when we use [the latter] in medical laboratories in order to establish our rights over them, and – at the same time – to maintain [that animals are different] as moral subjects, in order to avoid our duties (translated from Battaglia 1997: 124).

Battaglia's main point here is that philosophical speculation on animal behavior (as on other issues, as well) should coordinate scientific data and ethical reflections, in such a way that the latter support the former, when these are not able to provide an exhaustive response. When science is not able to give a 100% reliable answer on a given behavioral phenomenon, than ethics should be allowed to say the final word. And ethics, as conceptual basis for modern jurisprudence, is based on the principle that the weaker subjects, those who are not able to verbalise what they are doing and what they mean by that, should be favoured and protected. Such is the case with categories of subjects like infants or mentally-disadvantaged people.

Going back to the specifically scientific aspect, Battaglia's conclusion is that all the generally emotional manifestations previously considered dangerous for scientific research (empathy, most of all, as the generator of CHE par excellence), if handled with care, become useful and plausible clues for the whole theoretical apparatus. This is because:

1. By definition, empathy helps comprehension; hence it is easier to interpret a given phenomenon if we identify ourselves with it to some extent; and because
2. Animals are not only subjects of human scientific research: they are also the category humans belong to. Humans are not *more or less* similar to other animals: they *are* animals. To ignore this fact means to ignore a crucial part of the story.

To add a specific semiotic note to it, the concept of critical anthropomorphism, and empathy in particular, promote nothing other than Peircean abduction, as a valuable method for scientific inquiry.

In the light of all these points, a fair formulation of an alternative, purely zoosemiotic, Canon (ZC) is now possible. So, the rule: *In no case should actions or behaviors be interpreted as the result of an inferior psychic faculty, when it is possible to interpret them as a result of a superior faculty*. Comments:

1) Unlike Morgan's, *ZC approaches the problem of inferior and superior psychic faculties in terms of abductive, not rhetorical, possibility*. A behavioral phenomenon X (e.g., a beaver carries pieces of wood that weigh all the same) is observed in non-human animals; X can be explained with hypothesis A (the beaver is mentally estimating the weight of the pieces) or B (the beaver is driven by instinct); X resembles A on the basis of perceived similarity with A_H (i.e., A as performed by human beings); hence there is a reason to pursue A. Pursuing A does not mean automatically *stating* A as the final word on the matter. It means that our *research hypothesis* is now A, and the next step of our inquiry should be *oriented* towards A. If that proves to be a failure, that is if pursuing A is not *possible*, then further hypotheses shall be considered. In this sense,
2) *ZC aims at economic and ergonomic research*. Pursuing A is more economic and ergonomic than pursuing B, because A already provides clues and research tools, therefore a path to follow has already been indicated.
3) *ZC refuses the criteria for establishing differences between psychic faculties in the way Morgan's Canon does*. Such criteria, rather than anthropocentric, should focus on the notions of Umwelt, adaptation, free will and semiosis in general. Interpretive hypotheses that rely on concepts like instinct are by principle anti-semiotic, as they deny any interaction between the animal and her Umwelt.
4) As a semiotic-centred formulation, *ZC treats any behavioral phenomenon as complex and multi-layered and refuses on principle reductive one-sided interpretations*. An instance like the one involving the horse Hans represents a

paradigmatic example of a multi-faceted semiotically-rich subject for scientific inquiry, which scientists reduced to a quasi-joke by refusing to approach the episode in all its aspects. Once it had been established that the horse was not capable of mathematic computing, they ignored his still-astounding ability to interpret body-language, therefore missing a good chance to deepen the knowledge of horse communication in relation to their Umwelt.

5) *ZC fully supports the notion of critical anthropomorphism* (Battaglia 1997: 123–124), *and therefore rejects reductive hypotheses on an ethical basis, as well.* If the interpretation of the behavioral phenomenon X is complicated by the impossibility of making a totally reliable scientific choice between the hypotheses A and B, then – in addition to the motivation provided in point 1 of the present list – another reason to pursue A is the ethical one: "the presupposition of similarity, when there is no clear counterproof, should be interpreted in favour of the animals" (translated from Battaglia 1997: 124).
6) On a more strictly scientific level, *ZC promotes methodological empathy and,* more generally, *supports a more balanced and less radical use of anthropomorphism.* Anthropomorphism should be avoided when it applies human species-specific predicates to non-human species and when it produces distorted perceptions of the reality analyzed. It should not be avoided, in fact encouraged, when, through empathy, it helps comprehend the given phenomenon, when it favours biocentric approaches to non-human species, and when it is employed for didactic purposes, in order to facilitate attitudes of interspecific acceptance and compassion.

Chapter 4
A Glossary of People, Paths and Ideas

Abduction

Ancient method of investigation and reasoning, stemming from **Aristotle**'s logic, which gained scientific legitimation thanks to Charles S. **Peirce**. A surprising phenomenon, X, is observed. Among hypotheses A, B, and C, A is capable of explaining X. Hence, there is a reason to pursue A. Peirce ascribes the origin of the term to Aristotle, precisely to the word ἀπάγωγη, *apagoge*, translated as *abduction* (erroneously, according to Peirce, who thought the right translation should be "retroduction"; in his writings he uses both terms). The usage and contents of "abduction" are in conflict with "deduction" and "induction". If deduction means to proceed from a rule and a case to a result, and if induction proceeds from a result and a case to a rule, then abduction proceeds from a rule and a result to a case. Abduction is thus a "logic of discovery", a kind of critical thinking which opens doors of opportunity for scientific research and which, in the end, confirms the classifications provided by inductive and deductive procedures. In practice, according to Peirce, scientific investigation should go through the following phases: (1) Observation of an anomaly; (2) Abduction of hypotheses that explain the anomaly; (3) Inductive testing of the hypotheses in experiments; (4) Deductive confirmation that the selected hypothesis predicts the original anomaly.

In **>zoosemiotics**, abduction is not only a useful method of reasoning, but quite often the only one available. Particularly, it is an essential part of the paradigm of **>critical anthropomorphism**, as it tends towards an empathic, **>emic** perspective.

Disclaimer:
[1] The entries included in this glossary refer to words, concepts and scholars, which are considered (a) important in the development and systematization of zoosemiotics; and/or (b) applicable, to some extent, to zoosemiotic research. The glossary, therefore, does *not* include semiotic terms, even fundamental ones, that are already defined in other, more general, semiotic contexts (handbooks, encyclopaedias, monographs), and on which zoosemiotics, and this companion particularly, have nothing *else* to say. In that sense, one may compare the presence of a crucial notion like "code", that was slightly "adapted" to the zoosemiotic context, with the absence of the other crucial notion "sign", on whose "traditional" definitions this companion has nothing to add, except in those specific applications, like "icon", "index", "kinaesthetic signs" and others, which have an entry of their own.

D. Martinelli, *A Critical Companion to Zoosemiotics*, Biosemiotics 5,
DOI 10.1007/978-90-481-9249-6_4,

Abstraction

Abstraction, or generalization, is the ability of representing an event/entity in terms of general qualities/characteristics, independently from concrete realities, specific instances or actual objects. In semiotic terms, to abstract means to be able to consider an object as a token of a given type, to mentally represent that type, and finally, on the basis of that representation, to recognize other tokens of the same type. For example, abstracting *Hard Times* as part of the type "Dickens' novels" allows us also to recognize *A Tale of Two Cities* within the same whole.

The process applies also to the capacity of spotting the type even when the second token is merely a sign of the first one, or is manifest through another medium. In the previous example, not only an actual hard copy of *Hard Times* can be recognized as Dickens' novel, but also an electronic file of it, or even a picture of the cover appearing in a magazine. Evidently, thus, abstraction closely relates to such concepts as **>mind**, **>mental representation**, and **>symbol**ic **>semiosis**.

In psychological studies, abstraction is defined as a "higher brain function", and therefore has been long considered a **>species-specific** human feature. Already in the first decades of the twentieth century, however, and even more effectively starting from the 1950s, an increasing number of ethological studies (Hsiao 1929, Wolfe 1936, Cowles 1937, Köhler 1943, 1949, 1952, **>Eibl-Eibsfeldt** 1951, Rensch and Dücker 1959, Rensch 1965, Lehr 1967, **>Goodall** 1968, among others), in both natural and experimental contexts, found out that abstract representations are extremely common among non human animals. Subjects of such studies included pigeons, toads, elephants, parrots, ravens, squirrels, rats, monkeys, cats, and of course great apes.

The difference between the first and the second type of abstraction is pretty marked. If the mere ability to attribute a token to a type can be easily inducted with a simple behavioristic stimulus-response technique (after all, the infamous Pavlov's experiments were producing in the dogs nothing else than a capacity to generalize), at the same time, from a semiotic point of view, more relevant are those

---

[2] As it will be soon clear, most room in this glossary has been given to concepts and scholars that had no specific opportunity to be thoroughly discussed in the other parts of this companion, or for specific portions of general topics that *were* dealt with, which had to be overlooked despite their pertinence within the zoosemiotic discussion. The remaining entries, often shorter in length, are always provided with an indication of where, in the book, they were treated more in detail.

[3] A third reason for differences in the entries' length, or for presence/absence, are of course the personal scientific choices of the author. This companion, it shall be reminded, is "critical": it presents a view on zoosemiotics that is hopefully coherent in itself, but that is far from being complete, encyclopaedic and impartial. Priority was given to those concepts and scholars that were perceived to require more attention (because they were not previously given enough of it, or because they were given a *different* form of attention, that this companion claims was not fully appropriate).

[4] Omissions due to author's limitations in knowledge and competencies, finally, will certainly appear, and – regretfully – not only rarely. For these, the author can only apologize to the readers, and hope that his work will at least serve as a basis for a further, more accurate job.

instances that focus on the animals'capacity to recognize a given object via a completely different medium, with one only, or very few, characteristics in common with the first token (like the photograph of a person, related to the real person it portraits, only via a visual-iconic relation). Lehr (1967), who found that rhesus and capuchin monkeys were able to recognize pictures of insects and flowers, or the different instances of problem-solving in different species discussed by Köhler, and – in chimpanzees – by Goodall (1968), are among the most interesting studies on the topic.

Acoustic Channel

The sensory mode (>**channels**) connected with the production, emission and reception of sounds. Acoustic signs can be produced through the vocal apparatus, or by hitting parts of the body or the environment. The main pros of the acoustic channel are the broadcast transmission (i.e., the possibility of communicating in darkness, around corners, etc.), directional reception, rapid fading (which makes a fast continuation and response possible, thus increasing the communicative potential), immediate feedback, and a very high degree of specialization (as the example of language, developed first of all as a vocal-acoustic sign system, clearly demonstrates). The main limitation of the acoustic channel lies in the rapid fading, which – besides its aforementioned advantages – also implies that a message does not last in time, and must therefore be replaced by other means of communication (typically, the >**chemical channel**) in those cases where an enduring message is needed.

Adaptation

Adaptation can be defined as the result of a (normally gradual) process of phylogenetical and ontogenetical adjustment of an organism to its environment. Generally interpreted through the evolutionistic framework, adaptation is studied in zoosemiotics mostly using the interface of the >**Umwelt** theory.

In addition, as the concept of >**biosphere** is perceived as overlapping with that of >**semiosphere**, zoosemiotics is also interested in analysing how adaptation also involves the sign systems and repertoires of an organism, for example through processes of >**ritualization**.

Adoption

The process of raising one (or more) young specimen that is not offspring of the adoptive adult, or of any close member of the same community (a phenomenon known as "alloparental care"). Adopted and adoptive subject may belong to the same (>**intraspecific** adoption) or to different (>**interspecific** adoption) >**species**. From a zoosemiotic point of view, the phenomenon is interesting because it is based on a particular form of >**communication** (mostly visual) between the adult and the young. The interaction is normally started by the latter, who signals his/her status (particularly the age) to the adult, who then responds by accepting (i.e., manifesting parental care patterns) or refusing (sometimes aggressively) the new role. The phenomenon of adoption is mostly observed in birds and mammals.

Aesthetics

Generally speaking, Aesthetics stands for the study of the values of senses, emotions and taste, but in the modern sense (i.e., since the establishment, commonly attributed to Alexander Gottlieb Baumgarten, of "Aesthetics" as a branch of Philosophy), it is mainly defined as a philosophical reflection on art, culture and nature.

The application of aesthetic values and actions to non-human animals has been a leit-motif in western philosophy since the presocratics, and was more systematically studied since >**Darwin** 1871 and 1872 (and particularly in the mid twentieth century, thanks to studies like >**Morris** 1963). Within zoosemiotics, a groundbreaking research appeared in >**Sebeok** 1981 (chapter "Prefigurements of Art"), where the problem of the tension between the commonsensical ideas of art as a non-utilitarian form of behavior and of non-human behavior as strictly utilitarian is addressed:

> Over and over, we keep encountering the same pivotal aesthetic paradox: this emerges from a profound confusion about purpose; it drives us to ferret out compulsively any semblance of utility, usually defined as adaptive value. We find it difficult to conceive of art as a coherent part of animal life and can scarcely imagine it as an adornment of the creatures' leisure. All researchers in this field are stamped by a tension between a deeply felt conviction on the part of many distinguished and sensitive biologists that artistic activity indeed exists in the animal world, and the inability to face its presumed lack of importance, even uselessness. [. . .] the position assigned to the aesthetic life in Western culture, from Plato onwards, is imbued by an uneasy fluctuation between these two attitudes, that art is at once useless and fraught with significance, purposeless and yet important. (Sebeok 1981: 232)

In this companion (Chapter 2, Section 2.4.3), (zoo)aesthetics has been defined according to five major points:

1) The biological, not only or not too directly utilitarian, nature of aesthetic phenomena (as for instance argued in Hartshorne 1973);
2) The connection – also an etymological one – between the word "aesthetic" and senses, perception, plus concepts like "pleasure", "perception", "taste", sense of "beauty", etc;
3) The formulation of aesthetic messages, as suggested in Jakobson 1963, as more concerned with signifiers rather than with signifieds;
4) The aristotelian idea of tragedy (and by extension of all aesthetic messages) as *parà tèn dóxan* (i.e., unexpected, contrary to common beliefs) and *katà tò eikòs* (i.e., likely, believable);
5) The formulation of aesthetic information decoding, as suggested in Eco 1968, as consequent of three factors: (a) contextual interaction; (b) relative arbitrariness of the "matter" of signifiers as regards the signifieds (as in the

case of onomatopoeic words); (c) the levels of reality involved in the message articulation (physical matter of the signifiers, denotation, connotation, and so on).

Alarm Call

A sign or signal emitted by an animal in order to warn others of a danger that s/he has spotted or perceived. Scholars have emphasized that alarm calls may use the >**visual** (through specific bodily displays), the >**acoustic** (through the emission of vocalizations) or the >**olfactory channel** (through the production of chemical substances).

In zoosemiotics, the study of alarm calls, particularly in the monkeys of the genus *Cercopithecus* (see the seminal Cheney-Seyfarth 1990), has produced revealing findings in terms of the symbolic and syntactic use of signs among non-human animals. Vervet Monkeys produce three or four different kinds of alarm calls, depending on the specific kind of predator in the vicinity. A loud barking call is given for leopards, a short, double syllable cough stands for eagles, and a "chutter" sound stands for snakes (an additional call type appears in certain communities when baboons or human beings are in the vicinity). The calls are not similar to the sounds that those predators utter or produce, they are therefore of symbolic type, and the response of other monkeys to a given call is appropriate for escaping the corresponding predator.

Moreover, recent studies on alarm calls in the Greater Spot-nosed Monkeys (like Arnold-Zuberbühler 2006), have shown evidences of the existance of syntactic structures. This species produces two main alarm sounds, designating respectively a leopard, and a bird of prey. In addition to these designations, the monkeys form sequences with these sounds, whose meaning is not the simple sum of the parts (e.g., "A leopard and an eagle"), but something entirely different (generally interpreted as a suggestion to the rest of the group to move somewhere else).

This question is discussed in Chapter 2, Section 2.2.2.

Alarm Response

See **Alarm call**

Alex (1976–2007)

Short for *A*vian *L*earning *EX*periment, Alex was a male African Grey Parrot *Psittacus erithacus* trained in a rather successful >**Interspecific Communication Experiment**, by animal psychologist Dr. Irene Pepperberg. The training started in 1977 and was conducted first in University of Arizona, then at Harvard University and finally at Brandeis University, in Massachusetts, where Alex died of unknown causes at age 41 (against a life expectancy for his species of about 50 years).

The experiment started in a period when it became clear that, despite psychic faculties that are possibly the closest to the human ones, chimpanzees and other great apes do not have adequate anatomic and physiological characteristics to learn *spoken* human language. While these species were therefore trained with alternative methods (such as >**American Sign Language** or >**Lexigrams**), Irene Pepperberg thought that the idea of teaching to speak could be slightly modified and applied to a species that, in fact and notoriously, does not have these expressive limitations (if anything, it might have cognitive handicaps, as it is not a great ape, and not even a mammal). That was the case with the parrot Alex. The experiment was also meant to disprove the stereotype that parrots are just able to repeat things mechanically.

Alex proved to be able to understand and pronounce correctly about 150 English words, including several with multi-layered semantic functions, like "bigger", "different", "same", "zero", "over", "under", "smaller", and others. He could master quantities of objects up to six items, and also, not long before his death, started to learn reading, in order to relate written and spoken words (at the time of his death, he had been able to learn reading sequences of two letters). He was also, similarly to >**Koko** and >**Washoe**, capable of inventing new words, when his vocabulary did not allow him to properly refer to a given object. The most often reported example is the word "Banerry", a combination of the known words "Banana" and "Cherry" for naming an apple (similar to the former in consistency, and to the latter in colour). All in all, he was estimated to possess the intelligence of a 5 year old human being.

Alex's daily training consisted of different types of interviews, in which the trainers would ask him to name, request and describe objects and actions. When the answer was correct, instead of directly receiving a treat, Alex had to explicitly ask for it (e.g., "Want a banana", "Want a nut"). In order to avoid the >**Clever Hans Effect**, the experiments were structured in the so-called double-blind way. One trainer would ask the questions to Alex, and another one, who was unaware of the questions and thus impossible to be biased in his/her interpretation, would detect and transcribe the parrot's answers. Alex had a score of correct answers of 80% of the questions he was asked (including rather specific ones), and his mistakes were basically due to two factors: (1) *misunderstanding*: sometimes Alex would react to words that are similar in sound to the ones actually pronounced, (cable/table, fork/cork etc.); (2) >**Umwelt** differences: for instance, Alex would classify a piece of paper and a piece of leather in the same way. Possibly his criteria for categorization had more to do with the bi-dimensional and squared shape of the two objects, rather than their matter. In that sense, the request had to be considered inadequate as it contained more than one variable, and was thus subject to different categorizations of pertinence.

Shortly before Alex's death, the new "sensation", in the field of speaking parrots, became N'Kisi, also a *Psittacus erithacus*. By January 2004, N'kisi was reported to possess a vocabulary of nearly 1,000 words, frequently combined in complete sentences, with appropriate verb formes and tense, and with a variety of applications that go far beyond the information collected during the training.

American Sign Language

Often shortened as ASL or Ameslan, American Sign Language is the main sign language of the Hearing-impaiered community of English-speaking of United States, Canada, and Mexico. *Ad hoc* prepared versions of ASL were employed in some of the most successful **>Interspecific Communication Experiments**, particularly with the orangutan **>Chantek**, the gorilla **>Koko**, and the chimpanzee **>Washoe**. The main characteristic of ASL is that its **>semantic** and **>syntactic** aspects are totally comparable to the *normal* verbal language.

Ameslan

Short for **>American Sign Language**.

Analogies-Homologies

In biology, two structures or patterns are considered (1) *analogous*, if they are similar in function or constitution, although resulting from different evolutionary paths, or (2) *homologous*, if they share a common origin and the basic structure, even if they may evolved into substantially different functions. Homologies can be *phyletic* or innate (i.e., transmitted via genes), and *by tradition* or acquired (i.e., transmitted via culture). In zoosemiotics, as in other animal studies, the question anlogy-homology often emerges in relation to behavioral patterns (in this case, *semiosic* patterns) that show a certain degree of similarity between human and other animals.

The question is discussed in Chapter 2, Section 2.3.2.

Analogy

See **Analogies-Homologies**

Animal

In the denotative sense, an animal is a living organism of the kingdom *Animalia* bearing the following characteristics: (a) being multicellular; (b) being **>heterotroph**; (c) developing from an embryo (that derives from gametes produced in specialized organs). In most cases, animals are motile (in one or more stages of their life cycle) and provided with one or more sensory apparatuses. Every living organism possessing these characteristics is a potential subject for zoosemiotics research.

From a cultural point of view, the concept "animal" presents several connotations, and in this sense becomes an important topic for **>Anthropological zoosemiotics**. Among these connotations, one should mention at least the following:

1) Animal as "any other animal except humans"
2) Animal as referring to exclusively human characteristics

3) Animal as referring to a particularly uncivilized human being
4) Animal as referring to aggressive and/or violent
5) Animal as referring to natural and/or instinctive attitudes
6) Animal as referring to remarkable physical (mostly sexual) performances
7) Animal as referring to the ability of a human being to adapt to a certain context
8) Animal as referring to a zoomorphic non existent creature

A lengthy discussion on this topic appears in this companion in Chapter 3, Section 3.2.

Animal Ontology

Or Ontology of non-human animals. Theory, grounded in phenomenology, proposed in San Martin and Pintos 2001 as a continuation and extenstion of earlier reflections by Edmund Husserl on humans and other animals as living beings sharing a similar bodily experience:

> If animality is what we most originally experience, the common background from which we start before any division between "human animals" and the "other" animals is made, we need an ontology of animal or animate life as such, and, to begin with, animality, as it can also be called, is not to be identified with the soul but with "the concrete unity of body and soul". The "animality" of our being is where we are all partners, and from that experience we may outline an animal ontology. All animals, both humans and non humans, are transcendental subjectivities. Even though there are differences in level for nonhumans between the very elementary and those higher and closer to humans, we will ignore these obvious details here. (San Martin and Pintos 2001: 355)

The ontology is articulated in 11 points, and, in this companion, became part of the thesis on **>zoosemiotic universals**:

1. Every animal (EA) is a *Körper-Leib*;
2. EA is an *Ichilich* that rules its own body;
3. EA experiences a mental life articulated in time units;
4. EA experiences its own body in a direct way;
5. EA lives in the environment because it lives in itself; by consequence
6. EA is in material relation with the world;
7. EA perceptually feels itself as the 0 point of the world;
8. EA perceives the world from one point of view;
9. EA experiences a common animal world, and a social horizon of its own community/group/species;
10. EA – when interacting – has a set of semiotic problems to solve;
11. EA interacts with others primarily on an affective and emotional basis.

A more detailed explanation of this list and its context is in Chapter 2, Section 2.3.2.

Anthropocentrism

The philosophical or commonsensical attitude of considering the human animal at the center of and above other species (or other elements of reality). Anthropocentrism interprets Nature as (a) an entity existing *apart from* and *for the benefit of* humans, so that (b) nothing in Nature can be considered in itself, autonomously from humans; and (c) it is ethically acceptable for humans and non-humans to be treated in different ways. Nature is seen as lacking an *intrinsic value*, and instead carrying an *instrumental value*, i.e., the values it has for and to humans. In this sense, an investigation on anthropocentrism is crucial in **>zoosemiotics** for both methodological and ethical reasons.

In this companion (Chapter 4, see "Anthropocentrism"), and for the purposes of the human-other animal relationship, anthropocentrsism is classified in the following terms:

1) *Default*: an unavoidable form of anthropocentrism, related to the fact that the subject who observes a given animal species is evidently a human being, with all its resources, limits and modes of categorization.
2) *Binary*: a biased form of anthropocentrism. Here, the fact of being a different entity from the object observed (human, rather than another animal) produces a dualistic interpretation of reality, based on criteria of *difference* (*qualitative anthropocentrism*) and/or a strongly *hierarchical identity* (*quantitative anthropocentrism*), which puts the observer, and the group s/he belongs to, in a superior position in relation to the group observed.

Anthropomorphism

The representation of the Deity, or of a polytheistic deity, under a human form, or with human attributes and affections; or (what is the case in **>zoosemiotics** and other animal-related studies) the ascription of human characteristics to things not human. In animal studies, anthropomorphism represents an extremely problematic issue. In general, it is considered a scientific mistake or bias, which, historically, might have damaged or delayed a fair interpretation of animal behavior (as for instance in the typical case of the so-called **>clever Hans effect**). However, in the last few decades, particularly within philosophical discussions (see also **>Critical anthropomorphism**), there has been a comprehensive revision of the perception of the pros and cons of this attitude, as part of which its positive aspects have been emphasized.

Fisher 1990 presents an interesting methodological classification of anthropomorphism. A first separation is made between *interpretive anthropomorphism* ("all of the usual cases of ascribing mentalistic predicates to animals on the basis of their behavior" – Fisher 1990: 100), and *imaginative anthropomorphism* ("the productive activity of representing imaginary or fictional animals as similar to us" – Fisher 1990: 100). Interpretive anthropomorphism is in turn divided into *categorical* ("ascribing mentalistic predicates to creatures to which the predicates don't ever in fact apply" – Fisher 1990: 101) and *situational* (an animal's behavior is interpreted "in ways that could possibly apply to that animal in other circumstances, but which do not in the situation in question" – Fisher 1990: 101). Lastly, Fisher divides categorical anthropomorphism into *anthropomorphism by species* and *anthropomorphism by predicate*.

A lengthy discussion on anthropomorphism starts in this companion from Chapter 3, Section 3.3.2, and continues more or less until the end of the whole Chapter 3.

Anthropological Zoosemiotics

The branch of zoosemiotics that studies the semiotic interaction between human beings and other animals, including those of cultural and/or sociological type. Strongly anticipated by >**Sebeok** and by the zoo-biologist Heini >**Hediger**, this field is discussed at length in this companion in the entire Chapter 3.

One form of anthropological zoosemiotics, of a mostly biological orientation, is called *Communicational*, and regards areas of inquiry like applied zoosemiotics and interspecific communication. A second form is named *Significational/Representational*, and concerns the vast areas of myths, tales, allegories, ethics, and systematic classifications, in which non-human animals are for humans mostly a philosophical problem.

Anthroposemiotics

The field of semiotics that studies human >**species-specific** >**semiosis**. Though not problematic at all in its definition, the expression raises some questions when it comes to its "location" within the map of semiotic studies, particularly in relation with >**zoosemiotics**. Generally, there are two schools of thought: those who think that anthroposemiotics is part of zoosemiotics, exactly as the latter is part of biosemiotics (according to the logical scheme *human being = animal = living being*); and those who instead maintain that, because of the alleged species-specificness of such traits as >**language**, >**aesthetics** or >**culture**, the human beings deserve an entirely different niche within semiotic studies, *separated* from "natural" semiotics (zoosemiotics, >**biosemiotics**, >**ecosemiotics**, phytosemiotics, etc.), and autonomous from both the methodological and analytical point of view (musical semiotics, film semiotics, semiotics of marketing, etc.). A simplification of the situation might state that the first interpretation belongs to the semiotic tradition, while the second is more proper of the semiological one.

Curiously, the general tendency (at least in those environments that seem to have overcome the strictly anthropocentric semiological approach) is that of acknowledging the scientific and logical correctness of the first categorization (not only the human being *is* an animal, but also concepts like >**culture** and >**aesthetics**, and possibly >**language** too, are far from being inapplicable to non-human species), but *de facto* acting according to the second one (by, for instance, constantly keeping separate zoosemiotic from anthroposemiotic topics in congresses and publications, relegating the former in the company of other biosemiotic contributions, with the consequence that animal-related topics share their space with plants and micro-organisms, rather than with the human animals).

Anthrozoology

Partly an umbrella term, and partly a discipline with a specific paradigm, anthrozoology is an interdisciplinary field of inquiry originating within social sciences, and that nowadays comprises studies in anthropology, art, education, ethology, history, literature, philosophy, psychology, sociology and veterinary medicine. It is defined quite simply as the scientific study of the relationship between human and nonhuman animals, with the goal of creating "theoretical and conceptual bridges that not only link together widely separated disciplines but also span the gulf between the world of humans and the life of the rest of the planet" (Podberscek-Paul-Serpell 2000: 2). Such bridges are mainly constructed by looking at the symbolic, economic, ecological, and social consequences of human-animal relationship, thus encompassing topics from both human and natural sciences. According to the literature available, such topics include most frequently: human-pet relationships and their link with physical and psychological health; animal assisted therapy; different forms of zoonosis; abuse and cruelty to animals; and companion animals as social facilitators. Institutionally, anthrozoology moved its first steps in 1987, with the first issue of the journal *Anthrozoos*, and in 1991, with the creation of the International Society for Anthrozoology (ISAZ) at Cambridge University.

Anthrozoology is predictably an extremely close "relative" of >**anthropological zoosemiotics**. The latter may be interpreted as *part* of the former (that is, the specific *semiotic* addition to the long list of disciplines contributing to the field), or a different area of inquiry altogether (when one emphasizes the mainly socio-psychological identity of anthrozoology, as opposed to a study of the human-animal relationship as a phenomenon mediated by signs and codes).

Among the foundational literature on anthrozoology: Arluke-Clinton 1996, Serpell 1996 and Wolch 1998.

Anthro-Zoosemiotics

See **Anthropological zoosemiotics**

Appeasement

Semiosic pattern/s, generally of submissive type, displayed by an animal to another in conflict situations, in order to reduce the aggression shown by the latter to the former. Appeasement is normally used for avoiding the need to escape from the aggressor.

Applied Zoosemiotics

The application and exploitation of >**zoosemiotics** for human benefit (breeding, agriculture, veterinary medicine, etc.).

Aquinas, Thomas (1221–1274)

Italian philosopher and theologist. Possibly the highest philosophical authority of Catholic Church, Thomas Aquinas takes the principal responsibility for the penetration of Aristotelian ideas in the western Christian world: his own work can be read as an attempt to conjugate Aristotle (whom he calls "the Philosopher") and Christian precepts. This implies, among other things, a a clean break from overly mystical and spiritual philosophical formulations (as those of Ugo of San Vittore, for instance), to be replaced by a firmly establishment of dogmatism, ratio and hierarchies.

Although very little concerned with non-human animals in general, Aquinas' precepts about the human-animal relationship set decisive ethical standards for the Christian Church, producing a perception of such relationships that, to many extents, is still observed nowadays by fervent believers. In his *Summa theologiae*, indeed, Thomas makes a clear point on the unbridgeable difference between humans and other animals. The human being is an intellectual creature, master of his actions; all other animals are subordinated creatures, *functional* to the intellectual one. The Aristotelian framework soon becomes clear: in the natural world, just like the political one, there are masters and slaves, subjects and objects. The latter are at the disposal of the former. Humans are the only ones who know and perceive God, and therefore they are the sole beings created for their own sake. All other animals are created for human sake. The general invitation to kindness and pity that can be found in the Scriptures, says Thomas, is not to be intended as a duty. Humans should just be careful not to exceed in violence to animals, since, in future, that could turn into violence to other humans. If anything, in such cases when an animal is owned by a person, to kill that animal is an offence to the owner, exactly as killing a slave is an offence to his/her master. Rational creatures like humans cannot be friendly with irrational ones. Charity towards animals is exercised by God only, but this happens only because they are useful to the privileged creatures.

Architectural Signs

Architectural signs are one of the four "prefigurements of art" emphasized in >**Sebeok** 1981, along with >**pictorial signs**, >**musical signs**, and >**kinaesthetic signs**. Of all possible forms of art, architecture is the one most frequently interpreted within strictly utilitarian frameworks. While pictorial, musical and kinaesthetic signs are often claimed to bear a sort of intrinsic value which, at least to some extent, transcends the purposes of traditional biological functions, architecture on the contrary seems to be much more connected with them. In the first place, constructions fulfill a fundamental need for shelter and warmth (and, also, they may serve as prey traps, potential mate attraction, etc.). Hence, it becomes crucual to argue for the existence of aesthetic values in architectural signs (>**Aesthetics**). To Sebeok, this is a fundamental issue:

> [...] we must look for the artistic value that may be involved, although subordinated to the principal interest of the "survival machine" [...]. If there is such a subsidiary purpose, falling passively under the sway of "mere" biological advantage, or supplementing it, an effort must be made to ferret out this aesthetic component. Such a quest is far from trivial, for, in the end, it is tantamount to asking: what is art? (Sebeok 1981: 240)

The work of the Finnish architect Juhani Pallasmaa (1995) is extremely helpful in categorizing the functions of animals' constructions, which he orders as follows:

1. Protection from the physical environment (temperature control; water management; humidity control; gaseous exchange and ventilation; waste management);
2. Protection from predators (avoidance of detection or recognition; mechanical protection);
3. Food gathering (cultivation; feeding; leaf-rolling; prey capture; food storage);
4. Communication (structures which assist transmission; structures which are signals; male selection and reproduction);
5. Decoration

A so-called decorative function is thus included: "[...] some animal behavior related to nest construction looks like superfluous decoration, lacking [any] other obvious function in the structure" (Pallasmaa 1995: 55). In addition, within the so-called "communicative" function, one notices the presence of sexual selection, which related to aesthetic manifestations to many extents.

Frequently reported examples of animal architecture are the so-called >**Pergolates**, or bowers, built by the Satin Bowerbird *Ptilonorhynchus Violaceus*. The decoration of nests with diverse objects is a pattern displayed by numerous species, including the *Ploceus benghalensis* (who "plasters" its nest with mud, and then decorates it with bright-coloured flowers) and the *Ptiloris paradiseus* (who decorates its nest with snake skins).

Among insects, even the very "rational" bees show some sign of superfluous decorations. After building their cells in the traditional hexagonal shape, bees attach a small hexagonal decoration to the external surface of the queen's cell (which has a different shape from the other cells). Apparently, this small hexagon serves no purpose.

Worth considering are also those cases of architectural artifacts designed as a *support* to other artistic activities, as for instance music (> **Zoomusicology**):

> Animals have also invented devices to improve sound transmission. The male of the tree cricket cuts a hole in a leaf and, while singing, stands in the hole and presses its tegmen against the leaf, which acts as a resonator. The male mole cricket digs a double-mouthed burrow in the ground, which functions as a horn-like amplifier to amplify and direct sound; the invention looks something like an early gramophone. (Pallasmaa 1995: 51).

Architecture is however not solely a decorative activity. To speak of architecture in general, including that of humans, involves also (perhaps mostly) a number of behavioral patterns related with fabrication and construction of artifacts. This aspect is discussed at length in the entry >**Tools**.

Aristotle (384–322 b.c.)

Greek philosopher. One of the most important and influential figures in Western thought. His centrality is extended also at the zoosemiotic level, in terms of empirical observations (he was a forerunner in several zoological and ethological studies) and ethical reflections (establishing the motives of a most enduring perception of the human-animal relationship). His contribution is rather ambivalent. Aristotle is at the same time an attentive observer of Nature, and a biased interpreter of these observations, when they are somewhat in contrast with his whole, internally coherent, philosophical system. Paradigmatic is his definition of *Scala Naturae*, which served as a model for all classifications before Linnaeus. Though based on a principle of finalistic continuity of species and of correlations between organs, the scale aprioristically refuses the proto-evolutionary principles postulated by Anassimandrus: in the Aristotelian classification one simply finds a hierarchical scale from the most to the least perfect being.

Aristotle's works on animals number five: *History of animals, Parts of animals, Generation of animals, Movements of animals* and *Progression of animals*. As already mentioned, these texts alternate brilliant intuitions with rather glamorous mistakes: Aristotle seems incapable of understanding the function of muscles and the nervous system, of distinguishing between veins and arteries, and of fully comprehending the reproductive act (to mention one, semen is to Aristotle merely aimed at sexual excitement). Aristotle also rejects Alcmeon's opinion that the brain is the actual central organ of the body, maintaining that its sole function is to chill blood. To Aristotle, heart is the actual core of all organs.

Several other of his considerations are remarkable. The observations on the anatomy of octopus, cuttlefish, crustaceans, and many other marine invertebrates are detailed and accurate. Aristotle distinguishes cetaceans from fish, describes the embryological development of a chick, the chambered stomachs of ruminants and the social organization of bees. Most of his observations would only be confirmed many centuries later. Aristotle groups together animals with similar features into *genera* (although the term is used in a much broader sense than nowadays) and then distinguishes the >**species** within the genera. Animals are then divided into two types: those with blood, and those without blood (or at least without red blood): the distinction is closely correspondent to that between vertebrates and invertebrates. Blooded animals are organized into five genera: viviparous quadrupeds (mammals), birds, oviparous quadrupeds (reptiles and amphibians), fishes, and whales. Bloodless animals are cephalopods (such as the octopus), crustaceans, insects, shelled animals (such as most molluscs and echinoderms), and "zoophytes", or "plant-animals", supposedly resembling plants in their form, such as most cnidarians.

From the ethical point of view, as mentioned, the concept of *Scala Naturae* remains a crucial concept, particularly for its strong impact on Western philosophy.

To Aristotle, non-human animals, women and slaves are three categories subordinated to the free male human, as they are simply "useful" to the latter (Jowett 1885: 10). Aristotle's philosophical-political system demands for both natural and social hierarchies, the latter depending on the former. Of course, there is a difference between the human slave and the non-human animal: however, in terms of such a socio-natural setting, these differences are not really relevant. What matters is, instead, *utility*, and Nature seems to obey this principle ("the other animals exist for the sake of man, the tame for use and food, the wild, if not all at least the greater part of them, for food, and for the provision of clothing and various instruments" – Jowett 1885: 14)

In Aristotle's hierarchy, there is no real interaction between superior and inferior categories. The former act upon the latter, they impose an order and affect their nature and behavior, without being affected in any way.

ASL

Short for >**American Sign Language**.

Baer, Karl Ernst Von (1792–1876)

Estonian-German morphologist and embryologist. Possibly the founder of modern embriology, Von Baer courageously denied recapitulation theory (i.e., that the embryo of a more complex animals goes through a series of morphological stages similar to the adult forms of lower organisms), in a period in which everybody seemed to have accepted it, and instead formulated his own "laws of embriology":

1. The general characteristics shared within a large group of animals appear in the embryo earlier than specialized features.
2. The embryos' ontogeny develops from general to specialized characters.
3. A species' embryos do not go through other animals' definite, adult, forms, but rather separate themselves from the latter.
4. By consequence, the early embryo of a higher animal does not look like the adult of a lower animal, but simply like its early embryo. The human embryo, for instance, never assumed, in its early stage, the mature form of an invertebrate or of a fish.

Von Baer had a certain fascination for metaphysical ideas connected with Schelling's transcendental idealism, and allowed such ideas to permeate his scientific discourse, possibly damaging part of his otherwise solid structure and conclusions (e.g., he would maintain that the archetype of the organism, now a sort of transcendental entity, was in fact "guiding" the organism's morphological development: "The type of every animal both becomes fixed in the embryo at the beginning and governs its entire development" – 1828–1837 Vol. 1: 220). His rejection of recapitulation theory was in any case a crucial step in the later, "official", denial provided by the entire biologists' community.

Von Baer is known also, or perhaps mostly (at least in >**biosemiotics**) for his alternative, non-darwinian, way to evolutionism. He shared Georges Cuvier's taxonomy, maintaining that the animal kingdom is separated into four "archetypes": radiata (like starfish and sea urchins), mollusca (like clams and octopus), articulata (like insects and crabs), and vertebrata. Therefore, he found unacceptable that the different class of vertebrates were in fact descended from a common ancestor (funnily enough, >**Darwin** had supported this idea in his writings by also quoting von Baer's findings as a source: misuse or rhetorical strategy?). Von Baer's theory of natural history allowed for "limited" evolution, among closely related organisms, rather than large-scale transformation, as instead suggested by Darwin. By consequence he found it hard to believe in the mechanism of evolutionary causality.

Most of all, from a semiotic point of view, von Baer was not content with Darwin's conviction that species variation and natural selection were random processes, a stand which obviously was against his romantic teleological interpretation of life: "for a true understanding of nature, we cannot dispense with a governing intelligence" (1876, vol. 2: 473). This view, predictably rejected by modern biology, proved instrumental for two anti-darwinian schools: the quasi-creationist cause of the Intelligent Design and the sign-based biosemiotics. Particularly through Kalevi >**Kull**, von Baer's theories went through a consistent process of semiotic revisionism that placed him as a key-figure in evolutionism, to the detriment of Darwin:

> An explanation of the scarcity of semiotic biology in the twentieth century thus stems from the fact that Baerian biology has been in a suppressed minority position almost throughout of the century, except may be only for the first and last decades (i.e., in the periods of neovitalism or organicism, and post-Darwinism), when its supporters were slightly better known or noticed. However, the dialogue between these two lines of thinking in biology has never stopped, and has continuously enriched both views. The rapid growth of biosemiotics in the last decade can be seen as a parallel to the rise of post-Darwinism in evolutionary biology, which is also a result of that dialogue. What is needed, and what it may hopefully bring, is both the broadening and deepening of the views. This would include the better understanding and skillful interpretation of deep but forgotten investigations, grasping more from scientists of other views. (Kull 2005: 23–24)

This position, a program in fact, is one of the founding characters of a consistent portion of the biosemiotic community, which "not only presents a revolutionary idea of biology, but also announces that such a revolution comes from the heirs of the historical opposition to mechanism" (>**Barbieri** 2006: 102). The current situation, in biosemiotics, witnesses a rather heated discussion on this very position, as opposed to the "official" darwinian-mechanistic paradigm of biology. The debate is very much beyond the scopes of this companion, however it is mentioned in Chapter 1, Section 1.2.4.

Barbieri, Marcello (1940)

Italian biologist and biosemiotician, currently working at the Department of Morphology and Embryology, University of Ferrara (Italy). He is the leading figure in one of the two current main schools in biosemiotics, the so-called Code-based Biosemiotics (or Code Biosemiotics). His work is not primarily concerned with **>zoosemiotics**, if not at a very general level (see, for instance, his account on the different theories of evolution in 2003: 33–66), but his plea for a "Scientific biosemiotics" (discussed in Chapter 1, Section 1.2.4) creates an important bridge, in terms of scientific agenda, with the type of zoosemiotic paradigm that is being promoted in this companion.

About code biosemiotics, Barbieri says:

> [...] natural selection and natural conventions are two distinct mechanisms of evolution because information and meaning are two distinct components of life. Natural selection is the long term result of copying, whereas natural conventions are the long term result of coding, and the two mechanisms are distinct because coding cannot be reduced to copying (proteins cannot be copied). This is a version of biosemiotics that can be referred to as code-based biosemiotics, or more simply, as code biosemiotics because it assumes that semiosis is defined by coding not by interpretation. The main reason for this conclusion is that the rules of the genetic code have been virtually the same in all living systems and in all environments ever since the origin of life, which clearly shows that they do not depend on interpretation. (Barbieri 2009: 228).

Barbieri is also scientific director of the very book series in which this companion is published, and editor-in-chief of the journal *Biosemiotics*.

Bateson, Gregory (1904–1980)

Primarily an anthropologist, Bateson's work covered a range of interests that included biology, psychology, semiotics, social sciences and cybernetics. He offered several important insights for **>zoosemiotics** (and **>biosemiotics** in general), in terms of both methodology and analysis. Worth of mention are at least his reflections on such topics as **>Instinct**, **>Play**, **>Aesthetics**, **>Mind**, **>Communication** and Intelligence. His reflections, and writing style too, are often provocative and unconventional, and possibly best embodied by his famous metalogues, imaginary dialogues between a father and a daughter over controversial topics of the most diverse nature.

Particularly interesting, from a zoosemiotic perspective, is his view on communication. In the article "Problems in Cetacean and other mammalian communication" (Bateson 1972: 364–78), Bateson addresses the difficulty, as humans, in approaching communication processes in non-human species, with the hypothesis that the latter might use communication in a consistently different manner. The communication of relationship statuses (affection, disaffection, dependency and so on) is to Bateson the predominant communicative function in non-human (that is, to him, non-linguisitic) mammals. Among humans, the communication of relationship is less explicit when it comes to language, and mostly found in that proto-linguistic communication (such as gestures and proxemics) that is common to all mammals.

In Bateson, we also find a distinction between "analogic" and "digital" aspects of communication, the former concerning proto-linguistic communication (a muscle can be more or less tensed, a facial expression more or less intense, etc.), and the latter concerning possibly only human language (although the question is left open when it comes to dolphin communication).

Further contributions by Bateson to zoosemiotics are discussed in Chapter 3 of the present companion.

Bayle, Pierre (1647–1707)

French philosopher, particularly instrumental in the development of agnostic and secular ideas in Western thought, by arguing that faith cannot be justified by reason. In his *Dictionnaire historique et critique* there is a very interesting entry "Beasts (souls of)", which discusses matters regarding reasoning and other inner states in non-human animals (Bayle 1697/1825: 199–219), in an openly anti-cartesian fashion. The entry starts with a direct attack on mechanism ("a strange notion"), and then proceeds by defining the elements in perception and cognition in animals. Among the topics tackled (all gathered in the umbrella term "soul"), there is >**awareness**, *memory*, >**intentionality**, *sensations*, *reasoning* and *morality*. The general message conveyed is that of the biological continuity between humans and other animals, where differences are never of qualitative type ("it would not be more absurd to maintain, that the soul of man knows actually an object without knowing that it knows it; than it is absurd to say, that the soul of a dog sees a bird, without perceiving that she sees it"; "it is impossible for the school-philosophers to prove that the souls of men and those of beasts are of a different nature"; and finally "differences are only accidental, and are no marks of a specific difference").

Bee Dance

Also referred to as a form of >**language**, the dance of the honey bee is easily one of the most impressive examples of animal >**communication**. Its discovery is mostly due to the work performed, during the 1960s, by Karl von >**Frisch** and his then student Martin >**Lindauer**, who carefully observed and described the behavior of the *Apis mellifera*. By means of this dance, one or more bees inform the hive about the presence of food, nesting sites or water sources in the more or less immediate vicinity. The process is roughly organized in the following stages:

1. One bee (called scout) goes in perlustration of the neighbourhood of the hive, and finds, for example, a rich food source.
2. Impregnated with the pollen of the flowers found there, the scout flies back to the hive, starts acting frenetically and performs either a round or an eight-shaped redundant movement, very close to the other bees of the hive. The scout also produces sounds by vibrating its wings, therefore the other bees can receive

information via the **>tactile channel**, the **>olfactive channel** and the **>acoustic channel** (not the **>visual**, though, as the hive environment is too dark for allowing this type of communication).

3. By smelling the scout, the other bees get indexical (**>index**) information on the quality of the object found (in the example, the bees smell pollen, therefore it is a food source, and not – say – a nesting site that the scout has found). After **>Wenner** 1969, and the discussion followed, it became clear that the olfactive channel is playing a more relevant role than previously suspected (although it remains untrue, despite Wenner's beliefs, that odor is the *only* factor intervening in bees' communication)
4. By touching the scout, thus by perceiving what kind of dance it is performing, the bees get symbolic (**>symbol**) information on the location of the place (Fig. 4.1). If the dance is round, it means that the place is very close and does not exceed a distance of about 100 m. A dance in figure of eight represents a distance that ranges between circa 100 m and 13 km. When the distance is so big, it is clearly of great help to have precise indications on the route to follow, as well. That is provided by the orientation of the 8 figure in relation to the position of the sun. The middle part of the 8 (the one where the two lines cross each other) is the moment when the scout indicates the direction. That part is performed very carefully, and creates an ideal angle with the sun position: from that very angle, the other bees understand what direction to take. In the same part of the figure, the scout also provides information on the distance of the site.
5. By hearing the scout, the bees get additional (or reinforced and more precise) information on distance and direction of the site. For a long time, the presence of acoustic communication was not noticed by the researchers, but after Esch 1967 it became clear that the scout can create a tridimensional acoustic field through the wings' vibrations, and that the other bees can perceive the sound through their antennas (specifically on the so-called Johnston's organ).

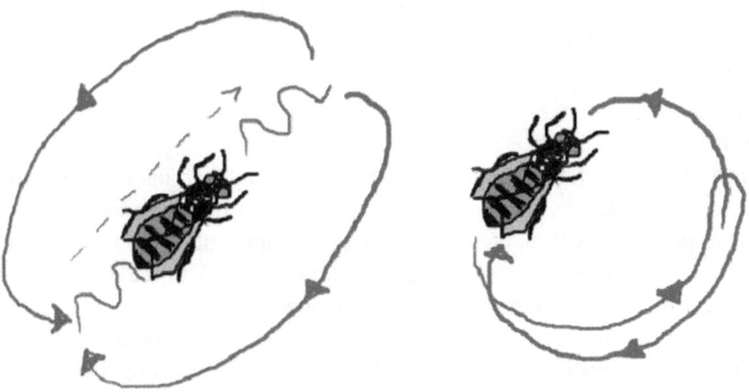

Fig. 4.1 The two types of bee dance

When showing the distance of the site, the scout – von Frisch noticed – provides an accurate and directly proportional correspondence (what in semiotics is called **>diagrammatic iconism**) between the time spent on the middle part and the distance of the site. In the community studied by von Frisch, the scout would spend a time between half and one second if the site was within 500 m, less than 2 s if the distance was within 1,500 m, up to 3 s when the site was within 3 km of reach, and so on.

Also, the axis of the very movement in the "8" middle part is progressively varied by the bee, As the directional aspect of the dance is crucial for saving time and energy to the other bees, the scout rotates approximately of 15 degrees per hour, in perfect accordance with the position of the Earth towards the Sun. This is possible thanks to the bees' endowment of the so-called solar chronometric orientation. Even in those weather conditions that make sun not visible, the bees equally succeed in conveying and learning the right information, as they are anyway sensitive or sensitised to polarized light.

Additional information on the type of route, and the travel conditions, are detected by the rhythm of the dance. From this element, the bees understand how much energy was spent in order to reach the place. Matched with the data received about the distance, this piece of information reveals whether the scout was flying against or in favour of wind, or if the route was particularly demanding.

Finally, the approximation in specifying the exact distance of a site requiring a round-dance (in point 4 of the above list) is not due to uncertain data, but to the existence of a local **>dialect**. Depending on their geographical origins, bees have a different conception of "close" and "far". For certain communities, it is important to perform a round-dance strictly if the site distance does not exceed the 50 m. Other communities seem to have a better sense of orientation, and can even afford a round-dance (that is, a dance without specific indications on the direction) if the site is 150 m far.

Lindauer 1961 and Frisch 1967 remain two "classics" on bee-dance, but the subject is still very popular among biologists and communication scholars, and new publications appear regularly: worth of mention are at least the works of Thomas Seeley (1995) and Fred Dyer (2002). For an insight on the discussion raised by Adrian Wenner, see Wenner 1969.

Behaviorism

Behaviorism is a theoretical trend (mainly in psychology), founded on the idea that all actions performed by an organism do not have to be interpreted as the result of internal physiological or mental processes, but are rather the result of a direct contextual conditioning. Key behaviourist scholars were Ivan Pavlov, Edward Lee Thorndike, John B. Watson and Burrhus F. Skinner.

Before being challenged (and partly overcome), in the second half of the twentieth century, by **>cognitive** approaches, behaviorism was widely popular in the study of animal behavior, affecting the early developments of **>ethology** and **>zoosemiotics** (the so-called Early **>Ethological Zoosemiotics**). For a short period, it was up to semiotics itself, through the work of Charles >**Morris**, to develop a specific "behaviorist" theory of signs: sign was to Morris "something that directs behavior with respect to something that is not at the moment a stimulus" (1946: 366).

Bekoff, Marc (1945)

American cognitive ethologist Currently Professor Emeritus of Ecology and Evolutionary Biology at the University of Colorado, Boulder, Bekoff is one of the most prominent figures in >**cognitive ethology**. His range of interests includes the study of mental and emotional aspects of animal behavior and different ethical subjects. Some of his publications (particularly Bekoff 1975, published in the journal *Semiotica*) constitute an important theoretical bridge between >**ethology** and >**zoosemiotics**. Interdisciplinarity, and particularly collaboration with human sciences, is to Bekoff an important goal to achieve in the process of building a theoretical paradigm for cognitive ethology (Bekoff 1995: 139). The question is discussed in Chapter 1, Section 1.1.2.

Bentham, Jeremy (1748–1842)

English philosopher, jurist and political radical. Heavily influenced by enlightenment thinkers, especially John >**Locke** and David >**Hume**, Bentham is primarily known today for his advocacy for utilitarianism and various forms of individual freedom (including women, homosexual and animal rights). His impact on the animal rights movement, particularly through the 1789 work *Introduction to the Principles of Morals and Legislation*, has been crucial, and therefore difficult to overlook also at a specific zoosemiotic level. Bentham argued that the ability to suffer, rather than the ability to reason, is the actual "insuperable line" to determine whether a living being is worth of ethical and juridical attention:

> The day has been, I grieve to say in many places it is not yet past, in which the greater part of the species, under the denomination of slaves, have been treated by the law exactly upon the same footing, as, in England for example, the inferior races of animals are still. The day may come when the rest of the animal creation may acquire those rights which never could have been witholden from them but by the hand of tyranny. The French have already discovered that the blackness of the skin is no reason a human being should be abandoned without redress to the caprice of a tormentor. It may one day come to be recognized that the number of the legs, the villosity of the skin, or the termination of the os sacrum are reasons equally insufficient for abandoning a sensitive being to the same fate. What else is it that should trace the insuperable line? Is it the faculty of reason or perhaps the faculty of discourse? But a full-grown horse or dog, is beyond comparison a more rational, as well as a more conversable animal, than an infant of a day or a week or even a month, old. But suppose the case were otherwise, what would it avail? the question is not, Can they reason?, nor Can they talk? but, Can they suffer? (Bentham 1789/1996: 283, footnote).

Bioacoustics

Interdisciplinary field of inquiry that mainly combines biology and physical acoustics, devoted to the investigation of the neurophysiological and anatomical mechanisms of sound production, spreading and reception in all animals. Although sharing a consistent range of interests with **>zoosemiotics**, bioacoustics (theoretically and empirically) rather proceeds on a parallel path, which does not cross but in few occasions with the former (e.g., **>Sebeok** 1968 and **>Tembrock** 1971). In that sense, the statement that "the branch of zoosemiotics studying acoustic systems of communication is also known as bioacoustics" (Nöth 1990: 161) should be considered misleading.

Biocentrism

Biocentrism is a ethical-philosophical orientation (not to be confused with a scientific theory of the same name, related to quantum physics) within human and natural sciences, theorized by several scholars, including Peter **>Singer** (1975), Tom **>Regan** (1983), Paul W. Taylor (1986), Sergio Bartolommei (1995), Leena Vilkka (1997), Dario Martinelli (2006a, in a specific semiotic application), and (not as a direct promoter, but as an acknowledged source) Charles **>Darwin**. It aims to the overcoming of **>anthropocentrism,** particularly in issues of research ethics and methodology. Unlike anthropocentrism, biocentrism stresses the centrality of the concept of "life", and deals with nature and animals as intrinsic, rather than instrumental, values.

Biocentrism is strictly linked with two other forms of critiques of anthropocentrism, i.e., zoocentrism and ecocentrism. The difference between the three concepts are described by Finnish philosopher Leena Vilkka:

> (1) Zoocentrism means a philosophy in which the issues, concepts and values of animals are central;
> (2) Biocentrism means a philosophy in which the issues, concepts and values of life are central;
> (3) Ecocentrism means a philosophy in which the issues, concepts and values of ecosystems are central (Vilkka 1997: 37)

And by the following scheme:

	Zoocentrism	*Biocentrism*	*Ecocentrism*
Key-concept	Consciousness	Livingness	Wholeness
Value	Well-being	Biodiversity	Beauty
Field	Animals	Life	Ecosystems

(from Vilkka 1997: 37)

In zoosemiotics, the notion is implemented by the use of Jakob von >**Uexküll**'s theories on >**Umwelt**, the concept of >**Zoosemiotic universals**, and a firm critique of >**Morgan's canon** and other similar notions. The first two elements in particular, with their general stress on biodiversity and life, plus the decision of rejecting "consciousness" as benchmark (as the concept is of risky application for a few animal species), have made "Biocentrism" a preferable term to "Zoocentrism" within the zoosemiotic discussion.

The question is discussed at length in Chapter 5, Section 5.5.

Biocommunication

Umbrella-term, defined in **Tembrock** 1971 and used in different fields of inquiry to describe certain types of >**intraspecific** and **interspecific semiosis** among different living (i.e., not only animal) >**species**. The term is adopted and theorized in a semiotic sense by Günther >**Witzany** (in 2007), particulalry according to the traditionally-semiotic classification of >**syntactic**, >**semantic**, and >**pragmatic** aspects.

Biorhetorics

Applied branch of classical rhetoric, introduced in >**biosemiotics** by Stephen >**Pain** in the 1990's (see Pain 2002). Pain maintains that human-animal relationship and nature conservation-related issues can be interpreted in terms of a "natural argumentation", where the force of a successful argument is equal to the distance from the position of the audience at X to the required position Y. Tropes, figures (sometimes meant in a metaphorical sense, sometimes in the traditional rhetoric sense) and the central components of the argument can be measured as factors of the force.

In a recent effort (Pain 2009), Stephen Pain thought also of a specific application of this concept to >**zoosemiotics**, using the term "Zoorhetorics".

Biosemiotics

Biosemiotics is an applied branch of semiotics whose main area of investigation is sign production and relationships in life forms. It consists of (1) the study of >**semiosis** in living organisms; (2) the interpretation of living systems as sign systems; and (3) the identification of >**semiosphere** with >**biosphere**.

Leaving aside the ancient "semiotics" of the medical observation of the body, carried out by the likes of Hippocrates or Galen of Pergamon, semiotics achieved an interest in extra-human and extra-cultural topics only in the nineteenth century, with Charles S. >**Peirce**, and in the early twentieth century with Jakob von >**Uexküll**. The Italian oncologist Giorgio >**Prodi** then came to denominate "Nature Semiotics" as the study of biological codes. It was however Friedrich S. Rothschild (1962: 777) who first used the term "biosemiotics" in a scientific context. In 1971 the Russian semiotician Yuri Stepanov extensively used the term "Biosemiotics" in his work, and – after Thomas **>Sebeok** used it in international contexts in 1986 – it started to spread all through the semiotic community, becoming nowadays one of the most important areas of expertise in semiotics.

Biosemiotics is also discussed in Chapter 1, Section 1.1, where a few definitions of the field are provided. In this entry, however, the following can be added in order to provide a more complete perspective:

> Sign processes penetrate the entire body of an organism. [...] Signification is the fundamental property of living systems that can be taken as a definition of life. Hence, biosemiotics can be viewed as a root of both biology and semiotics rather than a branch of semiotics. (Sharov 1998: 404–405)

Apart from the reaffirmation of the centrality of semiosis in biological processes, in this definition we also learn about the intimately interdisciplinary nature of biosemiotics. Such a concept is confirmed by Claus Emmeche, who also emphasizes the fact that biosemiotics focuses most of all in the "emergence of semiosis" in nature, and that this emergence may or may not coincide with the emergence of living forms.:

> Biosemiotics proper deals with sign processes in nature in all dimensions, including (1) the emergence of semiosis in nature, which may coincide with or anticipate the emergence of living cells; (2) the natural history of signs; (3) the 'horizontal' aspects of semiosis in the ontogeny of organisms, in plant and animal communication, and in inner sign functions in the immune and nervous systems; and (4) the semiotics of cognition and language. [. . .] Biosemiotics can be seen as a contribution to a general theory of evolution, involving a synthesis of different disciplines. It is a branch of general semiotics, but the existence of signs in its subject matter is not necessarily presupposed, insofar as the origin of semiosis in the universe is one of the riddles to be solved. (Emmeche 1992: 78)

As that branch of semiotics that deals with the living beings of the kingdom *Animalia*, **>zoosemiotics** must be considered part of biosemiotics.

Biosphere

The globality of the Earth's environments where living organisms live and operate, interacting with the lithosphere, the hydrosphere, and the atmosphere. The birth of the biosphere, by means of biopoesis, is dated at some 3.5 billion years ago. In the biosemiotic view, the concept of "biosphere" overlaps with that of the **>Semiosphere**.

Bower

See **Pergolate**

Camouflage

See **Deception**

Cassirer, Ernst (1874–1945)

German philosopher. A crucial figure in the post-Kantian idealist school of the early twentieth century. His major contribution to Western thought was the three-volume *Philosophy of Symbolic Forms* (published between 1923 and 1929). There, and in one of his subsequent works, *An Essay on Man* (1944), Cassirer defines the human being as a "symbolic animal" (*Animal symbolicum*). Non-human animals, says Cassirer, perceive their world by means of **>instinct** and senses, while the human being creates and shapes reality through symbolic meanings, that allow the conception of projects, hypotheses, forecasts and utopias. Symbolic forms are of linguistic, scholarly, scientific, and artistic types.

In a more semiotic-friendly form, the notion of "Symbolic animal" was partly re-elaborated in the concept of "Semiotic animal". For more, see Chapter 1, Section 1.2.3.

Channels

Channels are in semiotics the sensory modes used to transmit a message. The notion roughly corresponds to (but does not overlap with) that of the "senses", although attention should be paid not to limit the notion to the five senses possessed by humans. In **>zoosemiotics**, indeed, at least seven channels are classified: the olfactory, the gustatory (these two normally grouped into the definition of **>chemical channel**), the tactile, the thermic, the electric (these three gathered under **>tactile channel**), the **>acoustic channel**, and the **>visual channel** (not to mention the possibility of including also **>magnetoception**, as a channel of its own).

Chantek (1977)

From the Indonesian "Cantik" (beautiful, lovely), Chantek is a male orangutan of the species *Pongo abelii* raised at the Yerkes Regional Primate Research Center in Atlanta (Georgia) for a number of **>Interspecific Communication Experiments**, coordinated by anthropologist Dr. Lyn Miles. A program based on **>American Sign Language** was successfully attempted with Chantek starting in 1978. The distinctive feature of this program was that the trainee was put in a context in which all trainers would communicate with ASL, even among themselves, and nobody was making specific attempts to train him directly. Simply, when Chantek was told something, he was told in ASL, and when he wanted to communicate, he had to use ASL. Very soon, the orangutan learned to use the sign system spontaneously and efficiently, amounting to a vocabulary of several hundreds of signs, later enriched by a quasi-second language: the understanding of spoken English. His skills were so convincing that they had a political impact in the application of the ethic and juridical concept of **>personhood** in Great Apes (as for instance promoted by Paola Cavalieri's and Peter **>Singer**'s *Great Ape Project*)

Chantek is also known for his creativity, particularly expressed in **>tool-making**, painting (**>pictorial signs**), music-making (**>zoomusicology**) – alone or in collaboration with Lyn Miles herself, who is also a professional percussionist (the "duo" has also released some recordings under the stagename Animal Nation) – and several forms of handicrafting (especially necklaces); and for allowing advancements in the studies of >**self-consciousness** in non-human animals by displaying a great variety of revealing reactions to the mirror test.

Besides several documentaries and "popular science" articles and books, Chantek's efforts in ASL training are also academically documented in, a.o., Miles (1990, 1994).

Chemical Channel

The most elementary and common of the >**channels**. The chemical substances serving as sign vehicles between organisms within a shared environment are called semiochemicals (e.g., pheromones are those semiochemicals used in >**intraspecific** animal >**communication**). In animal >**semiosis**, there are species who have a clear distinction between taste and smell, and others who do not, and basically have the same receptor organ for both functions (as a matter of fact, there is always a close collaboration between these two senses: in humans the taste of food is always provided by a combination of the olfactory and the gustatory channels).

The main advantages of the chemical channel are three: it is relatively permanent, it can act at relatively big distances, and it can be decoded at a given time distance. In contrast, this channel has a slow variability and a high susceptibility to interferences. The territory signal emitted by dogs through their urine is a perfect example of these pros and cons: the message is quite permanent, it can be smelt at a certain distance, and can be decoded after several days by other dogs. At the same time, there is not much information spread (except a "this is my land!" kind of message), and the next dog will easily invalidate the message by urinating over it.

Cimatti, Felice (1959)

Italian zoosemiotician, currently Associate Professor in Philosophy of Mind at University of Calabria (Italy). A very important figure in the development of a cognitive approach to >**zoosemiotics**, through the seminal *Mente e linguaggio negli animali* (1998). Topics discussed in this work include the question of the >**semiotic threshold**, the >**functions of communication** in animals (according to Roman Jakobson's framework), the >**interspecific communication experiments**, and the >**zoosemiotic universals**.

Cimatti is also author of one of the few essays available on the figure and theories of >**biosemiotics** pioneer Giorgio >**Prodi** (Cimatti 2000).

Cleaning Symbiosis

See **Symbiosis**

Clever Hans Effect

Der Kluge Hans (German for "Clever Hans") was a horse owned by a math teacher and horse trainer named Wilhelm von Osten, between the nineteenth and the twentieth centuries, and who achieved a certain notoriety for allegedly being able to perform arithmetic and other intellectual tasks.

During increasingly popular shows, throughout the whole Germany, Hans would amaze the spectators by solving problems of different types by means of hitting his hoof on the ground (he was known for adding, subtracting, multiplying, dividing, working with fractions, telling time, keeping track of the calendar, distinguishing musical tones, and reading and understanding German). As a problem was presented to him (either orally or in written form), Hans would start tapping his foot and stop when the exact figure was reached.

A first investigation on Hans' skills was performed in 1904 by psychologist Carl Sumpf and 13 collaborators (known as the "Hans commission") from different fields, including veterinary doctors, circus managers and zoo directors, only to conclude that no trick was involved in the horse's performances. Psychologist Oskar Pfungst, in 1907, started a more thorough investigation on the case, introducing a few variables that turned out to be crucial (preventing Hans from seeing the questioner, making sure that the questioner did not him/herself know the answer to the problem, etc.). This way, Pfungst could demonstrate that Hans was not actually cognitively able to perform the tasks, but was instead very skilful in watching his observers' reactions to his "counting", until some involuntary cues in their body language would alert him that the task had been correctly performed.

The case of Clever Hans became paradigmatic for those experimental contexts where (a) the subject of the experiment ends up manipulating, rather than being manipulated by, its trainers, and (b) the trainers, possibly driven by the enthusiasm of an apparently remarkable performance of their trainee, fall victim of excessive **>anthropomorphism** in their interpretation. The phenomenon is known as "Clever Hans Effect", and became, among other things, the war ... *horse* of semioticians criticizing **>interspecific communication experiments**. The fear of falling prey of this methodological bias encouraged some of the researchers of these experiments to adopt the so-called **>emotion-free experimental contexts**, loosely based on Pfungst's principles of investigation in the Hans' case.

A detailed discussion on the Clever Hans Effect appears in Chapter 3, starting from the Section 3.3.

Code

The totality of the rules of transformation, shared by at least two subjects (and normally by an extended community), through which signs are more or less deliberately associated to one or more meanings. A code is at the same time a causal factor and a direct consequence of **>sociality**, in that it requires social interaction in order to be created, and – after being established – allows an increase and an improvement of such interaction. As a set of rules based on convention, **>arbitrariness** and **>ritualization**, codes tend mostly to create signs of symbolic type (**>Symbol**), however few are the cases where the sign-meaning association is purely and exclusively arbitrary. Codes can be established through any of the **>channels** available in each **>species**, and take different configurations according to the degree of familiarity and interaction within a community (up to the generally most difficult case of **>interspecific >semiosis**). Within the same community, let alone species, codes may be subject to progressive variations and adjustements, due to different factors: increasing efficiency of a new sign-meaning association, as compared to an old one of the same type; introduction of new subjects from a different community, who carry a different sign-repertoire (see also **>Dialect**), etc.

Using a well-known classification by Daniel Chandler (2002: 148–59), it can be said that non-human animal codes can be

1) *Social*, in at least the *bodily* and *behavioral* sense (but also in the *commodity* and *verbal* one, if these are meant in a broad sense);
2) *Textual*, when they result in an artificial or semi-artificial production (for instance in those specimens used in **>interspecific communication experiments**);
3) *Interpretative*, in at least the *perceptual* sense (but also in the *ideological* one, if we take to the letter Chandler's indication of ideology as a determinant for dominant, negotiated and oppositional rules for encoding and decoding).

In addition to Chandler's formulation, or perhaps as an extension of the category of Textual codes, one may also think of **>contextual codes**, that is, those sign repertoires that are typical (and somehow exclusive) of a specific situation and interaction.

Coetzee, John Maxwell (1940)

Writer and academic, 2003 Nobel Prize in Literature. Born in South-Africa, he is currently an Australian citizen. A passionate advocate for the animal rights movement, Coetzee tackled, in novels such as *Disgrace*, *The Lives of Animals* and *Elizabeth Costello*, a number of topics related to animal **>cognition**, **>communication** and **>ethics**, often offering reflections and interpretation of scientific relevance. A passage from *Elizabeth Costello* is used in Chapter 3, Section 3.3.4 of this Companion.

Cognition

The totality of mental activities and information processing occurring within an animal, in a fashion partly or totally untied from **>instinct**. It is the main target of **>cognitive ethology** and of **>modern ethozoomiotics**, the latter mostly focusing on how signs are constructed, interpreted and mediated. The study of animal cognition focuses on at least the following aspects:

1) *Attentive skills*: the capacity of focusing on and discriminating between different stimuli.
2) *Classification and categorization*: the capacity of creating semantic fields, and therefore making types out of tokens.
3) *Temporal cognition*: the capacity of learning, retaining and transferring past information for (immediate and/or remote) future use,
4) *Spatial Cognition*: the capacity of orienteering and navigating
5) Production and use of >**tools**
6) *Problem-solving*: the capacity of pursuing a goal by avoiding especially complex and unusual obstacles
7) >**Language**
8) >**Consciousness**
9) *Counting and computing*: the capacity of discriminating between quantities and perform basic mathematical tasks.

Cognitive

See **Cognition**

Cognitive Ethology

The evolutionary and comparative study of animal >**cognition** (thought processes, consciousness, beliefs, rationality). Cognitive ethology was pioneered by Charles >**Darwin** in his last works of of a *zoo*logical nature (1871 and 1872), and established by Donald >**Griffin**'s *The question of animal awareness*. It is a rapidly growing field with (rather unusually in a natural science) a strong inclination for interdisciplinary work. Currently, the list of eminent cognitive ethologists include Colin Allen, Marc >**Bekoff** (who also wrote articles of zoosemiotic contents), Frans >**De Waal,** Dorothy Cheney, Robert Seyfarth, Dale Jamieson, Remy Chauvin, and the great majority of scholars in >**interspecific communication experiments**. Cognitive ethology may easily be considered the closest relative to >**zoosemiotics**.

The main issues and problems which cognitive ethology deals with are efficiently summarized in Colin Allen's e-text "Philosophy of cognitive ethology":

> "There are two major planks [...] for cognitive ethology, both of which have come under attack. One plank is theoretical: it concerns the kinds of questions that scientists should be asking about animal minds. According to Griffin, scientists should no longer ignore questions about the intentionality, thoughts and conscious experiences of animals. This theoretical plank of Griffin's program has been attacked on the grounds that mentalistic, folk-psychological notions such as 'thought' and 'consciousness' are too imprecise or for serious scientific investigation. The other plank is methodological: it concerns the ways in which scientists should go about answering questions about animal minds. Here Griffin argues that our best chance for answering such questions comes from observing animals under natural conditions solving the kinds of problems for which their intelligence has become adapted by evolution. This methodological plank has been attacked on the grounds that such observations lack the controls required for proper hypothesis testing". (Colin Allen, at host.uniroma3.it/progetti/kant/field/ceth.htm).

Cognitive Map

See **Mental Representation.**

Comfort

Comfort, or comfort behavior, indicates a series of behavioral patterns related, to different extents, to the care, welfare and health of an animal's body. Such patterns include on the one hand activities like dustbathing, grooming, preening, rubbing, scratching, shaking, washing and others (pertaining, that is, to body-care), and on the other, activities like defecating, stretching, urinating, yawning and others (pertaining to the tension release of metabolic functions). Some of this activities, subject to a process of **>ritualization**, became essential components of social interaction (the typical example being grooming among great apes).

Within **>zoosemiotics**, a specialist in comfort behavior is Aleksei **>Turovski** (e.g., 2001 and 2002).

Communication

In general semiotics, communication is the process of transmission *and* reception of a message. For the purposes of this companion (within the framework of a tripartite definition of **>semiosis**, as discussed in Chapter 1, Section 1.1), communication is that particular semiosic process which, unlike **>representation** and **>signification**, requires *both* a source and a destination for the message. In communication, at least two animals (a sender and a receiver) take part in the semiosic phenomenon, and therefore the "sense" is exchanged, understood or misunderstood.

Communicational Anthrozoosemiotics

See **Anthropological zoosemiotics**

Conative

One of the six **>functions of communication** postulated by Roman Jakobson (1963). A message is conative when its focus is the receiver.

Condillac, Etienne Bonnot De (1715–1780)

French philosopher. A relevant exponent of a certain radical empiricism in the discussion around **>mind** (an approach that is often referred to as "sensationism"), a discussion that he took to a further extreme than John **>Locke**, by rejecting all forms of innate activities (while, in Locke, that rejection was limited to principles and ideas). In 1755, Condillac published a *Treatise on Animals*, meant as an attempt to explain the difference between human from animal souls. Free will, intelligence and individuality are the main discrimants: each human being forms him/herself on the basis of different models, assembled in a synthesis that creates the uniqueness of the individual. Animals, on the contrary, have a limited range of needs, and are fundamentally moved by **>instinct** and habit.

Moreover, as the philosopher later explains in the *Essay on the origin of human knowledge* (1756), they lack reason and symbolic **>language** (being able to express their feelings only through "inarticulate cries"), the latter activity consisting in "conventional signs" that require memory (another element missing in animals). Despite this consistent difference, Condillac is still in favour of a natural continuity between human beings and other species (openly rejecting **>Descartes**' mechanism), and – in the specific of semiotics – does not deny that animals can communicate with one another by using natural signs (such as the inarticulate cries, that become habitually associated with natural feelings):

> The similarity between animals and us proves that they have a soul; and the difference between us proves it is inferior to ours. The matter is made evident by my analyses, for the operations of the animal soul are limited to perception, consciousness, attention, reminiscence, and to an imagination which is not at their command, whereas ours possesses additional operations (Condillac 1756/2001: 39).

Condillac then pioneers the notion of "Semiotic animal" by asserting that the human ability to "attach ideas to arbitrary signs" creates a "mystery" in our imagination that allows to recall ideas from distant spaces and times, while animals are "stuck" in the immediacy of the present.

Consciousness

An umbrella-term that may refer to a vast scale of mental states, from subjective experience to simple awareness. When it comes to non-human animals, the data at scholars' disposal are not abundant, and must regularly be accompanied by philosophical conjectures. The range of hypothesis goes from the "pessimistic" Nagel 1974, who practically answered "No idea!" to his question "What is it like to be a bat?", to the optimistic Cheney-Seyfarth 1990, who instead offered a few hypothesis to their question (ironically referring to Nagel himself) "What is it like to be a monkey?".

The problem is largely part of the more general philosophy of **>mind**, and it is often discussed in terms of body-mind dualism and **>etic-emic** methodologies. It is obviously a central theme in both **>cognitive ethology** (it was, after all, the main issue in **>Griffin** 1976), and **>modern ethozoosemiotics** (where a thorough analysis can be found, a.o., in **>Cimatti** 1998). A convincing treatise on the subject is found in Mortenson 1987, where different anatomic-physiological-behavioral clues for consciousness are gathered as abductive indications for the presence of the phenomenon (Table 4.1).

Table 4.1 Indicators of consciousness, accordino to Mortenson 1987

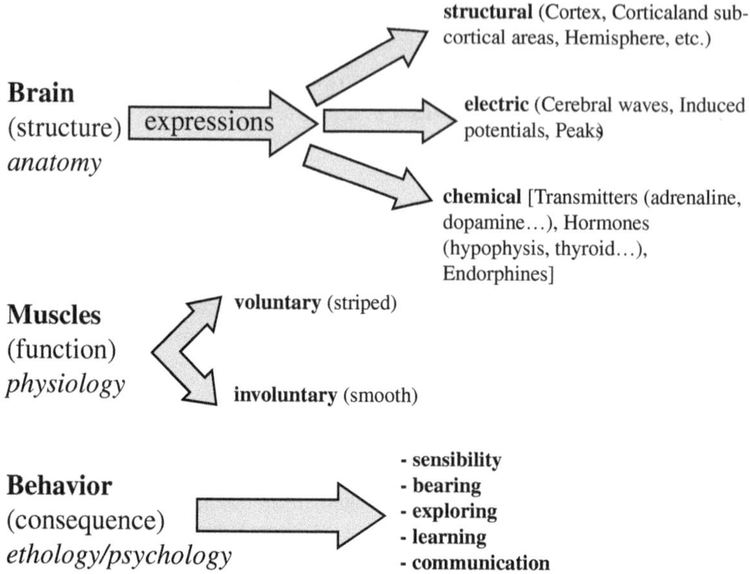

Contextual Codes

In this companion, a contextual code is defined as a type of >**code** developed specifically for a particular activity/interaction. It features signs that are used only in that context, and not in others (unless undergone a process of >**Ritualization**). Among the many examples of situations mediated by a contextual code, there are >**play**, courting rituals, >**food request,** cleaning >**symbiosis,** etc.

Convention

See **Codes, Symbol** *and* **Ritualization.**

Critical Anthropomorphism

Philosophical notion elaborated by Italian philosopher Luisella Battaglia in the essay *Etica e diritti degli animali* (Battaglia 1997). Within semiotics, it is applied as an attempt to overcome the methodological problems raised around >**Anthropomorphism** and >**Morgan's Canon.** As in both cases this companion suggests the adoption of a critical attitude, critical anthropomorphism constitutes a theoretical counter-proposal. It aims to use human experience as a means for recognizing given manifestations of emotional type, and integrate such experience with the theoretical and empirical findings of natural sciences (Battaglia 1997: 123–124). This way scientific and ethical reflections are not in contrast, but rather complementary to each other, avoiding certain, recurrent forms of methodological contradiction

("If human-animal similarities are accepted for scientific experimentation purposes, they should also be accepted in the field of emotional sensibility. In any case, any doubt about emotional sensibility should be gauged so as to benefit the weakest subject. In particular, the presupposition of similarity, when there is no clear counterproof, should be interpreted in favour of the animals" – from Battaglia 1997: 124).

Critical anthropomorphism is discussed at length in Chapter 3, Section 3.3.5.

Culture

The totality of information acquired and developed by a community and transmitted non-genetically from one generation to another. Phenomena of this kind have been observed in some insects, fishes, amphibians and reptiles, and in most birds and mammals. With few exceptions, the presence and the quantity of cultural processes are directly proportionate to the degrees of >**sociality** and social organization displayed by each >**species**.

At the core of any cultural process, there is the capacity of social >**learning**, which implies the transmission of some information from one or more "demonstrators" to one or more "observers/listeners". Several factors, at different (and changeable) stages, intervene in the process. These are (in no specific order): (1) >**imprinting**; (2) attention; (3) imitation (>**mimicry**); (4) memory; (5) social facilitation; (6) creativity; and of course (7) >**communication**. One famous case in which all of these factors are operating at the same time, and which was also a breakthrough in this topic's research, is that of the macaques *Macaca fuscata*, closely observed during the 1940s and 1950s by a group of researchers led by Masao Kawai in Koshima, Japan (Kawai 1965). Researchers had the idea of introducing a "novelty" in the life of a group of macaques, by leaving some sweet potatoes along the beach. One macaque began to wash the potatoes in the water, instead of brushing the sand off with her hand as other macaques did. In short time, this behavior became popular among the other members of the community and was passed on from generation to generation. In addition, the macaques proved to make a cognitive use of the innovation, as further modifications to this pattern were eventually provided (for instance, they started washing the potatoes in salt, rather than fresh, water, in order to enahnce their flavor).

Although the >**intraspecific** examples of cultural transmission are certainly more recurrent in nature, it must be underlined that >**interspecific** cases are far from rare, a most peculiar example being that of the so-called slave-ism performed by certain species of ants on weaker species of insects. Among the vast literature on animal culture, worthy of attention are Thorpe 1961 and 1963, Gardner et al. 1994, Bonner 1980, and Box 1973 could be singled out.

Dance

See **Kinaesthetic signs**

Darwin, Charles (1809–1882)

English naturalist. Quite simply, the most important animal scholar of all times. All fields of inquiry pertaining to the study of animals owe an incalculable debt to his theories and observations, and there is certainly no point in listing all of them, besides those that have been instrumental for the development of **>zoosemiotics** (and still considering that Darwin'ss many other contributions to animal studies were instrumental to this field anyway, although possibly in a less direct way).

If the theory of Natural Selection is what gave Darwin most of his fame, it is in his later works that one finds the basis for the development of **>cognitive ethology**, zoosemiotics, **>sociobiology**, and other fields that characterized animal studies in the twentieth century (including elements of research ethics: Darwin's methodology, notoriously, relied upon personal observations and descriptions of an anecdotal and non-invasive nature). In *The Descent of Man* (1871), Darwin introduces reflections and observations about animal **>cognition**, reasoning, use and making of **>tools**, **>consciousness**, **>learning**, **>aesthetics**, **>sociality** and morality (and even spirituality, when one considers those remarks on the supernatural in animals). In the subsequent *The Expression of Emotions of in the Man and Animals* (1872) Darwin follows up the arguments of the previous book, and focuses on topics that are even more pertinent to modern zoosemiotic research: signalling behaviour, **>representation** and interpretation. Darwin also meticulously describes the main features of what is nowadays known as **>multimodality**.

The importance of Darwin's writings in the zoosemiotic (as in any other) context, certainly resides in their modernity. Accurate descriptions of signalling behaviour, important methodological distinctions (**>intraspecific semiosis** and **>interspecific semiosis**, or "intercommunication", as he calls it; innate, learned and ritualized sign repertoires, etc.), and specific theoretical formulations.

It is safe to affirm that Darwin is to animal studies as Bach is to music and Shakespeare to theatre.

Dear Enemy

Ethological phenomenon that describes an animal's particular reaction to the invasion of its territory on the part of a neighbor. The aggressive response, in this case, is shorter and less intense, as compared to the case when a total stranger is invading the territory. The "dear enemy" phenomenon, thus, implies the animal's capacity for individual recognition, and it is observed especially in birds and mammals, although relatively recent studies on the coral reef fish *Pomacentrus partitus* (e.g., Myrberg and Riggio 1985) showed that fishes are also less aggressive towards their own "dear enemies".

Deception

In semiotic terms, deception is a communicative action where a sender (1) breaks the association of a sign-meaning established in one or more codes, (2) mentally represents the receiver's mind (for instance, mapping his/her expectations), and (3) takes advantage of the broken association.

It is an extremely common practice in the entire animal kingdom, and includes such phenomena/patterns as **>mimicry** and **>imitation**, through to linguistic lies, created by the likes of **>Koko**, **>Washoe** and other subjects of **>interspecific communication experiments**. A classification of the different typologies of deception was provided in Bouvet 2000, and is articulated in seven cases:

1) *performing* – as in several courting rituals implied different forms of **>exhibition**;
2) *hiding* – as in mimicry;
3) *simulating* – as in **>play**;
4) *dissimulating* – as in those cases where a certain pattern is purposively displayed ("staged") in order to keep the receiver's attention off from another pattern;
5) *deceiving* – as in those cases where a sign associated to a given meaning is deliberately emitted for achieving a different goal than the one pertinent to that meaning;
6) *disguising* – as in cases of body-decoration or transvestitism;
7) *pure lying* – as in the mentioned "linguistic lies".
8) A thorough discussion on deception appears in Chapter 2, Section 2.4.1.

Deely, John (1942)

American semiotician and philosopher. Currently Professor of Philosophy at the Center for Thomistic Studies of the University of St. Thomas (Houston, Texas). Deely is a key-figure in contemporary American semiotics. His work is particularly concerned with general theory of semiotics, and its impact on philosophy (an activity resulting in the volume *Four Ages of Understanding*, a monumental history of philosophy from a semiotic perspective). Deely is also very active in **>zoosemiotics**, having several times discussed the notions of **>Umwelt** (e.g., Deely 2001), Semiotic Animal (e.g., Deely 2005), and others, often following in the footsteps of Thomas **>Sebeok**, of whom he was close friend, and to some extent heir.

Descartes, René (1596–1650)

French philosopher and mathematician. Descartes' theories on animals are probably among the very few on the subject to be well-known. Animal-rights activists have always been very critical of Cartesian conceptions, mentioning them as the typical example of human prejudices towards other animals. What is certain is that the Cartesian idea of the animal-machine was a major breakthrough in a discussion that, until then, was mostly animated by theological, ethical and political reflections.

Descartes' philosophy, as always happens to the most influential thinkers, is to a large extent controversial. While it is generally acknowledged that he marked the beginning of modern philosophy, one cannot avoid remarking that many of his reflections, starting from his metaphysics, were in fact animated by a truly medieval spirit.

The opinion expressed by Descartes on animals is quite difficult to misunderstand: animals are totally thoughtless and conscienceless. They are simple mechanisms. To maintain that a cat thinks is to Descartes as silly as stating that a clock does. Descartes does not distinguish between mechanism and organism: his view is a form of reductionism that replaces organic structures with mechanic components. The main points in Descartes' argumentation on animal cognition, as illustrated in his *Discourse on the method* (1637, here edited and translated in Veitch 2008) are the following:

1) Descartes does not deny the existence of emotions in animals: he simply maintains that they have no awareness of them. All bodies are machines, in the Cartesian system, including human bodies (Veitch 2008: 44–5);
2) The difference between humans and other animals concerns the human possession of a soul, expressed through ratio and language (Veitch 2008: 45–6. "There are no men so dull and stupid, not even idiots, as to be incapable of joining together different words")
3) The "problem" with animals is thus the absence of soul. They are *res extensa*, but definitely lack *res cogitans*. The existence of soul in humans is to Descartes an undisputable fact: to deny this fact is as big a mistake as denying the existence of God (*sic*). Similarly, to consider "beasts" as provided with soul is also a mistake, for it means to put up for discussion well-established theological and moral dogmas, which the whole civilization is based on (Veitch 2008: 48).

In this respect, Descartes' philosophy has a strong ideological component and guarantees a scientist's support to a conception highly promoted by Christianity: the absolute domain of humans on Nature. This is probably one of the reasons why Cartesianism was so successful in the following centuries, and – to some extents – is still alive in certain scientific environments. Together with >**Aristotle** and Christianity, Descartes constitutes the most important step in the formation of the western anthropocentric attitude towards non-human animals.

The notion of Cartesian mechanism was already being criticized in the eighteenth century, >**Hume** being one of the most illustrious examples. The main weakness, as was remarked on several occasions, was the use of the living being-machine comparison as a full explanation of life, rather than as effective metaphor. As philosopher Gino Ditadi remarks:

> Machines are built, organisms grow. This means that a comprehension of organisms must focus on processes. Cells, for instance, can be understood only on the basis of relational processes that reflect the dynamism of an organic system. While the activities of a machine are determined by its structure, such a relation is reversed in organisms, for their structure is determined by processes. Organisms display a high degree of flexibility

and plasticity. Machines work according to linear cause/effect chains, organisms work with retro-action and are an open system, in a constantly dynamic balance (translated from Ditadi 1994: 116).

Descriptive Zoosemiotics

The portion of zoosemiotic research that studies >**pragmatics**, >**semantics** and >**syntactics** in zoosemiosis.

Design-Features of Language

See **Hockett**, Charles.

De Waal, Frans (1948)

Dutch ethologist and primatologist. Professor of Primate Behavior in the Emory University psychology department in Atlanta, Georgia. A major expert in primate behavior, De Waal plays a central role, in modern ethology, in the study of morality and >**sociality** in great apes, offering an approach that he openly ascribes to the recent tradition of >**cognitive ethology**, of which he is one of the leading exponents. It is particularly through studies such as De Waal 1996 that his research achieved world notoriety. A central thesis in that work is that most animals, and particularly great apes are not only, and fully, "moral patients", but also provide evidence of "moral agency", becoming thus, like humans, *insiders* of the ethical discourse.

De Waal's research is not of semiotic type, however it offers several insights for the ethical program proposed in the Chapter 5 to this Companion.

Diagrammatic Iconism

A particular case of iconism (>**Icon**). It occurs for instance when a scout bee relates the distance of the food source to the speed of its dance (see **Bee dance**): the greater the distance, the faster the dance. This way a form of qualitative relation between sign and object is created, even if the speed of a movement is obviously not a "feature" of a spatial distance: what matters in this case is the correspondence (almost metaphorical, yet clearly iconic) between the increase of a value (distance) and the proportional increase of another value (motion).

Dialect

A variety of a sign system that is distinguished from other varieties of the same system by features of sign-object association, structure, social use, and by its use by a community of users that are set off from others geographically or socially. Normally, the structures and characteristics of semiosis among animals are species-specific, i.e., they vary from species to species. As in humans, however, there are individual nuances and, most of all, differences from one population group to another within

Fig. 4.2 The songs of the humpback whale present a wide range of local and regional dialects

the same species. These are called dialects, and are basically divided in mycrogeographical (or local) and macrogeographical (or regional) dialects. In the first case, variations are very subtle, and mostly concern very close-knit populations which interact with much frequency, so that their communicative codes mutually affect each other. In the second case, variations are quite evident, being put into action by communities which have very little, if any, occasion for communicative exchange, their respective codes thus growing in total autonomy. Present in a wide number of species, dialects are prototypical in birds (see, a.o., Armstrong 1963) and cetaceans (a.o., Payne 1996) (Fig. 4.2), and are closely connected with the development of >**culture**.

Discontinuity

The idea that, from an evolutionary point of view, human beings constitutes a radical deviation from the rest of the animal kingdom, and therefore the differences between the human and the other animals are of qualitative type.

It consitutes one of three general methodological frameworks, along with >**gradualism** and >**Umwelt** theory, to classify the different scholarly approaches to nature and animals, as explained in detail in Chapter 1, Section 1.1.4.

Distant Space-Time Semiosis

The capacity, typical of very complex semiotic systems like human >**language**, to construct or refer to entities or situations that are not spatially or temporally tangible. Distant space-time semiosis includes description/account of non-experienced past (like historical events), planning/forecast of not immediate future, and the dialectic/rhetorical/discoursive creation of possible worlds (non-existent situations or characters, abstract reflections, what-if scenarios, etc.).

The notion generally corresponds to Charles >**Hockett**'s design-feature of "Displacement", and is discussed in Chapter 3, from Section "3..

Domestication

Generally speaking, domestication is the process of taming an **>animal** (usually a non-human one), normally by generations of breeding, to live in close association with another animal (usually, a human one), in such a way that the former becomes dependent on the latter's care and, in most cases, unable to live in its original conditions again. According to Thomas **>Sebeok** (e.g., 1998: 70), a pre-requisite to domestication is the "reduction or possibly total elimination of the reaction of escape in a non-human animal from a human one". Domestication implies a consistent introduction of new signs and codes (**>code**) in the interaction between the subjects involved, up to the point that the **>ethogram** of the tamed **>species** is radically altered.

Early Etho-Zoosemiotics

See **Ethological Zoosemiotics**

Echolocation

Echolocation is the most typical example of **>proprioceptive semiosis**. It is the animal's capacity to locate objects of different type (obstacles, preys, co-specimens) and locate him/herself through the emission of sounds (especially ultrasounds) and the reception of their feedback, once they hit the various elements of the environment. Echolocation is predictably characteristic of those species that have to deal with environments of limited visibility (e.g. in the night, or underwater). Although not empirically defined, echolocation was first "guessed" by the Italian scientist Lazzaro Spallanzani in 1793, as he found out that blind bats were still fully able to find their way among obstacles of all kind. It was Donald **>Griffin**, in 1944, to coin the term and elaborate a theoretical model on the subject (Griffin 1958). Started in 1938, his research on echolocation was crucially supported by the world's first ultrasound microphone, devised by the American physicist George W. Pierce.

Echolocation has reached a remarkable specialization in mammals, in the bats *Chiroptera* and in the cetaceans *Odontoceta*, but less refined systems can be found also among bird species like the *Steatornis caripensis* or the genus *Collocalia*, and among insects, like the moths of the family *Arctiidae*.

Ecocentrism

See **Biocentrism**

Ecosemiotics

Term introduced in Nöth 1998 to define the semiotic study of the human-nature (or culture-nature) relationship, mostly focusing on how human being relates to and make sense out of the natural environment.

Eibl-Eibesfeldt, Irenäus (1928)

Austrian ethologist, particularly known for founding Human >**ethology**. Currently Professor at, and Head of, the Max-Planck-Institute for Behavioural Physiology, in Andechs (Germany). Pupil of Konrad >**Lorenz**, Eibl-Eibesfeldt is one of the key-figures in the development of the ethological disciplines, and produced several works of (indirect) zoosemiotic interest, focusing on such topics as morality (>**Ethics**), >**sociality** and >**communication** (a.o., 1970 and 1971), plus specific contributions to the study of human behavior.

Electric Channel

See **Tactile Channel**

Emic-Etic

Terms derived from the linguistic words "phonemic" and "phonetic", and introduced by Kenneth Pike (1954). They indicate respectively the description of one or more behavioral patterns from the inside perspective of the actor (emic); and the description of one or more behavioral patterns from the outside ("culturally neutral") perspective of the observer (etic).

Methodologically useful in several fields of inquiry, particularly ethnology and anthropology, the concept is essential also in >**zoosemiotics**, and is here discussed in Chapter 2, Section 2.3.1.

Emotion-Free Experimental Context

In >**interspecific communication experiments**, an emotion-free context is a type of experimental setting arranged in order to avoid any possible unwanted input from the experimenters that might result in a >**Clever Hans effect**. Such inputs include facial expressions that might unconsciously suggest to the subject of the experiment the solution for a specific task, affective – therefore, allegedly, not anymore scientific – relations established with the subject of the ICE, and others. In an emotion-free context, researchers might wear masks, observe the experiment unseen from another room, leave the subject alone in the laboratory, avoiding any kind of interaction, double- or triple-check data, and so forth.

The majority of interspecific communication experiments constructed within an emotion-free context throughout the last decades ended up producing poor results (i.e., the attempts of teaching human language to specimens of non-human species failed), in contrast with those contexts where emotional and personal relationships with the animals were encouraged (as the >**Washoe**, >**Kanzi**, >**Chantek** and >**Koko** programs), which were reportedly successful. Such discrepancy activated a heated discussion, animated by two schools. On the one hand, those (including many semioticians) who think that the failure of emotion-free experiments is the ultimate proof that: (a) non-human animals are not able to learn human language; and (b) the Clever

Hans effect is the real core of the problem, and therefore interspecific communication experiments are in principle misleading non-scientific practices. On the other hand, those who maintain that the attempt to teach language in a context that eliminates, rather than encourages, emotional and personal interaction is a contradiction in terms, as language is both caused by and focused on interaction. In favour of this orientation, Felice **>Cimatti** (1998: 147–151) argued by presenting a typical experimental situation (using Washoe and **>American Sign Language** as example). According to the detractors of interspecific communication experiments, at the very moment Washoe is asked to perform the ASL gesture corresponding to a given object, she does not have the slightest idea of how to perform such a task, since she is not able to understand the relation between signifier (the ASL gesture) and signified (the object). Thus, she proceeds by attempts, moving her arms at random, until – unconsciously or not – her trainers display apparent satisfaction, which lets her know what the right gesture is. In the detractors' opinion, this is a typical example of Clever Hans effect (and, more generally, a form of anti-scientific **>anthropomorphism**): unless one is able to invalidate any risk of this type, the experiment will be a waste of time, and the risk can be avoided only by creating an "emotion-free" context. However, says Cimatti, to consider the use of a communication system as unrelated to social and emotional interaction is a self-evident contradiction and a more serious mistake than anthropomorphism. What is the point in learning a communication system if one has no one to use it with, does not receive adequate stimuli, and, in the end, does not have any reason to learn it? In such a case, says Cimatti, the mistake is not only methodological, but also theoretical: "if language is a system based on emotional interaction [. . .], then it is simply impossible to study it without considering this aspect" (translated from Cimatti 1998: 149).

Endosemiotics

See **Endosemiotics-Exosemiotics**

Endosemiotics-Exosemiotics

Biosemiosis can take place either *within* a living being and/or *between* two or more of them. Discussions around the former case do not yet include the field of **>zoosemiotics**, and instead consists of a branch of biosemiotics named *endosemiotics*, or – according to the specific cases – protosemiotics, microsemiotics, cytosemiotics, etc. Endosemiotics involves the message exchange among cellular organelles, cells, tissues, organs and organ systems. The interpretation of the symptoms of a given disease as signs of the disease itself, as suggested by Hippocrates, can be considered an early example of endosemiotic analysis.

Along with endosemiotics, there is *exosemiotics*, which regards the entire spectrum of message exchange between two or more complex organisms. Fields of exosemiotics are mainly *phytosemiotics* (whose object is the semiosis among plants), *micosemiotics* (semiosis among fungi) and finally *zoosemiotics* (semiosis among animals). A relevant branch of zoosemiotics is obviously **>anthroposemiotics**, i.e., the semiosis among the human animals.

Finally, the globalistic conception of life as a whole of sign processes, that is, the interpretation of the fully and intimately semiotic nature of life, mostly promoted by the American semiotic tradition (**>Peirce** and **>Sebeok**, first of all), led a few biosemioticians to the belief that the entire universe can be analysed and interpreted semiotically (see **>Pansemiotism**).

Ethics

Branch of philosophy that deals with the values underlying, and relating to, human conduct, with a focus on the (social, legal, etc.) rightness and wrongness of actions and to the (culturally-specific) "goodness" and "badness" of the motives and ends of such actions.

The possibility that animal ethics (animal rights, **>biocentrism,** conservation, ecology, etc.) can and/or should be part of the zoosemiotic agenda is discussed in the Chapter 5 to this companion.

Ethogram

In **>Ethology**, an ethogram is the behavioral catalogue of a species, that is, the totality of activities and patterns that tipify it. The compilation of an ethogram is preceded by a stage of careful observation (in real time, and through footage) of the animal in different circumstances, while it normally avoids experimental contexts (as captive animals tend to present a varied and incomplete repertoire of behavioral patterns, as compared to free ones). Sebeok suggested a semiotic interpretation for this concept:

> In semiotic terms, this concept encompasses an animal's species-specific communicative code, in confrontation with which the human observer's role necessarily becomes that of a cryptanalyst, of someone who receives messages not destined for him and is initially ignorant of the applicable transformation rules. (Sebeok 1990: 40)

Ethology

A branch of zoology, ethology is the scientific study of animal behavior. The term stems from the Ancient Greek "ethos" ("custom", "behavior"), and was first coined in 1762, by the Academie Française des Sciences. John Stuart Mill employed the word in his *System of Logic* (1843), but in a different sense, proposing the establishment of a field of inquiry, "ethology" indeed, that would account for individual and national differences in character. It was the American Myrmecologist William Morton Wheeler, in 1902, to popularize the term in the modern sense. Leaving aside the word, however, traces of, so to speak, proto-ethology can be already found in studies on **>instinct** and innate behavior, such as Pernau 1716, Reimarus 1773 and Spalding 1873. Not to mention the acknowledged pioneering contribution by **>Uexküll**.

Before ethology, the main field devoted to the study of animal behavior was comparative psychology. The differences between the two disciplines are both epistemological (obviously, as the scientific context is very different) and methodological: comparative psychology prefers to perform empirical observation in artificial experimental contexts, while ethology concentrates on behavior in natural situations. Because of this, the two approaches may be considered either complementary or competitive, and so they acted all through the twentieth century (collaborating in some cases, and disputing over their respective findings in others). Another difference is of geographical type, as comparative psychology developed most strongly in North America, while ethology was more followed in Europe (this distinction not being alien from philosophical and ideological consequences).

One of the main focuses in ethology is the study of the evolution of behavior and the interpretation of behavior in terms of natural selection, and that of course reveals the instrumentality of Charles **>Darwin** to the foundation of modern ethology (particularly, the book The expression of the emotions in animals and men, 1872). Early ethologists, like Oskar Heinroth and Julian Huxley, focused primarily on the notion of instinct, as an explanation for animal behavior. An important step was the construction of the concept (and research related) of **>ethogram**, which allowed both a qualitative and quantitative data-collection about behavior.

The turning point was the generation of Konrad **>Lorenz** (whose name is still the most immediate association with the term "ethology"), Karl von **>Frisch** and Nikolas **>Tinbergen**, who (1) established unmistakeably ethology as an autonomous discipline; (2) set a number of methodological standards that are still used nowadays (like Lorenz's identification of the so-called Fixed Action Patterns, Tinbergen's famous **>Four Questions**, or the notion of **>imprinting**), and (3) engaged in sophisticated investigations of animal behavior that led to important, and sometimes revolutionary, discoveries (like von Frisch with **>bee dance**).

Through the work of this generation, ethology became strong particularly in Europe before World War II, and also in UK right after, when Tinbergen moved to Oxford University, and figures like William Thorpe, Robert Hinde and Patrick Bateson came onto the scene. In continental Europe, meanwhile, another generation of ethologists, pupils of the old masters, was growing fast, providing new names of historical importance, such as Irenäus **>Eibl-Eibesfeldt** and Martin **>Lindauer**.

The appearance of **>zoosemiotics** in 1963 had a small but significant impact on ethology, and, because the early stage of **>Sebeok**'s conception of the field aimed to an umbrella term that would unify different approaches on animal **>communication**, collaborations between semioticians and ethologists were not rare (e.g., Sebeok 1968). This collaboration decreased from the 1980s onwards after Sebeok's biosemiotic turn, and the conception of the entire **>biosemiotics** as a field more rooted in **>philosophy** and in the Peircean tradition, rather than in natural sciences (in fact, often dismissing them as inadequate to accurately describe life phenomena).

During the 1970s, ethology underwent a certain theoretical "restyling", witnessing the birth of two branches that eventually replaced classical ethology in many of its methodologies and general conceptions. These were >**sociobiology** and, most of all, >**cognitive ethology**. It was precisely the impact of a cognitive approach to animal studies that created the conditions for a *entente cordiale* between ethologists and zoosemioticians, on the basis of a common paradigm and a common research ethics (see Martinelli 2009).

At present, it is safe to say that ethology consists of at least the following branches:

1) *Field ethology*: the classical investigation of animal behavior in natural conditions;
2) *Applied ethology*: the practical exploitation of ethological knowledge in activities like breeding, veterinary medicine, companion animals, etc.;
3) *Cognitive ethology*: the study of the mental processes of acquisition, representation and use of information in animals;
4) *Quantitative ethology*: the study of animal behavior via the employment of quantitative data analysis (to be considered a "continuation" and a refinement, in a statistically-oriented and nowadays informatized environment, of the research started on the field);
5) *Human ethology*: the application of ethological theories to the study of human behavior (in the way pioneered by Darwin 1872);
6) *Social ethology*, or *sociobiology*: the systematic study of the biological bases of social behavior.

An excellent introduction to classical ethology, before the cognitive breakthrough, is Eibl-Eibesfeldt 1970. The relationship between ethology and zoosemiotics is discussed in Chapter 1, Section 1.1.2.

Ethological Zoosemiotics

The most classical branch of >**zoosemiotics**, that studies the semiosic processes occurring among animals. Introduced by Thomas >**Sebeok** in 1963 and developed by him and several other scholars, this field is discussed at length in this companion in the entire Chapter 2.

Ethological zoosemiotics can be divided into an early stage (eEZ) and a modern one (mEZ). The first case includes the period when zoosemiotics was mostly conceived as an umbrella term for gathering different approaches on animal >**communication** (that is, not >**semiosis** altogether). In this stage, zoosemiotics was mainly influenced by classical >**ethology** and >**behaviorism**. The second stage, started in the late 1970s, witnesses some important innovations: the extension of zoosemioticians' interests to the entire spectrum of semiosis; the (attempt to a) formation of an autonomous paradigm for zoosemiotics; and the introduction of >**cognitive** approaches.

Etho-zoosemiotics

See **Ethological Zoosemiotics**

Exosemiotics

See **Endosemiotics-Exosemiotics**

Expressive

One of the six >**Functions of communication** postulated by Roman Jakobson (1963). A message is expressive when its focus is the sender.

Exhibition

In a semiotic sense, exhibition consists of a set of displays which, through a process of >**ritualization**, become simplified, stereotyped and exaggerated. Generally, but not exclusively, exhibitions are species-specific and occur in courting- fighting- or feeling-related contexts. Forms of exhibition can be detected in most species, particularly birds and mammals, and – for what we are able to perceive – seem to occur mostly through the >**visual channel** and/or the >**acoustic channel**. The biological function of some exhibition patterns is often not clear, and a typical case is that of >**kinaesthetic signs**.

Food Request

A form of >**intraspecific** (and occasionally >**interspecific**) interaction, aimed at obtaining food. It usually occurs between parents and offspring, but also in other forms of relationship, such as between mates or between pets and human beings. Food request is one of those situations that require a specific >**contextual code**: young chicks of nesting birds >**species** open their beak wide in vertical position; other species (like the *Larus argentatus*) request food by repeatedly hitting the parents' beak on a specific spot; various *canidae* (including jackals and wolves) regurgitate their food after being rubbed on their mouth by the young specimens; apes of the species *Hylobates*, *Macaca*, *Papio* and others may request food by opening up the palm of their hands (like "beggars"), and so on. The semiotic nature of these signs is evidently of indexical type (>**index**). However, a very interesting phenomenon occurs when these signs (or entire codes) are subject to a process of >**ritualization**. In these cases, the signs do not assolve their original function anymore, but are abstracted to a symbolic function (>**symbol**). The jackal *Canis aureus*, for instance, extended the mouth-rubbing sign not only to the simple, indexical, request for food, but also to a general situation of affection and friendly recognition (a "greeting" or a "kiss"). In the sea lion *Otaria byronia* the same sign is also used to pacify conflicts among females. In these cases, as in others, it is clear that the sign goes through different stages of redefinition and abstraction of its meaning (indeed: ritualization), and assumes a symbolic connotation: I rub your mouth to request food, *therefore* I am in a close relationship with you, *therefore* I use this sign also to friendly greet you, *therefore* (in the sea lions' case) I use this sign also for re-establishing a friendly situation.

Four Questions

Expression employed to define Nikolas >**Tinbergen**'s influential methodological model for ethology, articulated into four fundamental points (Tinbergen 1963). The questions are grouped into two categories (including two questions each): *Proximate mechanisms* and *Ultimate mechanisms*. Proximate mechanisms include

1) *Causation* (or *Mechanism*), investigating the stimuli eliciting the response, and the modification of the latter after recent learning.
2) *Development* (or *Ontogeny*), investigating the modes of behavioral change in time units (age), in relation with earlier experiences.

As for ultimate mechanisms, the questions concern

1) *Evolution* (or *Phylogeny*), investigating the animal behavior in relation with other species performing similar patterns, and in general the phylogenetic aspects of such behavior.
2) *Function* (or *Adaptation*), investigating the impact of the given behavioral pattern/s on the animal's general survival and reproduction chances.

The four questions, clearly framed in a behavioristic approach (>**Behaviorism**) had a remarkable impact at interdisciplinary level as well, being employed also in >**Sociobiology** and Social sciences.

Fouts, Roger (1943)

American primatologist and psychologist. Currently Professor of Psychology at the Central Washington University and co-director of the Chimpanzee and Human Communication Institute (CHCI) in Washington. At first assistant, then leader of the >**Washoe** program, Fouts has been very instrumental in the development of what is commonly regarded as one of the most succesful >**interspecific communication experiments**. Encouraged by a special bond with Washoe (he was hired in 1967, after an unimpressive interview, for the sole reason that the chimpanzee showed immediate sympathy to him), he accompanied her and the other chimpanzees eventually added to the community throughout her entire life, following their improvements and popularizing the training program via different media (including the publication of the very succesful and informative Fouts 1999).

As with many colleagues in primatology, Fouts too became an animal rights advocate, fighting particularly for the extension of human rights to Great Apes.

Fossey, Dian (1932–1985)

American zoologist and primatologist. Together with Jane >**Goodall** and Birutė >**Galdikas** (the three are also known as Leakey's angels, for being all pupils of famed anthropologist Louis Leakey), the most important scholar in >**interspecific semiosis** with great apes in the wild. Inspired by Prof. Leakey, Fossey became fascinated with the rare mountain gorillas of the Rwandan wilderness, and decided to go studying closely their lives and habits. Fossey achieved the gorillas' acceptance in their community by developing means of >**communicaton** with them (a.o., Fossey 1972 and 1974). Her study interest developed into a passionate commitment for the gorillas' welfare and preservation, up to become a *persona non grata* to Rwandan natives, who extensively hunt gorillas for their skin (sold to rich tourists in form of various gadgets). After years of campaigns and fights with local institutions, Fossey was found murdered in her tent. in officially mysterious circumstances. Her legacy remains big, and her autobiography (Fossey 1983) is still the best-selling book about gorillas.

Frisch, Karl Von (1886–1982)

Austrian ethologist and Nobel Prize winner in Physiology or Medicine in 1973. Along with Konrad >**Lorenz** and Nikolas >**Tinbergen**, with whom he shared the prize, von Frisch must be considered the most important figure in classical >**ethology**. His main contribution to animal studies remains the extensive research conducted on the >**language** of honey bees (>**Bee dance**), together with his then student Martin >**Lindauer** (as documented, a.o., in Frisch 1967). Their work included studies on the bees' sensorial mechanisms, their various communication systems, their sensitivity to polarized light, and the pheromones emitted by the Queen bee and her daughters (these topics are also discussed in the specific entry on "Bee dance", and in various points of the companion).

Less known but still important, are von Frisch's studies on >**architectural signs** in animals (Frisch 1974).

Functions of Communication

According to Russian linguist Roman Jakobson (1960: 353–7), there are six main functions within a communication system, each concerned with a particular element of the semiotic process: >**expressive** (referring to the *sender*), >**conative** (referring to the *receiver*), >**phatic** (referring to the establishment of *contact* between sender and receiver), >**referential** (referring to the *context* of the message), *metalinguistic* (referring to the *code* used in a given communication act), and >**aesthetic** or *poetic* (referring to the form of the *message*). These functions do not work in isolation and are not mutually exclusive. A message typically is the intersection of two or more functions, where one of those is simply (but not always) the dominant and more evident one.

Felice >**Cimatti** (1998: 59–105) proposes a systematic application of Jakobson's model within >**zoosemiotics**. The issue is dealt with in Chapter 2, Section 2.2.3.

Gabriel, Peter (1950)

World-famous singer-songwriter, former member of the progressive rock band Genesis. "Traditional" musical merits apart, Gabriel is a promoter of >**interspecific** musical interaction: in 2001, he visited the Georgia State University's Language Research Center, leading a keyboard-based jam session with some bonobo apes. Later, more importantly, Gabriel founded the project "ApeNet" (since 2007 renamed "InterSpecies Internet", and currently inactive), with the intention of linking great apes through the internet:

> ApeNet is a consortium of foundations and individuals who support interconnecting great apes with each other, as well as with humans, through enculturation and technology. ApeNet was founded by British musician Peter Gabriel and American philanthropist Steven Woodruff to provide new living and communication solutions for enculturated great apes.

[...] Mission: To link enculturated great apes with each other through the internet, establish culture-based great ape preserves, establish an internet-based Journal, and to encourage and support great ape welfare and conservation. (from www.apenet.org, no longer online)

Gaius Plinius Secundus (23–79)

Roman philosopher and naturalist, also known as Pliny the Elder. His presence in a zoosemiotic companion is self-evident by his massive encyclopaedic work *Naturalis historia*, which arguably comprises the entire knowledge on Nature and animals (and medicine, mathematics, geography, etc.) available at the time. It consists of 37 books, of which the 8th, the 9th, the 10th and the 11th are entitled "Zoology", and contain several topics of ethological and zoosemiotic interest. Among these, the detailed (and fairly complete) description of a few species (particularly the honey-bees), a few hypotheses (later found out to be correct) on zoological phenomena (as in his interpretation of the "origin of amber", accurately described as a product of fossilized resin of pine trees), and a rather poetic (and still scientifically interesting) description of the nightingale's singing activity.

Galdikas, Birutė (1946)

German primatologist and ethologist. The youngest of the three >**Leakey's angels**, together with Dian >**Fossey** and Jane >**Goodall**, Galdikas is currently Full Professor at Simon Fraser University in Burnaby, British Columbia. As per anthropologist Louis Leakey's plan (he wanted to perform field research on the three closest "relatives" of the human being, the chimpanzees, the gorillas and the orangutans), Galdikas focused her work on the orangutans of the Tanjung Puting Reserve, in Indonesian Borneo, expanding scientific knowledge on their behavior, habitat and diet (Galdikas 1995). Like the other two "angels", she then devoted most of her attention to the advocacy of social and ethical causes related to her research subject, achieving several recognitions and founding the Orangutan Foundation International, of which she is currently president.

Gardner, Allen and Beatrix

See **Washoe**

Generalization

See **Abstraction**

Goodall, Jane (1934)

English primatologist and ethologist, and founder of the Jane Goodall Institute. Pupil of anthropologist Louis Leakey (along with the other two **>Leakey's angels**, Dian **>Fossey** and Birutė **>Galdikas**), Goodall studied for nearly 50 years chimpanzees' **>sociality, >cognition**, **>tool** making and **>semiosis** in Gombe Stream National Park, Tanzania (a.o., Goodall 1964, 1968, 1971, 1986). Similarly to the other two "angels", Goodall too accompanied her scientific research with a strong ethical commitment for primates' conservation and awareness. In this activity, she also managed to achieve specifically scientific results, as for instance the taxonomic inclusion of chimpanzees, bonobos and gorillas in the Hominids' family, and the promotion of **>critical anthropomorphism** (she deliberately chose to give names to her chimpanzees, challenging the contemporary conventions that would regard this practice as anti-scientific). This and other methodological aspects of her work generated some criticism, and accusations of biased research.

Goodall's environmental and humanitarian work was recognized in several ways by several institutions: most notably, she is a UN Messenger of Peace.

Gradualism

Term that refers to the idea of an evolutionary adaptive continuum among animal **>species**, in which, depending on the contexts, a given species displays the most refined of a given (physical, behavioural, etc.) trait, and in which, position after position, the characteristics of the diverse animals are less and less complex and refined, although, in most cases, adequate for ensuring the survival of the species in question. In this sense, differences across species are of purely quantitative, and not qualitative, type.

It consitutes one of three general methodological frameworks, along with **>discontinuity** and **>Umwelt** theory, to classify the different scholarly approaches to nature and animals, as explained in detail in Chapter 1, Section 1.1.4.

Griffin, Donald (1915–2003)

American zoologist. He is generally considered the founder of **>cognitive ethology**, or rather his 1976 book *The Question of Animal Awareness* is considered the main transition point from classical to cognitive ethology. His "historical" role apart, Griffin performed extensive research on several topics of zoosemiotic interest, including animal **>communication**, navigation, acoustic orientation and sensory biophysics, delivering a major contribution to the discovery of **>echolocation** in bats (see Griffin 1958).

Gustatory Channel

See **Chemical Channel**.

Hediger, Heini (1908–1992)

Swiss zoologist, director of Zürich zoo and founder of the field of "zoo biology" (i.e., the ethological observation of captive animals). Considered by his friend Thomas **>Sebeok** as one of the main influences for the whole **>zoosemiotics**, Hediger had a crucial impact on the study of human-animal interaction (**>Anthropological zoosemiotics**) in contexts of captivity. He studied closely the proxemic, social and territorial aspects of animal behaviour, combining **>Uexküll**'s **>Umwelt**-theory with modern **>ethology**. Among his many focuses:

1) Animal's *expression*, defined as "variable nonpathological phenomena of the animal, which may help to an understanding of their situation" (Hediger 1968: 143): face-displays, general posture, kinesics are organized into acoustic, optic, olfactory and "internal" expressions (the latter including reactions of secreting tract, vomiting, urinating, level of adrenaline etc.);
2) Animal's managing of *interpersonal distances* (e.g., 1968) in social or generally interactional contexts (among other things, this research inspired Edward Hall's theories on social distance).
3) Different forms of human-animal relationship (e.g., 1965), setting the basis of later reflections by Sebeok, and in general opening the door to a very important portion of anthropological zoosemiotics;
4) The **>clever Hans effect** (e.g., 1981), this, too, extremely influential for Sebeok, and for most of zoosemioticians' reviews on **>interspecific communication experiments**.

Among the several tributes given by zoosemioticians to Hediger, one should at least mention Sebeok 2001b and **>Turovski** 2001.

Heterotroph

A living organism that is not able to chemically produce its own food, and must therefore consume other organisms. Being heterotrophic is one of the conditions that separates an >**animal** from other organisms like plants, which are on the contrary autotrophic.

Hockett, Charles (1916–2000)

American linguist. One of the key-figures in structural linguistics, Hockett made a very important contribution to the study of >**language**, by elaborating a so-called design-feature approach, aimed to emphasize the differences between non-human animal >**communication** and human language. Such approach led to the definition of 13 design-features of communication (Hockett 1960a), all of which available in human language, while only some can be traced in other animal communication systems. These are:

1) Vocal-Auditory Channel (>**Acoustic channel**).
2) Broadcast transmission and directional reception (>**Acoustic channel**).
3) Rapid Fading (>**Acoustic channel**)
4) 4Interchangeability: A subject can both speak and hear the same signal, therefore anything that can be heard can be also spoken.
5) Total Feedback: A subject can hear him/herself speak, therefore s/he can control speech production.
6) Specialization: Hockett meant specialization mostly as that capacity to "produce information" (in other words, what Jakobson called the >**referential function** of communication), arguing that this is typical of human beings, while other animals mostly tend to communicate their own emotional state (>**expressive function**)
7) Semanticity: A signal can be matched with a specific meaning.
8) Arbitrariness: Signs can be symbolic, and do not need to have a specific/natural connection with the objects they represent (>**symbol**).
9) Discreteness: Phonemes can be distincted from each other in a digital sense.
10) Displacement (see >**Distant space/time semiosis**).
11) Productivity: New and unique meanings can be created from previously existing utterances (associated to different meanings).
12) Traditional Transmission: languages are also acquired through >**learning**.
13) Duality of patterning: Phonemes are combined into words, and words are combined into sentences.

While leaving no doubt that only human language is the very communication system that is able to incorporate *all* the 13 features, Hockett also discussed the presence of some of them into other animals, particularly focusing on honeybees (> **Bee dance**), whose communication implies at least the existence of Broadcast Transmission and Directional Reception, Semanticity, Displacement and Productivity; and gibbons, who display all features except Displacement, Productivity, Traditional Transmission, and Duality of Patterning.

Hoffmeyer, Jesper (1942)

Danish biosemiotician. Currently associate professor at the Institute of Biological Chemistry, Copenhagen University, he is president of the International Society for Biosemiotics Studies, and a leading figure in one of the two main schools in biosemiotics, the so-called Sign-based Biosemiotics (see also >**Barbieri**, Marcello). His work is not primarily concerned with zoosemiotics, but his contribution remains important in the framework of a general systematics of the semiotic study of living organisms (see 1996), a field where also his colleague Claus Emmeche deserves a mention.

Homology

See **Analogies-Homologies**

Human-Other Animal Relation

See **Athropological zoosemiotics** and **Anthrozoology**

Hume, David (1711–1776)

Scottish philosopher. Key figure of the Enlightenment, Hume represents a radical cut with the "sick metaphysicians" (as he calls them) of the seventeenth century, i.e. Spinoza, Hobbes, and most of all >**Descartes**. His points of reference are rather >**Montaigne**, >**Locke**, Bacon, >**Bayle** and Newton. By consequence, his considerations of non-human animals follow more closely a proto-evolutionary approach. Similarly to Locke, Hume makes it clear that the differences between humans and other animals are simply a matter of degree. His attack to Cartesianism, as appearing in the *Treatise on human nature* (1740/2003) is pretty direct and to the point:

> Next to the ridicule of denying an evident truth, is that of taking much pains to defend it; and no truth appears to me more evident, than that beasts are endowed with thought and reason as well as men. The arguments are in this case so obvious, that they never escape the most stupid and ignorant. (Hume 1740/2003: 126)

The main points of his argumentation are the following:

1) Similarities between humans and other animals concern both the emotional and the intellectual area: all animals aim at seeking pleasure and avoiding pain, all animals care about their own life, all animals share the same principles at the basis of reasoning (Hume 1740/2003: 127–8);
2) Exactly on the issue of reasoning, Hume most radically challenges the philosophical tradition. Reasoning, in Hume, departs from the senses, imagination and experience: they allow the act of deducing and believing that future (whether immediate or not) will conform to given expectations. Hume argues against one of the very foundations of traditional philosophy: mathematic thought is not

the expression of reason, but simply a consequence of senses and imagination. In fact, it is expressively the search of the perfect science that makes humans imperfect beings (Hume 1740/2003: 295);

3) Although reasoning, in humans and other animals, is claimed to be caused by passions, Hume does not mean to be an irrationalist: his goal is simply that of establishing an adequate causal relation between emotions, experience and feelings on the one hand and intellect, thought and act on the other (Hume 1740/2003: 233).
4) Finally, although he excludes them from the idea of justice, Hume includes the other animals in the idea of morality and >**ethics**, and in fact asserts very clearly that the exclusion from justice has nothing to do with taking care of them. Morality, as founded on feelings, is a primary virtue: justice is an artificial one. It is no coincidence that modern philosophical utilitarianism – inspired by Hume – includes many of the philosophers most concerned with the animal rights case (Peter >**Singer** above all). Hume seems to have no doubts about the fact that all animals are moral beings (Hume 1740/2003: 283)

Icon

In Charles S. >**Peirce**'s theory of signs, an Icon is a sign that bears qualitative features of the object. An iconic sign is a sign that looks/sounds/smells etc. like the object.

Iconic >**semiosis** is rather common among animals, the various phenomena of >**mimicry** (visual and vocal) being possibly a classic example. Several forms of >**deception**, also, are of an iconic type. Finally, on a more general level, prey-predator interactions and >**adaptation** processes often imply the production and the exploitation of iconic signs. A particular case of iconism is the so-called >**diagrammatic iconism**.

Icons are further discussed in Chapter 2, Section 2.2.1.

Imitation

A paricular case of >**mimicry** consisting of taking from a model (i.e., another individual or >**species**) one or more patterns not present in the mimic (i.e., the imitating individual/spiecies). Imitation may occur through any of the >**channels** available to a species, although it is normally associated with the >**acoustic channel**, and may be displayed for various purposes, including courtship, deception, defence or aggression.

Imprinting

One of Konrad >**Lorenz**'s most important contributions to >**ethology** (although generically anticipated by Douglas Spalding in 1873). Imprinting can be defined as the capacity to learn specific types of information at certain critical periods in development. Lorenz had an insight of this phenomenon while famously observing young geese and chickens spontaneously following their mothers from almost the first day after they were hatched. He then reproduced the procedure with eggs that were incubated artificially, discovering that the stimulus can come from any source and still serve as imprinting, as long as it is presented during a critical period (the few days after hatching).

Index

In Charles S. **>Peirce**'s theory of signs, an Index is a sign that bears a physical connection with the object. An indexical sign is thus a natural cause or consequence of the object. Its categorization requires a semantic memory, in that one has to recognize the relation between the referent and event/state.

Indexes (or indices) may safely be defined as the most recurrent signs in zoosemiosis. A large number of markers, symptoms, indicators and signals produced through any of the available **>channels** are in one way or another indexical, from the tail-wagging of a dog (index of its excitement) to the wheel display of a peacock (index of its health conditions). By consequence, the significance of indexicality, within zoosemiotic study, is to be considered more central than in the case of **>anthroposemiotics**, particularly when it comes to its cultural aspects, whose logocentric nature predictably manifest a greater interest for symbols (**>symbol**) and icons (**>icon**).

Indexes are further discussed in Chapter 2, Section 2.2.1.

Ingold, Tim (1948)

British anthropologist. Currently Chair of Social Anthropology at the University of Aberdeen. In 1988, he edited the collection *What is an animal?*, one of the most thorough inquiries on the many aspects of the concept of "animal" in cultural and anthropological sense (without diregarding **>zoosemiotics**, and hosting one essay by Thomas **>Sebeok**). The book is a crucial source for **>anthropological zoosemiotics**. Ingold contributes two texts, both very relevant for the context of this companion. The first one, an extensive introduction to the whole book, is an overview of the subjects of the volume, as dealt with by the various contributors (a.o., Mary Midgley, Stephen Clark and Brian Goodwin). In his second contribution, entitled "The animal in the study of humanity", Ingold goes through some of the "hot" topics in the humanity-animality debate: **>culture**, **>language**, symbolic communication (**>symbol**) and others. Many of the recent definitions of culture, Ingold observes, have been taking for granted its human species-specificity, but without really proving it. That applies also to the often-made distinction between a supposedly *true* human culture and a non human *proto-culture*, or to the use of language symbolic quality as a prove that only human information is transmitted in ways that are not only genetic: "What most anthropologists have failed to realize... is that the opposite of symbolically encoded behavior is *not* genetically transmitted behavior, or crudely, 'instinctive' behavior" (Ingold 1988: 84–5). Although not disputing the human uniqueness of symbolic thought, Ingold rejects the idea of using it as a passepartout for explaining humanity, the latter presenting more elements than not of continuity with animality, and rather concludes that the main human achievement, and witness of its uniqueness, expressed through language (and its implications at cognitive level), is the fact of being the only species that was able to make a world for itself.

Instinct

A critical notion *par excellance*, instinct is generally defined as a genetically acquired force that drives animals to react to a stimulus in certain fixed ways. A central notion in classical >**ethology**, in fact – at least, partly – the very notion *around which* the whole field moved its first steps, instinct is becoming more and more an obsolete, if not trivialized, concept, and has been almost completely abandoned by modern >**cognitive ethology**. The main critiques of the notion, and its supporters, came from at least three sides: (1) from a behavioristic point of view, it makes no allowance for environmental influences upon behavioral patterns; (2) from a cognitive point of view, it long prevented from an accurate description of several patterns, originally thought of as "instinctive"; but later found out to be the result of different categories of motivation and cognition; (3) from a methodological point of view, it became a banal "black box" (or "explanatory principle", as it is called in >**Bateson** 1969) where scholars put most (or all) of the observations they were not able to explain with the necessary simplicity required by >**Morgan's canon**, or to explain in principle.

The notion of instinct is thoroughly discussed in Chapter 3, Section 3.3.5.

Intentionality

In a general sense, intentionality represents the conscious planning or performance of a given action, aimed to a goal. In philosophy, particularly in the scholastic and gnoseologic traditions, the expression has been used in a broader sense, designating any mental state provided with a content (e.g., thinking, willing, longing), and nearly overlapping with the notion of >**consciousness**. In >**ethology**, the concept has been practically overlooked (and replaced with the notion of >**instinct**), until the coming of >**cognitive ethology**, when – on the contrary – it became central. In semiotics, finally, the concept is often used in relation to the so-called >**semiotic threshold**, as a watershed between semiotic and non-semiotic acts (this aspect is discussed in Chapter 1, Section 1.1).

Interspecific

Any form of >**semiosis** occurring between two or more different animal species. It is an expression largely relying upon the notion of >**species**, in itself highly problematic, in that (a) it is a human elaboration, and (b) it is constantly under revision and update. Having said that, and taken into account the multitude of grey areas where the boundary between one species and another is far from being clear, many are the examples of semiotic interaction where the specific difference goes beyond any doubt. Among these, the prey-predator interaction, several forms of >**mimicry**, and >**symbiosis**.

From a semiotic point of view, a difference between species (or, more simply, groups or communities) implies that the two parties involved share only a limited number of patterns of perception and codification of the world (i.e., very possibly, they live each in their own >**Umwelt**), and in that sense the interaction is either very basic, or largely subject to interference and misunderstanding. A simple >**code** is normally established via a trial-and-error process, and, depending on the situation, the parties may strictly stick to it (e.g., in the case of cleaning symbiosis), or highly manipulate it (e.g., in prey-predator interaction).

A particular case of interspecific semiosis is the >**communication** occurring, in experimental contexts, between human beings and other species, particularly the attempt of the former to teach their own language to the latter (> **Interspecific communication experiments**). This instance is widely discussed in Chapter 3, from Section 3.3 of the present companion.

Interspecific Communication Experiments

The expression "Interspecific Communication Experiments" (ICE, from now on) refers to a range of experimental programs (conducted usually by biologists, psychologists or linguists), aimed at testing the possibility of teaching human >**language** to other animal >**species**, particularly great apes.

Generally, the scholars working on ICE advanced the basic claim that there is no unbridgeable gap between human language and other >**communication** systems. Such a claim relied on the Darwinian idea of evolutionary continuity between *homo sapiens* and other species, a continuity where language plays only the role of a more refined system for communication. Together with the Darwinian one, another stand, of behaviouristic type, would define language as a >**conative** tool, whose function is basically that of transmitting thoughts from a sender to a receiver. In this sense, language was seen as a "normal", albeit complex, behaviour, that could be learnt as any other.

The first experiments were affected by the (eventually proven) wrong assumption that the vocal and >**acoustic channel** was the only reliable source for a communication system to be considered true language. The idea probably originated from writings by Samuel Pepys (*The diary of Samuel Pepys*, 1661), Johann Conrad Amman (*Surdus loquens*, 1692 and *Dissertatio de loquela*, 1700) and Julien Offray de La Mattrie (*L'homme machine*, 1748), who all shared the opinion that monkeys could easily be taught to speak. Among the first experiments of this type, the chimpanzees Peter, raised by clinical psychologist Lightner Witmer around 1909, and Joni, raised by ante-litteram cognitive primatologist Nadia >**Kohts** between 1913 and 1916. Later attempts to teach spoken language to chimpanzees were performed on Vicki, raised by Keith and Catherine Hayes between 1952 and 1958, and Gua, raised by Winthrop and Luella Kellog during the late 1960s. All the chimpanzees – after years of intensive training – learned a maximum amount of 3–4 words, of the likes of *mama*, *papa*, *cup* and *up*. All of them, however, proved to understand dozens of words. Similar attempts were also made with other primate species, like the orangutans trained by William Furness around 1916, and the gorilla Toto (Maria Hoyt, in 1942). The real problem – as demonstrated in Liebermann 1968 and 1972 – lay in the neurological and anatomical incapacity of non-human bodies to articulate sounds in the way humans do. The difficulties of these apes were *expressive*, not *cognitive*.

Later, in 1977, Irene Pepperberg applied the idea to a species that does not have these expressive limitations, the parrot >**Alex**. This experiment was more successful, and Alex proved to be able to understand and pronounce about 100 English words.

As it became clear that not only verbal communication had to be considered language, the interest in ICE increased, also because of the growing fame of behaviorism. In 1966, Allen and Beatrice Gardner started a program based on >**American Sign Language** (i.e., gestures, not spoken words) with the chimpanzee >**Washoe,** with remarkable results. Other succesful ASL-based programs started in 1972 (by Francine Patterson with a gorilla named >**Koko**), and in 1978 (by anthropologist Lyn Miles, with an orangutan named >**Chantek**). ASL was also used in Herbert Terrace's >**Nim Chimpsky** program. After few years of apparent success, Terrace had a very famous second thought: he realized that the whole training was heavily affected by the >**clever Hans effect**, and – with rare professional honesty – he admitted it.

ASL was not the sole alternative to speech. David Premack, whose project started in 1966, trained the chimpanzee Sarah by using colored plastic symbols that had no iconic or indexical relation with the object represented (e.g. an apple was represented by a blue triangle). The project achieved mixed results: apart from concrete objects, whose acquisition was fast and efficient, Sarah showed ability to learn abstract concepts like "name of" (used as "what's the name of x?" when she did not know how to call a certain object), colors, big-small, equal-different, square-round, etc. She could also answer questions like "what is the color of. . .". However, suspicions that her training was again affected by the Clever Hans effect remained high.

Scholars then elaborated ICE that excluded, as much as possible, any interaction between trainer and trainee (>**emotion-free experimental contexts**). Duane Rumbaugh, in the mid 1970s, attempted to teach language by employing symbols. The trainee was a chimpanzee named Lana. Rumbaugh felt that computer automation could prevent anyone from cueing the animal. The machine was programmed to perform certain tasks, like dispensing food or displaying an image in response to pressing the proper symbol on the keyboard. Lana was then given a special keyboard which had symbols (called >**lexigrams**) in place of regular keys. The trainer was connected to Lana with another, equally equipped, computer. The interaction would thus occur only through the respective monitors. Results were again controversial: Lana learned to formulate requests and to engage in relatively complex conversations, but she would also do so only in response to a specific need (usually, hunger or thirst). Rumbaugh was followed by his wife, Sue Savage-Rumbaugh, who carried out one of the most interesting and successful ICE programs, with the bonobo >**Kanzi,** and with a technique that was still based on the use of lexigrams.

Another ICE worth mentioning, among many that were and are performed, is the program conducted by Lou Herman on two dolphins, Phoenix and Akeakamai, in the 1980s. The training consisted of two simple sign systems: Phoenix learned an acoustic language generated by an underwater speaker, and Akeakamai an adapted version of ASL. In both cases, the words produced consisted mostly of concrete entities, actions, modifiers (like "right", "left", "bottom", "surface") plus >**metalinguistic** signs, such as "yes", "no" and "erase" (used as a signal to interrupt an activity). In order to avoid the Clever Hans effect, the trainers wore a mask that prevented them to show facial expressions. Moreover, one trainer would give the message, and another one, without knowing the message, would record the dolphins' answers. The >**syntax** of the messages was organized in sequences, like "surface frisbee bring-to basket" (i.e., on the surface there is a frisbee: bring it into the basket). When a task could not be performed, the dolphins would touch a panel correspondent to "no"; when it was, then, after performing it, they would touch a panel correspondent to "yes". This project, too, achieved interesting results. Phoenix and Akeakamai also proved to be able to understand messages that referred to a distant time. They would memorize the instructions, and – as the conditions became favourable – perform the task.

Intraspecific

Any form of >**semiosis** occurring within the same animal >**species**. It is the "space" where most of zoosemiosis occurs, and the most refined and complex forms of >**communication** can be found. Members of the same species (or group or community) share the same >**Umwelt**, therefore share information on a basis that can be at the same time objective and subjective, and that relies on well-established codes (>**Code**), and, not rarely, cultural innovations and exchanges (>**Culture**).

Joni

See **Koths**, Nadia.

Kant, Immanuel (1724–1803)

German philosopher, one of the most influential in modern Western thought. Among his most prominent works, one must mention the *Critique of Pure Reason*, the

Critique of Practical Reason, and the *Critique of Judgment*. It is in the latter work that Kant expresses his finalistic ideas about human beings, and their role in Nature. Human beings, and human beings alone, are the ultimate purpose of creation. No being can actually claim to be the final goal of creation, but since humankind is the final goal of Nature and is a moral species, therefore it can be considered the final goal of creation, as well. This moral character is founded on intellect and ratio, which – says Kant – are superior in humans than in other animals (Kant 1790/1978: 77). The principle is extended also to **>aesthetics** and aesthetic sense (the main subject of the *Critique of Judgment*), which are issues that may concern exclusively a being that is at the same time animal and rational:

> The agreeable, the beautiful, and the good thus denote three different relations of representations to the feeling of pleasure and displeasure, as a feeling in respect of which we distinguish different objects or modes of representation. Also, the corresponding expressions which indicate our satisfaction in them are different The agreeable is what GRATIFIES a man; the beautiful what simply PLEASES him; the good what is ESTEEMED (approved), i.e., that on which he sets an objective worth. Agreeableness is a significant factor even with irrational animals; beauty has purport and significance only for human beings, i.e., for beings at once animal and rational (but not merely for them as rational-intelligent beings-but only for them as at once animal and rational); whereas the good is good for every rational being in general-a proposition which can only receive its complete justification and explanation in the sequel. Of all these three kinds of delight, that of taste in the beautiful may be said to be the one and only disinterested and free delight; for, with it, no interest, whether of sense or reason, extorts approval. And so we may say that delight, in the three cases mentioned, is related to inclination, to favour, or to respect. For FAVOUR is the only free liking. An object of inclination, and one which a law of reason imposes upon our desire, leaves us no freedom to turn anything into an object of pleasure. All interest presupposes a want, or calls one forth; and, being a ground determining approval, deprives the judgement on the object of its freedom (Kant 1790/1978: 49).

The consequences of such statements, in Kantian philosophy, are mostly of an ethical type: human beings have no real obligation or duty towards other animals. What they have is a moral duty towards humanity, in order not to damage it in any form, and this may occasionally include the exploitation of other animals as well. Kant is not opposed to it when it is "useful" to humankind, but he is rather clear on the fact that none of these exploitations should be excessive or unmotivated. The reason, as already emerging from **>Plato**, **>Locke** and most of all **>Porphyry**, is: those who are cruel towards animals are more likely to be cruel towards other humans. However, the difference in the spirit animating Porphyry and Kant is enormous: ferocious attack versus wars, violence, Christianity and meat-eating in Porphyry's case, simple invitation to humane attitudes in Kant's.

Kant's reflections have been of great influence for modern **>anthropocentrism**, because of their seemingly moderate and reasonable approach. They add kindness and some ethical conscience to the Aristotelian and Cartesian frameworks. The non-human animal's life is clearly subordinate to that of the human being, and the intellective capacities of the former have nothing to do with those of the latter. Still, a certain respect and the avoidance of gratuitous and unnecessary exploitation are due.

Kanzi (1980)

A male bonobo *Pan paniscus* trained in human **>language** through the use of **>lexigrams**. The program was (and still is) conducted at the Language Research Center at Georgia State University. Sue Savage-Rumbaugh, leader of the project, felt that the core-question for interspecific communication was to establish whether apes understand the signs they produce, rather than finding out about their grammatical capabilities. Therefore, the entire methodology of the program deviated from the trends of that time (for instance by moving away from some of **>Hockett**'s design features of language – productivity and displacement, most of all – which were so popular in other studies). Savage-Rumbaugh's work departed from a quite critical attitude towards other **>interspecific communication experiments**. In all previous programs, she maintained, the primates were not using signs symbolically (**>symbol**), but rather indexically (**>index**): they learned to associate certain behavioural patterns with certain consequences. In her view, indeed, the main point of human language is that it uses symbols, according to a three-steps process: (1) the physical external substance of a word, e.g. "door", as either written down or spoken with a given linguistic pattern; (2) the relationship between that and a real door (when the word "door" is pronounced one conjures up a mental representation of a door); and (3) the capability of symbols to make one think about things that are not present, or even not existent.

The experiment, performed on a small community of bonobos, among which Kanzi is the most famous, achieved excellent results. Kanzi, who was born when the project had already begun with other apes, took everybody by surprise from the very beginning of his own training, showing he already knew some lexigrams and could already understand a remarkable amount of spoken words (evidently, this competence was achieved by observing his fellow apes, especially his adoptive mother Matata, who was part of the initial stage of the project). Kanzi's competences grew rapidly. Once it became clear that he did not need a "basic course" to grasp the meaning of spoken words and written symbols, the trainers abandoned the usual reward-based training, and went straightaway to the "advanced" level, by adding new symbols to his keyboard. Kanzi was not trained to associate the lexigrams with specific objects, but rather the trainers would naturally employ the symbols in conversations with him, generally helping him to understand the general sense (as it is often done in certain techniques for learning a foreign language). This way, Kanzi's vocabulary steadily increased to over 200 produced and 500 understood words. Kanzi, particularly, showed an unquestionable comprehension of spoken words, something that had always been a weak point of the other interspecific communication programs. In the various tests he was given (including recorded words played to him through headphones), he was nearly 100% accurate on all words that were part of his vocabulary at any given age. He was also able to respond to speakers with different accents as well as to artificially-produced words. Far from succeeding only with single words, Kanzi showed comprehension of sentence structures, word order and grammar rules. Similar results were achieved with two other bonobos of Kanzi's community, Panbanisha and Panzee.

Kaplan, Gisela (1944)

Australian cognitive ethologist, with a specific interest in animal **>communication** (with an impact on **>zoosemiotics** and **>zoomusicology**, too). Kaplan is professor in Animal Behaviour at the University of New England. Her education (she has one PhD in Arts and one in Animal Behaviour) led her to animate her approach to animal studies in that fully interdisciplinary manner that both zoosemiotics and **>cognitive ethology** envision. A prolific writer of over 300 research articles and 20 books (and multiple award-winner for her research publications), Kaplan is a specialist in vertebrate cognition, primate and avian behaviour, particularly birdsong (benefiting also from her first career as a musician). Since the 1990s, she specialized on in Australian songbirds' vocal communication, **>cognition**, and the functions of song and **>mimicry** (focusing mostly on the Australian magpie *Gymnorhina tibicen*). Such topics have been tackled also within specific zoosemiotic contexts (as in Kaplan 2009).

Also worthy of mention is her long collaboration with Prof. Lesley Rogers, a very relevant figure to the study of animal cognition and brain development (a.o., Rogers-Kaplan 1998 and 2004, particularly the first – an innovative research on animal communication).

Kinaesthetic Signs

One of the four "prefigurements of art" emphasized in Thomas **>Sebeok**'s *The Play of Musement*, (1981: 210–259), along with **>architectural signs**, **>musical signs**, and **>pictorial signs**. The word "kinaesthetic" designates any use of bodily movement for aesthetic purposes (**>aesthetics**), and focuses on the vast area of "dancing" rituals performed in front of a potential partner, or with/against a contendant. As such, dance may or may not be an implement to musical signs (**>zoomusicology**). Kinaesthetic performances have been observed in a great number of **>species** and configurations. Among the many examples:

1) The particular "postures" assumed as visual displays or in relation to a musical performance (as, a.o. the 45-degree body inclination of humpback whales before their notorious songs, see Payne 1982: 472).
2) Particular group performances serving as social rituals (like the chimpanzees' "rain-dance" described in Goodall 1971: 54);
3) Courting rituals of different types (to a certain extent recurrent in nearly all species, particularly birds, but also in insects like the Nymphalid butterflies described in Akimushkin 1988: 16–7)
4) Ritualized or semi-ritualized challenges to mating or territory rivals (another extremely recurrent pattern in birds);
5) Duet dances that reinforce a monogamic bond (as the so-called ceremony of triumph in geese of the genus *Anser* – see Mainardi 1992: 122–3);

Fig. 4.3 A dancing ritual between two cranes

6) Performances involving role-playing (such as those rituals based on body-decoration occurring among cichlid fish like *Tilapia* or *Haplochromis* – see Akimushkin 1988: 113–4)
7) Performances imitating "real" life – see Akimushkin 1988: 107–8)

Anthropologists like Curt Sachs (1937), Anya Royce (1977) or Gertrude Kurath (1977) also showed a certain interest in animal kinaesthetic signs, particularly in relation to their search for the biological origins of dance (as in the case of Sandhill Cranes portraited in Fig. 4.3). Aspects like the recurrence of the circling choreographies (Kurath 1977: 47 and Sachs 1937: 10), the presence of spectator (Royce 1977: 4) or the social transfiguration/identification (Kurath 1977: 43) were often mentioned.

Kircher, Athanasius (1601–1680)

German scholar, "master of a hundred arts". Often referred to as "the last Renaissance man", Kircher cultivated the most diverse interests in the most diverse fields of knowledge, discussing with equal competence egyptology and geology, medicine and music. It is in this latter area that one finds a very intriguing contribution to the study of animal acoustic >**communication**. In 1650, Kircher published the twelve books of *Musurgia universalis* (a work nowadays known as the forerunner of the so-called "theory of passions"). However, the first volume (*Anatomicus de natura soni et vocis*, Chapter "De vocibus naturalibus in animalibus eorumque anatomia") is, considering the times, one of the most complete dissertations on the subject of animal vocalizations of aesthetic type (>**zoomusicology**).

Kircher works essentially on three levels: (1) direct investigation of the anatomical and physiological aspects of sound production; (2) personal observation, which points out those features that are not detectable on empirical basis; and (3) anecdotes, collecting facts related to the subjects discussed (a practice that more recently has been accused of >**anthropomorphism** and general scientific inaccuracy, but is now experiencing a "second youth" in >**cognitive ethology**).

Kircher elaborates a classification of singing species, using the (very limited) zoological knowledge of that time (Kircher 1650: 25), and noticing that the definition of "voice" is something that fits best to humankind and some birds, while nature has gifted other animals with rough vocalizations ("Boues mugiunt, balant Oues, Equi hinniunt, barriunt Elephantes" and so on – Kircher 1650: 26), that are nevertheless still suitable for displaying emotions and feelings.

Kircher transcribes different birdsongs on score, and does the same also with the American animal *Pigritia* (the sloth *Choloepus didactylus*), of whose night call Kircher emphasizes the ascending and descending diatonic scale, covering the interval of a major 6th.

Birds remain however the main focus, and are taken into consideration in sections 3–5 of the chapter. Species analyzed, and trascnribed on a score, are the parrot (with their "amazing imitations of human speech"), the nightingale (with their "beautiful chant", and their capacity to sing "planus, gravis, acutus, creber, extensus" – Kircher 1650: 29), the magpie, the wren, the rooster and the hoopoe. The chapter is concluded with a small dissertation on insects (cicadas, locusts and crickets) and amphibians (frogs), of which Kircher provides only anatomical descriptions.

Kluge Hans, Der

See **Clever Hans Effect**

Kohts, Nadia >(1890–1963, full name: Nadezhda Ladygina-Kohts)

Russian primatologist, she can be considered a forerunner of cognitive primatology. Between 1913 and 1916, she raised a young chimpanzee, Joni (see Kohts 1935), in her home in Moscow, conducting different types of cognitive test (including tool, mirror, art and discrimination tasks). She invented the matching-to-sample paradigm, and worked specifically on emotional responses in Joni, providing in-depth studies of such feelings as jealousy, guilt, empathy and loyalty, and describing facial expressions in detail. She would then compare Joni's behavior to her own infant son, Roody. Among her experiments, of particular semiotic interest is her failed attempt to teach spoken human language to the chimpanzee.

Koko (1971)

Short for Hanabi-Ko, Koko is a female lowland gorilla (*Gorilla gorilla*) trained in **>American Sign Language** (in fact, to an *ad hoc* version of it called "Gorilla Sign Language") by psychologist Dr. Francine **>Patterson**. Along with **>Washoe**, **>Kanzi** and others, Koko provided one of the most successful results of **>interspecific communication experiments**.

Trained since 1972, first in Woodside, California, then in Maui, Hawaii, Koko is reported to understand and use more than 1,000 ASL (or GSL) signs, and to understand more than 2,000 spoken words in English. Within the same training program, there were and are also other gorillas, namely Michael (Mike), a long-term "buddy" for Koko (the latter repeatedly signing GSL grief expressions when the former died in 2000), who reached a proficiency of about 600 signs (and showed very strong inclination towards painting – >**Pictorial signs**), and the younger Ndume, who was chosen by Koko herself in a kind of audition (Koko was exposed to some videos, and asked to choose her favourite companion from the gorillas displayed). In 1998, Koko was also the first non-human animal to "chat" through the Internet, in a specific event hosted by AOL.

One of the most important characteristics of Koko's training program was a direct challenge to >**Clever Hans effect**: Dr. Patterson wanted to establish a solid and emotionally-close relationship with the ape. Her theory was that children gain motivation in learning a >**language** most of all because of the interaction with parents. To deprive them of such a fundamental element would mean to deprive them of language itself. Results with Koko proved very satisfying, just like Washoe's case (who received similar affective treatement). Koko could comprehend spoken words and use the correspondent symbols (which means that the signified was a semiotic entity in her mind). Moreover, she could create new signs to describe objects she did not know the name of (among the most revealing cases: "white-tiger" for "zebra", "finger-bracelet" for "ring", "drink-fruit" for "melon", "animal-person" for "gorilla" and "water-bird" for "swan"). Such a skill is crucial in distinguishing a cognitive use of the signs from a mechanical one.

The >**referential** use of the signs was also very important. Koko did not make just simple requests, but she would talk to Patterson about what she could see around her. Basically, she would engage into casual conversation, once more dismissing the widespread belief that language-trained apes could not go beyond a basic, >**conative** use of the signs.

Well-documented, as with Washoe, is also Koko's extensive use of lies (> **Deception**), as with her recurrent habit of stealing candies and then blaming her friend Mike in front of Dr. Patterson.

Koko's life and training is documented in both scientific and semi-scientific literature. Among the former: Patterson 1978, 1981, and Patterson-Linden 1981. Interesting resources are also available at The Gorilla Foundation website (www.koko.org).

Kull, Kalevi (1952)

Estonian biologist and biosemiotician. Currently Professor of Biosemiotics and Head of the Department of Semiotics at Tartu University, Kull is a relevant exponent of the so-called Tartu-Copenaghen school of >**biosemiotics**. His interests occasionally cross with >**zoosemiotics**, but his importance is mostly to be found in the effort to systematize the biosemiotic disciplines in a general, also historical, sense (see, a.o., Kull 1999 and 2005). Further important contributions include a thorough analysis, from a semiotic viewpoint, of the notion of >**Umwelt** (Kull 1998 and Kull-Torop

2003), >**semiosphere** (as corresponding to >**biosphere**), and an attempt to emancipate the figure of Karl Ernst von >**Baer** and his alternative non-darwinian theory of evolution.

Lana

See **Interspecific communication experiments**.

Language

With the possible exception of >**instinct**, there is hardly a more critical topic in animal studies than the notion of language. The definition itself is already a problem, and is very far from producing a compromise among scholars (*across* different disciplines, or even *within* the same one). In semiotics, a certain agreement has been reached (especially after >**Sebeok**'s repeated efforts in the last 20–30 years of his life) in considering language (a) a secondary and tertiary >**modelling** system (as opposed to non-verbal sign systems, that can only be primary); and anyway, quite simply, (b) the "name" to give in principle to the *human* modelling system, regardless of similarities or differences with other systems (making therefore the concept a founding element of human identity as such). As Sebeok firmly pointed out, it is not even *allowed* to use the word "language" as a colorful metaphor to describe complex, but non-human, semiosic processes ("As a rule of thumb, picturesque conjunctions of the word 'language' with the generic word 'animal' . . ., ape or dolphin, or a category of domestic pets. . ., or in phrases like 'the language of flowers', are unscientific nonsense, rhetorical tricks designed to mislead by assuming as part of the premise the conclusion that is supposed to be demonstrated" Sebeok 1995: 10).

The problematicity of such stands is however very high, and calls into question the possibility that these affirmations lack an empirical basis (becoming, thus, exactly what Sebeok accused other scholars to practice: *petitio principii*), and – not least – the fact that plenty of other disciplines (directly or indirectly involved in the study of language) seem not to agree *at all* with this position. The risk, it is feared, is to make this terminological choice (language as the human modelling system, *a priori*) semantically as irrelevant as the choice (exercised only in nine world countries) of calling "soccer" what everywhere else is known as "football", without changing a single comma in the rules of the game.

As a general rule, it seems that human sciences are very eager to consider language as a >**species-specific** human feature (and, as mentioned, as the very characterization of human identity), radically separated from other animals, while natural sciences are often in favour of a >**transpecific** extension of the concept, suggesting that, if anything, a difference between the human and other >**communication** systems is only a matter of degrees. To start with, natural sciences (and some human sciences, too) hardly share the opinion that, in language, the "modelling" function precedes in importance the "communicative" one, as instead suggested in semiotics. Secondly, most natural sciences do not agree that >**syntax** and symbolicness (>**symbol**), which characterize respectively secondary and tertiary modelling systems, are exclusive property of (human) language. Counterproofs offered include phenomena like birdsong, >**bee dance** (indeed, often referred to as "language"), >**alarm calls**, and the results of >**Interspecific communication experiments**, all extensively (and empirically) studied in >**ethology** and other sciences. Thirdly, if it is possible, as it *is* possible, to study the biological evolution of language, then it becomes hard to maintain the hypothesis of a "radical", clear-cut, separation between human language and other modelling (and communicative) systems.

The discussion remains lively and far from concluded, and it is probably significant to notice that, despite their firm interdiction of the use of the word "language" even as a simple metaphor within a non-human and non-verbal context, semioticians themselves (including Sebeok) hardly manage to resist the temptation to talk about "language" in the most diverse (non-human and/or non-verbal) contexts (for instance when referring to primary modelling systems as typical of "natural languages", an expression which, from *that* semiotic point of view, should be an oxymoron).

The notion of language is discussed in Chapter 2, Section 2.2.2 and in Chapter 3, from Section 3.3.

Leakey's Angels

Nickname given to the three main pupils of anthropologist Louis Leakey who performed research on great apes' >**interspecific** >**semiosis** in the wild. The "angels" are/were Dian >**Fossey** (researcher on gorillas), Birute >**Galdikas** (researcher on orangutans), and Jane >**Goodall** (researcher on chimpanzees).

Learning

The term defines the process of acquiring and retaining information through ontogenesis, i.e., the consequence of interaction between an animal and its environment, rather than through phylogenesis. While in classical >**ethology** it was often claimed that almost all non-human behavior is due to instinctive predispositions (>**instinct**), the current scholarly position (a.o., Bekoff and Jamieson 1990, Hamilton and Marler 1966, Mainardi 1975) generally holds that:

1) All animal >**species** (including protozoa) are provided with so-called "non-associative" learning abilities (i.e., learning based on constant exposure to a unique event or stimulus);
2) Most of them (except for protozoa) are provided with "associative" learning abilities (i.e., learning based on constant exposure to numerous causally related events or stimuli);
3) In many species, particularly the so-called higher ones, there is "complex" learning (i.e., learning based on >**cognition**, rather than simple association).

Further categorizations of the concept concern (a) the specific learning processes associated with singing abilities in bird and mammal species (*song learning*); (b) the form of learning that occurs without an explicit stimulus, with the animal acquiring information on his/her own environment in an exploratory way (*latent learning*); (c) the learning processes specifically tied to an organism's >**Umwelt** (*perceptive learning*); (d) the information received and stored by an organism in its prenatal life (*prenatal learning*); (e) the acquisition of a (subjectively or objectively) "correct" reaction to a stimulus, after repeated attempts (*trial-and-error learning*); and (f) the type of learning specifically emerging from social interaction and parent-to-offspring transmission (*social* or *cultural learning*).

Leclerc De Buffon, George-Louis (1707–1788)

French naturalist and mathematician. His *Histoire naturelle*, published between 1749 and 1788 in 35 volumes, is considered the most important work by a naturalist in the entire eighteenth century, and was extremely influential for those such as Jean-Baptiste Lamarck and Georges Cuvier. The books encompass most of the knowledge of the time on the animal and mineral kingdoms, fully reflecting the encyclopaedic spirit of the Enlightenment. Because of its essentially secular nature, the *Natural History* had a few problems with religious institutions, and was for instance condemned by the Faculty of Theology at the Sorbonne (an event that forced Buffon to publish a retraction).

Innovative in many respects, there are several aspects of Buffon's work that make it particularly worthy of note:

1) The so-called Buffon's Law, i.e., the notion that, despite similar environments, different geographical areas display distinct plants and animals (an anticipation of biogeography);
2) A rough anticipation of >**Darwin**'s theories, appearing in volume 14, where Buffon maintains that all quadrupeds developed from 38 species only;
3) The considerations of similarity between humans and apes (although not corroborated by the hypothesis that they might have a common origin);
4) A general inclination to a systematization of animal >**species**, that makes him a pioneer of modern taxonomy.

Lenneberg, Eric

See **Morgan's Canon**

Lestel, Dominique (1961)

French social scientist with a strong interest in >**zoosemiotics**. Maître de Conférence at the Department of Cognitive Sciences at Paris' École normale supérieure, Lestel is internationally renowned for his works in the field of human-animal relationship

(1994, 1995, **>Anthropological zoosemiotics**), particularly **>interspecific communication experiments** (1996, 1998, 1999) and animal **>culture** (2001, 2002). Dominique Lestel contributes to zoosemiotics with insights from social sciences. Social scientists, to him, have been rather insensible to such issues as animal cultures (and particularly the communication processes involved), and a scientific interaction with **>biosemiotics** may be of benefit for both parties. Lestel maintains that "the semiotic dimension of animal cultures has been totally neglected by ethologists" (2002: 40), as animal communications are in fact more complex than normally believed. In a fashion similar to **>Griffin**, he shows that most of the supposedly distinctive features of human **>language** are in fact present in other animals as well, particularly the notions of tradition, meta-communication, innovation, flexibility, multiple articulation and dialogues.

Lexigrams

Symbols placed on a keyboard, used in **>interspecific communication experiments**, that represents words in a purely symbolic (**>Symbol**), non-iconic (**>Icon**) and non-indexical (**>Index**), manner. Lexigrams must also be combined according to certain rules, therefore reproducing an elmentary grammar. The term was coined in 1971, thanks to Ernst von Glasersfeld (who designed the first 120 symbols and the first draft for a grammar), and first employed in interspecific communication in the occasion of the Lana project, directed by Robert M. Yerkes (**>Interspecific communication experiments**). The system was then extended to 384 signs, and the grammer upgraded, when the **>Kanzi** project started in 1980.

Lindauer, Martin (1918–2008)

German ethologist, specializing in animal **>communication**. A pupil of Karl von **>Frisch**, with whom he collaborated upon extensive research on the **>bee dance**, Lindauer is an important figure in **>zoosemiotics** as well, for being not only a field researcher on communication, but also a remarkable theoretician on the subject. Works like Lindauer 1990 are crucial for the meticolous systematization of **>biocommunication** research, including a definition of it, a classification of the **>channels**, and an investigation of its social dimension.

Locke, John (1632–1704)

English philosopher. Apart from being one of the most important proto-semioticians, John Locke showed consistent interest in medical, chemical and biological sciences, and his work on animal **>cognition**, as displayed in his *Essay Concerning Human Understanding* (1690, here in the 1825 edition) is of particular importance for **>zoosemiotics** too.

His main arguments are:

1) In contrast to Descartes, Locke makes no division comparable to the Cartesian *rex extensa* and *res cogitans*. To him "matter" is able to think. Neither is any suggestion made that animal life is actually comparable to the functioning of a machine. The sole realm that can possibly be interpreted in terms of mechanism is to Locke the vegetal one: there, and there only, the subject is incapable of sensations and ideas. What marks the difference between animals and plants is perception (Locke 1690/1825: 100).
2) Perception varies in grades according to the single capacities of each species. Animals are thus organisms provided with sense, memory, and ability to make plans and comparisons. In the *Essay*, Locke discusses both human and non-human memory and its capacity to activate reasoning, and uses birdsongs as main example (Locke 1690/1825: 104).
3) Locke's consideration of birdsong is also interesting from an aesthetic point of view (>**Aesthetics**), as he stresses non-utilitarian characteristics in this activity: Locke describes birds as "wasting their time" by re-composing their sound models, without any apparent evolutionary advantage being secured for themselves or their own species. These birds are consequently able to sing "just for the sake of singing", expending the same effort as they would if it were a matter of life or death (Locke 1690/1825: 104). Another reflection on the subject is rather sarcastic about Cartesianism

    > For, though I should grant sound may mechanically cause a certain motion of the animal spirits in the brains of those birds, whilst the tune is actually playing; and that motion may be continued on to the muscles of the wings, and so the bird mechanically be driven away by certain noises, because this may tend to the bird's preservation; yet that can never be supposed a reason why it should cause mechanically- either whilst the tune is playing, much less after it has ceased - such a motion of the organs in the bird's voice as should conform it to the notes of a foreign sound, which imitation can be of no use to the bird's preservation. But, which is more, it cannot with any appearance of reason be supposed (much less proved) that birds, without sense and memory, can approach their notes nearer and nearer by degrees to a tune played yesterday; which if they have no idea of in their memory, is now nowhere, nor can be a pattern for them to imitate, or which any repeated essays can bring them nearer to. Since there is no reason why the sound of a pipe should leave traces in their brains, which, not at first, but by their after-endeavours, should produce the like sounds; and why the sounds they make themselves, should not make traces which they should follow, as well as those of the pipe, is impossible to conceive. (Locke 1690/1825: 104–5)

4) In any case, the mental faculties of non-human animals remain to Locke inferior, in degrees, to those of human beings. Ideas are composed and compared, but not at such a complex level as in humans. Animals do not count, do not easily distinguish and finally lack the ability of abstraction (Locke 1690/1825: 107–8). Yet, with Locke there is no doubt that non-human animals "are not bare machines", and that "we cannot deny them to have some reason. It seems as evident to me, that they do some of them in certain instances reason, as that they have sense".
5) Finally, in another of his works, *Some thoughts concerning education* (1692, here in the 1823 edition), Locke anticipates a topic that will be also discussed

in Kant's *Critique of Judgement*, that is, our ethical obligations as humans to respect other animals and treat them kindly (Locke 1692/1823: 112–3)

Lorenz, Konrad (1903–1989)

Austrian ethologist and Nobel Prize winner "for discoveries in individual and social behavior patterns". Possibly the most influential figure in classical **>ethology,** and certainly the most popular one. Following the steps of his teacher and mentor, German biologist Oskar Heinroth, Lorenz contributed crucially to the foundation of ethology, in its modern form. Among the (many) concepts he helped to define and/or develop, one must mention at least **>instinct** (a.o., in Lorenz 1949, 1963, 1965) , **>imprinting** (a.o., 1949 and 1979), human-dog co-evolution (1950), aggression (a.o., 1965), plus an extensive (and positively ethical-philosophical) discussion on humanity and its role within and towards Nature, especially in the last years of his life (a.o., 1973 and 1974). Among the **>species** he observed closely, and always in a non-invasive way (something that animal-right supporters always like to point out in relation to the category in which he won his Nobel prize – "Physiology or Medicine"), there are the jackdaw *Corvus monedula*, the greylag goose *Anser anser* and other nidifugous birds.

Magnetoception

The capacity to use magnetic fields to gather information about direction, altitude or location of given objects or places. Magnetoception is important, if not central, in the navigational skills of many animal **>species**, particularly migratory ones. It has been observed in many birds (first and foremost), and in flies of the family *Tephritidae*, honeybees *Apis mellifera*, turtles of the species *Caretta caretta* and *Dermochelys coriacea*, lobsters of the family *Palinuridae*, crocodiles of the family *Crocodylidae*, sharks of the superorder *Selachimorpha*, cartilaginous fish of the order *Chimaeriformes*, stingrays of the family *Dasyatidae*, common cows *Bos primigenius*, and deers of the family *Cervidae*. In humans, too, the presence of magnetic deposits of ferric iron in the ethmoid bone, allows, to a certain extent, magnetoceptive abilities.

As discussed in Chapter 2, Section 2.1, the question remains open whether magnetoception, a sensorial ability that seems to function only in **>signification** processes, should or should not be considered a **>channel** in all respects.

Maran, Timo (1975)

Estonian zoosemiotician. Currently Senior Research Fellow at the Department of Semiotics of Tartu University. One of the most important figures in the new generation of zoosemioticians, and certainly one of the most active. Maran is a specialist in **>mimicry** research (2005, 2007a, b, 2008a), but also tackled general issues in the theory and history of **>zoosemiotics** (2005 and 2009), extending his interests to **>ecosemiotics**, as well (a.o., 2008b). Maran regularly gives lecture courses in zoosemiotics and ecosemiotics for the University of Tartu.

Marchesini, Roberto

(1959) Italian biologist and epistemologist. One of the world leading figures in **>anthrozoology** (or Zooantrhropology, as he reformulated the discipline, by introducing new and specific theoretical and methodological elements – see Marchesini 2000b, 2004 and 2005) and animal **>ethics** (a.o., 1996, 1999, 2000a). He is President of the Italian society for applied behavioral sciences (Società italiana di Scienze Comportamentali Applicate) and founder and director of a School in Human-Animal Interaction (Scuola di Interazione Uomo Animale).

Marler, Peter (1928)

English ethologist, with a strong interest in **>zoosemiotics**. Currently Professor Emeritus at the Animal Communication Laboratory of University of California, Davis. Initially trained as a botanist, Marler turned quite early in his carreer to ornithology, giving a major contribution to the research on the social functions of birdsongs (Marler-Slabbekoorn 2004). He also studied extensively animal acoustic **>communication** (particularly primates'vocalizations, as in Marler 1977). His approach has always been keen to zoosemiotics, by both adopting explicitly its paradigm (a.o., Marler 1978), or – perhaps more valuably – by "hinting" at it in biological contexts. In this sense, it is certainly remarkable that Marler had used semiotic theories in the study of animal communication even before **>Sebeok**'s introduction of zoosemiotics in 1963. In Marler 1961, indeed, there is an interesting attempt to combine ethological and semiotic viewpoints by using Charles **>Morris**' theories (1948), particularly his distinction between identifiers, designators, appraisors and prescriptors, as applied to the analysis of birdsong.

Martinelli's Canon

See **Zoosemiotic Canon**

Maynard Smith, John (1920–2004)

Evolutionary biologist with a strong interest in animal **>communication**. Maynard Smith's use of (and contribution to) **>zoosemiotics** must be regarded as one of the most significant points of contact between traditional biological sciences (particolarly evolutionary ones) and semiotics, in that way envisioned by **>Sebeok** (1972 and 1990) and practised by the likes of Peter **>Marler** (1961 and 1978). Timo **>Maran** (2009) summarizes very efficiently the most relevant deployments of semiotic terminology and methodology in Maynard Smith's work. They include **>Peirce**'s sign typology (particularly the **>icon->index->symbol** triad, as in Maynard Smith-Harper 1995 and 2003), the concepts of "signal"and "meaning" (Maynard Smith-Harper 2003), the concept of "information" (Maynard Smith 1982, 1999 and 2000), and several others of less frequent recurrence.

Mechanism

See **Descartes**, René.

Mental Map

A mental, or cognitive, map is the capacity of orienting oneself in space and heading towards a certain destination, by using signs available in the environment (and belonging to any of the possible >**channels**). To possess a cognitive map means to be able to solve a consistent number of spatial and temporal problems: finding the shortest way for a familiar destination, finding an alternative way when the usual one is not available, going back to familiar places from unknown ones, etc.

Many experiments were made in order to verify the existence of mental maps in animals, on a basis that was cognitive and not simply instinctive (>**instinct**). One of the most interesting concerns the honey bees *Apis mellifica* and was illustrated by Stephen Gould (1986: 861–3). The bees were trained to reach a place (P1), north from the hive (H), and after a while were suddenly closed in a dark box, and moved to a completely unknown place (P2), south from H and from which P1 was not visible. The hypothesis was: if the map is a simple instinctive reflex, then all that bees will be able to do is reaching H from P2. If, on the contrary, bees can develop a map regarding places they are exposed to eventually (as P1 was), then they should be able to reach also those places. The result was, indeed, that the bees *could* reach P1 from P2, despite the fact that this was the first time ever they were attempting that path. In other words, the spatial representation in the bee's **mind** is not behavioristically ruled by external stimuli, but by an inner source, gathered with experience and with a mental processing of the information available.

Mental Representation

A mental representation is an inner sign of an external object (which functioned as its stimulus), and that can be reactivated also in absence of the original stimulus. It is defined at length in Hebb 1949 and, with a specific zoosemiotic approach, in >**Cimatti** 1998. The most typical example of mental representation is the >**mental map**.

Metalinguistic

One of the six >**Functions of communication** postulated by Roman Jakobson (1963). A message is metalinguistic when its focus is the >**code**.

Metasemiosis

The capacity of using signs being aware that they are signs, as in >**deception** or >**play**.

Miles, Lyn

See **Chantek**

Mimicry

Mimicry, or more precisely Biological Mimicry, is an umbrella-term that covers several forms of iconic similarity (appearance, behavior, messages...) of one >**species** to another. It involves all >**channels** (not only the visual, as is sometimes implied), and occurs largely (but, by far, not only) in insects. Mimicry occurs when an organism (called the *mimic*) evolves to share common characteristics with another organism (the *model*), while a third party (the *dupe*) is the receiver of the mimicry sign (e.g., a predator common to the mimic and the model). Sometimes, model and dupe are the same organism. Mimicry is, in the majority of cases, advantageous to the mimic and harmful to the receiver. Usually, models and mimics are closely related organisms, but sometimes the phenomenon occurs among biologically "distant" species.

A particular case is *Camouflage*, occurring when the model is a non-living or abiotic part of the environment (as in the famous case of chameleons). Occasionally, mimics may have multiple models during the stages of their life, or they may be polymorphic, i.e., different individuals imitate different models. Models, also, may have more than one mimic.

The main typologies of mimicry are the following:

1) *Defensive* (or protective): it occurs when mimics avoid/deceive predators or other enemies by resembling other species or (in the *Camouflage* case) inanimate parts of the environment. Forms of defensive mimicry include the *Batesian* (after English naturalist Henry Walter Bates: the mimic shares signals similar to the model, without being as unpalatable as the latter. It is observed, a.o., in *Lepidoptera*, *Hymenoptera*, some snakes, and Octopuses of the genus *Thaumoctopus*); the *Müllerian* (after German naturalist Fritz Müller: two or more species have very similar signals, and both are "unpalatable". Observed in *Lepidoptera* and *Hymenoptera*); the *Mertensian* (after German herpetologist Robert Mertens: an unpalatable prey mimics a less dangerous species. Observed in some snakes species); the *Browerian* (after biologist Lincoln P. Brower: an intraspecific case of Batesian. Observed in *Lepidoptera*).
2) *Aggressive*: it occurs when the mimics are predators or parasites that deceive/lure the prey/host by appearing like a harmless species, or (in the *Camouflage* case) inanimate parts of the environment. It is observed in many different orders.
3) *Auto-* (or *intraspecific*) *mimicry*: mimic and model belong to the same species, although the dupe could easily belong to another. It occurs typically (but not only) in acoustic >**imitation**, especially among birds (as in the case of the Superb Lyrebird portraited in Fig. 4.4).

Fig. 4.4 Superb lyrebirds are among the most skilful mimics of natural and artificial sounds

A fourth form of mimicry, called *Reproductive*, occurs among plants only.

Within >**zoosemiotics**, the major expert in mimicry is undisputably Timo >**Maran** (2005, 2007a, b, 2008a), while general classics on the subject include Brower 1988, Owen 1980 and Wickler 1968.

The concept of mimicry is closely related to (but does not overlap with) that of >**deception** and therefore constitutes part of the discussion in Chapter 2, Section 2.4.1.

Mind

Leaving aside the endless philosophical, psychological and neurological implications of such a concept, and focusing only on its relevance within the zoosemiotic context, a "mind" can be defined as a network of biological and physiological activities and relations (including brain functioning, sensorial perception, nervous system, etc.) that allow an organism to acquire, store and elaborate the information it is, was, or might be exposed to, and produce an adequate response (which includes not producing any response whatsoever), after a consideration of the variables involved at any stage of this process.

A network of this kind is essential in a great number of semiosic processes, including >**modelling**, >**deception**, >**play**, >**aesthetics**, >**mental representations**, >**metasemiosis**, etc. As such, it must be considered an integral part of any member of the Animal Kingdom (>**Zoosemiotic universals**).

The concept is also discussed in Chapter 2, Section 2.3.2.

Mirror-Test

See **Self-consciousness**

Modelling

A process through which a phenomenon/entity is executed and/or multiplied on the basis of a virtual or real general model. Such a model serves, thus, as a cognitive input for the given phenomenon to be activated and to expand in complexity. Such a quality is for instance typical of sign systems, which are not only a *result* of perception, but in fact support and extend it at many levels. The expression "modelling system" has been indeed used in semiotics for describing the primary quality of natural and verbal **>language**. Originally Yuri Lotman's idea (1977), the "Modelling system theory" was taken up by Thomas **>Sebeok** (e.g., Sebeok-Danesi 2000). Sebeok made a clear distinction among

1) "primary modelling systems", typical of natural languages (gestures, facial expressions, body posture, proximity, and so forth), which consist in an instinctive ability to model reality in an "iconic" manner;
2) "secondary modelling systems", typical of (human) verbal languages, where the modelling process occur in an indexical and syntactical way; and finally
3) "tertiary modeling system", again typical of verbal languages, which *extend* the properties of secondary models to the symbolic extent (as for instance may happen in certain forms of **>culture**, or in specific conventional codes as mathematics or religions).

Modern Etho-zoosemiotics

See **Ethological Zoosemiotics**

Morgan, Conwy Lloyd (1852–1936)

English psychologist who developed the so-called **>Morgan's Canon**.

Morgan's Canon

Named after Conwy Lloyd **>Morgan**, the Morgan's Canon is a particular case of the possibly more famous **>Ockham's Razor**. It asserts that "in no case should actions or behaviors be interpreted as the result of a superior psychic faculty, when it is possible to interpret them as a result of an inferior faculty" (Morgan 1903: 59). The canon was conceived as a reminder to scholars of the dangers of **>anthropomorphism** and what later became known as **>Clever Hans effect**, and still is a "classic" in some animal studies in that role.

Within semiotics, the Canon has been mentioned in relation to the case of **>interspecific communication experiments** by the likes of Thomas **>Sebeok** and

Eric Lenneberg to warn other scholars about the risk of over-interpreting anthropomorphically the results of such experiments. In the book *Speaking of Apes* (Sebeok–Umiker-Sebeok 1980) the Canon is quoted while evaluating chimpanzee **Washoe**'s allegedly spontaneous >**American Sign Language** sign productions. If Allen and Beatrice Gardner, trainers of Washoe, looked favourably at the creation of new signs in Washoe, as symptoms of enthusiasm towards, and cognitive appropriation of, the new communication tool, Lenneberg does not hesitate to apply the Razor and the Canon to such instances, and finds that "we are simply testing our own ingenuity to assign interpretations to productions that might, for all we know, have been emitted randomly" (in Sebeok–Umiker-Sebeok 1980: 80).

Despite these instances, the position held in the present companion (Chapter 3, from the Section 3.3) aims to be rather critical towards the possibility of finding any reasonable application of the Morgan's Canon in semiotic contexts. Taking a similar position to that taken on >**Instinct**, the opinion here is that, in fact, the Canon is in principle an anti-semiotic concept (for this discussion, see the entry >**Zoosemiotic canon**).

Morris, Charles (1901–1979)

American semiotician. His role, within the zoosemiotic context, was instrumental for completing an ideal transition from >**Peirce**'s general idea of semiotics as something "not just concerned with the human being", to >**Sebeok**'s introduction of >**zoosemiotics**. In Morris' semiotics there is a clear description of the sign as a concept encompassing life altogether, and whose extents can be described in biological, rather than anthropological, terms. Strongly influenced by >**behaviorism**, Morris' "behavioristic biospychology" (as his friend Ferruccio Rossi-Landi called it) is exemplified in Morris 1946.

Morris, Desmond (1928)

British sociobiologist and popular science writer. His work has been influential for >**zoosemiotics** at more than one level, particularly in the field animal >**aesthetics** and >**anthropological zoosemiotics**. Pupil of Nikolas >**Tinbergen**, Morris started in the late 1950s a thorough investigation on the painting skills of chimpanzees (>**pictorial signs**), resulting in Morris 1963. The research led also to the curious extravaganza of an exhibition *in incognito* of chimpanzees' paintings, mixed with human ones, that was very favourably reviewed by the unaware critics. Morris, himself a painter, intended to make a point that – when assessed unbiasedly – animal art can stand the comparison with the human one. Later, his work focused on the animal components of human behaviour, resulting in his best-known (and most controversial) works, including *The naked ape* (1967). Morris is still active as writer and TV personality.

Multimodality

Multimodality refers to those instances of >**semiosis** when different patterns cooperate (or compete) to display one or more texts. It operates in almost every semiosic context (particularly in communicative ones), and it can be safely said that only the most elementary occurrences constitute an exception. It is however in such instances like >**play**, linguistic or pseudo-linguistic interaction (>**language**), >**aesthetics**, >**deception** and others that multimodality becomes a primary and unavoidable semiotic strategy.

Zoosemioticians – and communication researchers in general – tend to categorize signals by the primary sensory >**channel** involved, but the reality is rarely so straightforward as multiple channels are regularly engaged simultaneously. This is especially true in highly social, group-living animals. As anticipated by >**Darwin** (1872), multiple, concurrent stimuli are very important in order for a signal to be fully efficient. However, the influence of such stimuli on the signal and its meaning has only recently been closely analyzed (see, for instance, Horn 1983). Behavioral neuroscientists have found that many species (and particuarly primates, birds, and insects) are concerned with the processing of information from multiple sensory channels, in the fields of both attention and perception. In particular, posture and movement are the stimuli most frequently associated with the production of the signal. Partan and Marler (1999) have provided a synthetic but exhaustive account of the most important points in multimodal >**semiosis**. When the diverse signs are redundant (i.e., they all concur to convey the same text), there can be:

- Equivalence: The multimodal signal provokes the same exact reaction as the signs emitted separately.
- Enhancement: The multimodal signal produces a reaction that is increased in intensity.

  When the multimodal signal is based on the combination of nonredundant signs (i.e., they convey different texts), there can be a wider range of possibilities:
- Independence: The different signs are independent and produce distinct reactions that are not in relation with each other, although they are combined.
- Dominance: One of the signs prevails on the other(s). It is a typical occurrence in play behavior.
- Modulation: One nonredundant sign affects the other(s), by modulating its/their effect.
- Emergence: The multimodal signal provokes an entirely new reaction that has nothing to do with the separate signs.

  Multimodality is discussed more in detail in Chapter 2, Section 2.3.3.

Musical Signs

One of the four "prefigurements of art" emphasised in Thomas **Sebeok**'s *The Play of Musement*, (1981: 210–259), along with **Pictorial signs**, **Architectural signs**, and **Kinaesthetic signs**.

See **Zoomusicology**.

Nim Chimpsky (1973–2000)

An obvious pun with Noam Chomsky, Nim was a chimpanzee *Pan troglodytes* trained in >**American Sign Language** by Dr. Herbert Terrace at Columbia University. The project, which – in terms of the data collected – was moderately succesful (Nim acquired the respectable knowledge of 125 signs, and communicative interactions between him and the trainers were reportedly satisfying), achieved a unique notoriety for Terrace's eventual statement that the whole program was heavily affected by the >**Clever Hans effect**. Going against his own interests, and therefore showing rare professional ethics, Terrace admitted that all what initially looked like Nim's cognitive acquisition of linguistic skills, was simply the result of active manipulation from the chimpanzee's part, and self-deception from the trainers' part. Terrace thought that none of Nim's achievements could be barely compared to what true >**language** should be, and that no more than 25 signs could be *really* considered to be cognitively understood by Nim.

Terrace's statement opened a huge (still on-going) debate on the validity of >**interspecific communication experiments**, resulting in at least three factions: (1) those who thought that Terrace had underrated Nim's cognitive capabilities, and therefore the program was far from being a failure; (2) those who agreed that Terrace had failed, but on the basis of a badly-structured program, not because of the chimpanzees' general impossibility to learn language; and finally (3) those, like most semioticians, who admired Terrace's honesty and agreed that in principle no species other than the human one can acquire a specifically linguistic competence. Thomas >**Sebeok**, in particular, was very firm in considering Nim's case as paradigmatic of all interspecific communication experiments, and praised Terrace for being the only one who did not fall into self-deception, or did not perform an "outright fraud" (as he called it) to the scientific community.

N'kisi

See **Alex**

Objects

See **Things-Objects**

Occam's Razor

See **Ockham's Razor**

Ockham, William Of (1288–1348)

English logician and Franciscan friar, to whom the so-called > **Ockham's Razor** is attributed.

Ockham's Razor

Ockham's, or Occam's, razor is a principle attributed to William of >**Ockham**. It says: *entia non sunt multiplicanda praeter necessitatem*, that is "entities must not be multiplied beyond necessity" (or, as it is often referred to, "plurality should not be posited without necessity"). It is the main source of inspiration for the more radical >**Morgan's canon**, widely discussed in this companion, both basically conveying the message that, in interpretive (empirical or not) environments, complication is not a preferrable solution, if a simpler option is available and justifiable. Such *lex parsimoniae* was instrumental in the paradigms of many disciplines, including biological sciences, but it was (and is) not immune to criticism, particularly when it comes to a rather common (mis)use that transformed a plea against unnecessary complication into a plea in favour of compulsory simplification (Einstein reportedly said: "Everything should be made as simple as possible, but not simpler"). Historical detractors of Ockham's razor include Karl Menger, Leibniz and >**Kant**, who also went as far as to postulate a *counter*-razor: "The variety of beings should not rashly be diminished".

Most of the discussion on these issues focuses on the Morgan's canon, whose entry is here suggested to consult, together with >**zoosemiotic canon**, and the more detailed discussion available in Chapter 3, Section 3.3.5.

Olfactive Channel

See **Chemical Channel**.

Olfactory Channel

See **Chemical Channel**.

Osten, Wilhelm Von

See **Clever Hans Effect**

Pain, Stephen (1956)

English biosemiotician. His main field of specialization is natural argumentation systems, an area he pioneered by founding a biosemiotic branch called >**biorhetorics** (a.o., Pain 2002). Through biorhetorics, Pain had often the chance to deal with animal >**communication** and >**cognition**, as, for instance, in Pain 2006 and 2009.

Pansemiotism

Holistic conception of life as a whole as sign processes, developed within >**biosemiotics** especially since the 1990s. Pansemiotism interprets the nature of life as fully and intimately semiotic, following thus a path that was mostly traced by the American semiotic tradition (e.g., >**Peirce** and >**Sebeok**), and by teleological biology (>**Baer**). This led a few biosemioticians, particularly those belonging to the so-called sign-based biosemiotics (>**Hoffmeyer**), to the belief that every single process in life (and possibly not only in life) can be analyzed and interpreted semiotically. Pansemiotism is highly problematic, and is heavily criticized by those scholars who think that biosemiotics and >**zoosemiotics** should exclusively rely upon empirical bases.

In this companion, the question is discussed in Chapter 1, Section 1.1.1 and Section 1.2.1.

Patterson, Francine (1947)

American psychologist, leader of one of the most successful >**interspecific communication experiments**, conducted with the lowland gorillas >**Koko**, Mike and (after Mike's death) Ndume, at the premises of The Gorilla Foundation in Woodside, California. Dr. Patterson trained the apes with a personally modified version of the >**American Sign Language**, called Gorilla Sign Language. She documented her work with scientific and semi-scientific essays and books (among the former: Patterson 1978, 1981, and Patterson-Linden 1981), and via different media, including a 1978 documentary called *Koko, A Talking Gorilla* (directed by Barbet Schroeder), several TV programs, and even a Internet chat event hosted by AOL, featuring Koko herself.

For more on this Interspecific Communication Experiment, see **Koko**.

Peirce, Charles Sanders (1839–1914)

American philosopher and logician. He regarded logic as a branch of semiotics, and contributed enormously to the latter, by establishing a specific paradigm and line of research in the field, particularly in the elaboration of the notion of "sign". His semiotics, open to metaphysics, biology and indeed logic, became in the late twentieth century the most serious alternative to the Saussurean "linguistic" tradition, up to the establishment of two major branches: the American "semiotics" (Peirce, >**Morris**, >**Sebeok**...), focused on the entire sign-production of living forms (>**semiosis**) and the European "semiology" (Saussure, Greimas, Barthes...), of an intimately logocentric nature. It is certainly thanks to Peirce that semiotics became interested *at all* in non-human semiosis, and this fact alone justifies his inclusion in the olympus of >**biosemiotics**, particularly, but also of >**zoosemiotics**.

Peirce's impact on modern semiotics has been so big and "charismatic" that a few suspicions of misinterpretation/exaggeration of his theories have arisen, particularly when it comes to his metaphysics, heavily relying upon the assumption of God's existence, and employed as straight philosophy of science in more than one case. Another critical remark that has been advanced is of methodological type: the publication, in two main stages, of the *Collected Papers* (1935–1966), created an omnicomprehensive texture of Peirce's writings that include several fragments, revisions and private communications that he himself – it is argued – would have probably refused to publish. In that sense, it is not rare for arguments that take completely different directions to both cite Peirce (in two different moments of his life) as an authority, regardless of the fact that the philosopher himself evidently and very legitimately changed his mind at some point.

Despite these reservations, however, his contributions to semiotic theory, and particularly to the study of non-human semiosis, have been, and must be considered, of the utmost importance.

Pepperberg, Irene

See **Alex**

Pergola

See **Pergolate**

Pergolate

Pergolates, or bowers, or pergolas, are constructions built for mating purposes by the passeriform Satin Bowerbird *Ptilonorhynchus violaceus*, that are of remarkable interest from a semiotic point of view, and particularly in the field of aestehtic signs (> **Aesthetics**). Karl von **>Frisch** argued that "[. . .] those who consider life on earth to be the result of a long evolutionary process will always search for the beginning of thought processes and aesthetic feelings in animals, and I believe that significant traces can be found in the bowerbirds" (Frisch 1974: 244).

During the mating period of the year, the male specimens (different from females because of a more uniform, dark blue, plumage) build a construction, only vaguely reminiscent of a nest, whose function is that of hosting the mating encounter. The pergolate is extended vertically, in form of two concentric arcs, although other species of the same genus erect cone-shaped maypoles around a tree (famous was the case of the Italian naturalist Ottaviano Beccari, in the end of the nineteenth century, who thought he had found dollhouses built by creative young girls).

Apart from building techniques, pergolas are notable for the meticulousness with which the bird decorates the inside and the outside of his construction. All types of objects are utilized, from those already available in Nature (twigs, shells, flowers, etc.) up to small human artefacts (fragments of bottles, pens, pieces of clothing, etc.). The objects are usually brightly-coloured, preferably blue, like the male's plumage colour. Blue is also the colour of a special paint spread in different places of the bower (especially the entrance), that the bird obtains by chewing some berries, and distributed with a chip of wood held in his beak.

The placing of the objects seems not to be random:

> Every time the bird returns from one of his collecting forays, he studies the over-all colour effect. He seems to wonder how he could improve on it and at once sets out to do so. He

> picks up a flower in his beak, places it into the mosaic, and retreats to an optimum viewing distance. He behaves exactly like a painter critically reviewing his own canvas. He paints with flowers; that is the only way I can put it. A yellow orchid does not seem to him to be in the right place. He moves it slightly to the left and puts it between some blue flowers. With his head on one side he then contemplates the general effect once more, and seems satisfied. (Frisch 1974: 243–244)

Importantly, none of the decorative objects have structural importance to the whole construction: the presence of decorations is thus "unnecessary".

After finishing his job, the male stands awaiting the female, who, if attracted by the pergola, heads towards him. While she "evaluates" the construction, the male welcomes the female by holding a symbolic object in his beak (usually, a flower, a leaf or a berry) and by performing a courtship dance.

Personhood

Legal status generally assigned to individuals who are acknowledged as possessing rights and duties under a country's Constitution. During various historical periods (but predominantly in the twentieth and twenty-first centuries), several philosophers, ethologists and animal rights activists (as individuals or organized in specific groups/projects) have claimed the status of personhood for non-human animals, particularly Great Apes, cetaceans and elephants, considered the closest **>species** to human being, in terms of intelligence, emotional life, **>sociality** and morality. In more than one case, the status was granted, most recently in 2008, when the Spanish Parliament's environmental committee approved the resolutions proposed by the Great Ape Project, an organization founded by philosophers Paola Cavalieri and Peter **Singer**.

Pfungst, Oskar

See **Clever Hans Effect**

Phatic

One of the six **>Functions of communication** postulated by Roman Jakobson (1963). A message is phatic when its focus is the **>channel**.

Phenomenology

See **Animal ontology**

Phyletic Homology

See **Analogies-Homologies**

Pictorial Signs

Pictorial signs are one of the four "prefigurements of art" emphasised in Thomas **>Sebeok**'s *The Play of Musement*, (1981: 210–259), along with **>architectural signs**, **>musical signs**, and **>kinaesthetic signs**. Since the 1950s, the study of pictorial skills among non-human animals has gone hand in hand with **>interspecific communication experiments**, both types of research often performed by the same scholars on the same subjects. It is thus predictable that most attention was paid to chimpanzees, the species most commonly used in these experimental programs. Compared to other forms of animal art, pictorial signs are mostly studied in captivity. Among the few situations *also* studied in nature, is the research on visual aesthetic preferences in animals, which, if not a pictorial context *tout court*, constitute at least a premise to visual arts (Sebeok 1981: 233).

There are nine categories of perception that seem to provoke aesthetic pleasure in animals, particularly in primates:

1. Saturated colours (preferred to unsaturated ones);
2. Primary colours (preferred to mixed ones);
3. Brilliant colours (preferred to non-brilliant ones);
4. Rhythmical repetition of equal components (probably because it facilitates comprehension and produces "pleasure of recurrence");
5. Bilateral and radial symmetry;
6. Steady curves, like circles, spirals, wave-lines, and so on (preferred to irregular curves);
7. Conspicuous lines or shapes (preferred to indistinct ones);
8. A certain balance between the left and right halves of a picture (preferred to an unbalanced arrangement);
9. The same colours or conspicuously different colours (preferred to nearly equal colours), when two objects of different colour have to be combined.

As for the mere pictorial activity, researchers have put into evidence some interesting cases, mostly within Great Apes, noticing that (a) apes seem to display a *wish* to draw, and not only to *play* with paper and pencils; (b) apes are not conditioned or forced to draw: drawing seems to be their own choice. Comparative psychologist Paul Schiller, who worked with several chimpanzees during the 1950s, pointed out that the chimpanzee Alpha would not use a pencil as part of a game, but would use it only for what it is made for, that is, to write (Schiller 1951: 110).

Particular attention was given to the self-rewarding elements involved in pictorial activity. Scholars tried to show that (a) drawing in chimpanzees has a clear aesthetic component; and (b) that the experiments were not affected by the **>clever Hans effect**: "In all cases the animals received no assistance or guidance from the experimenters, except for the provision of and [. . .] the familiarisation with the drawing or painting equipment. Attempts to influence the kind of picture produced by provoking imitative responses were always most unsuccessful" (Morris, 1963: 141).

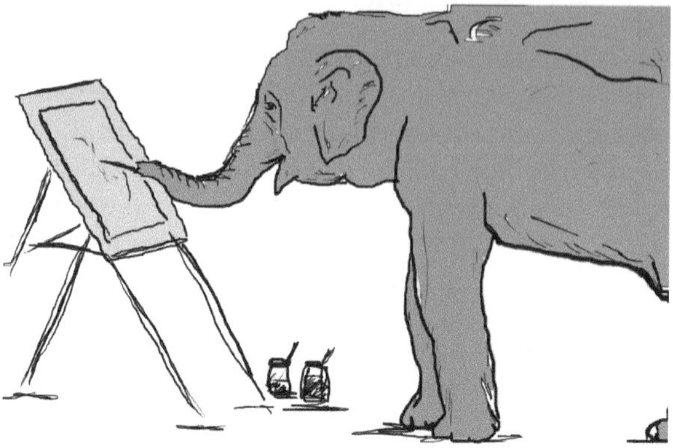

Fig. 4.5 Painting elephants, an increasing (and, most of the times, unethical) attraction in zoos

Desmond >**Morris** (1963: 144) considered the self-rewarding element as the focal point of his theories on chimpanzees' art ("Actions that are usually referred to as play, curiosity, self-expression, investigation, and so forth, come into this category of self-rewarding activities"). Morris systematically analyzed chimpanzees' drawings, emphasizing the constants emerged by comparing the drawings available and the respective processes of realization. This led him to postulate the existence of six basic >**principles of picture-making**: Self-rewarding activation, Compositional control, Calligraphic differentiation, Thematic variation, Optimum heterogeneity, Universal imagery.

Interesting studies were also made on elephants (whose artistic skills – too often evinced through unethical, invasive methods – are nowadays a circus-like attraction in several zoos – see Fig. 4.5). Masson and McCarthy (1996: 311–315) report a few examples from captivity contexts, the best-known being that of the female Indian elephant Siri. Her pictorial abilities went totally unnoticed until 1980, when a new keeper, named David Gucwa, was employed to look after her. Gucwa noticed that Siri would trace signs on the sand, using sharp-shaped stones as pencils, and then seemingly pointing the final product with her proboscis. The keeper thought of providing Siri with a pencil and some big sheets of paper that he would hold steady. The result was a long series of abstract drawings (reportedly praised by artist William de Kooning), for which the elephant received no reward. Once the news spread out, other zookeepers showed no surprise, claiming to have often seen their own elephants drawing the same way.

Plato

(ca. 427–347 b.c.) Greek philosopher and mathematician, one of the most important figures in Western thought. Plato's contribution to topics keen to >**zoosemiotics** departs from his reflections on Nature and reality. Plato does not speak of reality as it is, but rather as it should be: his main speculations (the Demiurge above all) aim, generally speaking, at the constitution of a better world, founded on intelligence and not on chance. This concept fits perfectly with Nature and its constitutive elements: it is a human duty to refine the natural order, "collaborating" with the Demiurge in order for Beauty to prevail on *Kaos*.

Plato believes that life began as happy, without any such concepts like property, hunting, war and other types of violence. He describes this original life as a time when human beings and other animals established relations and conversations of mostly philosophical matter, exchanging information on each other's knowledge. The work wherein Plato mostly reflects upon non-human animals is the *Timaeus*, i.e. – most probably – the most metaphysical (if not occasionally delirious) of his writings, an off-the-limits attempt to mathematize reality at all costs. According to Plato, each body is provided with a soul, plants included, although the latter are much more limited than other beings (Lee 1965: 105).

The soul is strictly related to the body: the former determines the appearance of the latter, in form of expressions of guilt or merit. Indeed, human and non-human beings were born with faults to be punished for. To be male or female, to belong to a species instead of another, are all exact consequences of such faults or merits. Particularly, almost all non-human animals were in the past human beings that wasted their life in some way.

The animal world has two dimensions. One is the *divine*: animals are not divine, but they anyway "tend" towards the Divine, the Light, the Perfection and the Good. The other dimension is that of *ipseity* (i.e. reflected on the animal itself): this is an obscure, imperfect and fallacious dimension. All living beings have this double face, and their Light depends on how close they are to the Divine. The maximum distance from Divine implies the presence of just a single weak fragment of light.

A peculiar attempt to explain the origin of animals follows, in which – among other things – it is suggested that human beings were constantly remodelled into other organisms in relation to their qualities and actions in their previous life (e.g., innocent light-minded men became birds, pedestrian animals derived from rude, non-philosophical, people, and so forth).

Play

"Playing" is commonly defined as a set of behavioural patterns that non-seriously imitate otherwise serious patterns and situations. Animals tend to play when primary needs are already attended to, and therefore energy can be expended for other purposes than utilitarian ones. Juvenile specimens tend to be the most interested in this category of activity, which probably supports the hypothesis that the two basic functions of this behavior are amusement and the learning of rules of social behaviour.

Playing is one of those situations that require a **>contextual code**, i.e., signs that are specific for that context and not used in others (unless re-articulated in different forms, as in **>ritualization** processes). The communicative patterns employed

– that can be visual, chemical, acoustic and tactile (>**Channels**) – can be either innate or learned, and in most cases are metasemiotic (since they have the dual purpose of pointing out the unserious intention of the sender, and referring to the serious pattern they are imitating), and symbolic (since they have no physical resemblance to the object represented, and they were shaped in consequence of a code agreement).

Play can be distinguished between *autotelic* (when the rewarding element of the activity is the activity itself) and *simulative* (when imitative patterns appear); between *individual* and *social*; and between *kinetic* (when the players do not use any object) and *instrumental* (when the players do). Social play can also be divided into >**intraspecific** (the most obvious case) and – the not rare – >**interspecific**.

A thorough analysis of the issue appears in Chapter 2, Section 2.4.2.

Poetic

One of the six **Functions of communication** postulated by Roman Jakobson (1963). A message is poetic when its focus is the message itself.

Porphyry of Tyre (234–circa 305)

Neoplatonic philosopher born in Tyre (nowadays Lebanon). He is one of the most important forerunners of animal rights philosophy, supported with rational arguments that are also interesting for >**zoosemiotics**. *On abstinence from animal food* (*Perì Apokhês Empsykhon*, in English also known as *On abstinence from killing animals*) is a passionate ethical defence of non-human animals. Although firmly opposed to Christianity (he also wrote a pamphlet named *Katà Khristianon, Against the Christians*, survived in fragments only), Porphyry was an attentive reader of the Bible and other Christian texts, and knew Hebrew language. Even Augustine, who perfectly knew Porphyry's aversion to Christianity, could not help defining him as an "eminent philosopher".

Already in the *Katà Khristianon*, written in 268, Porphyry attacks Paul of Tarsus on the issue of meat-eating. In the First Letter to Corinths, Paul had stated that humans should eat "everything the butcher sells" without feeling guilty, for God owns every living being, and every living being is at human disposal. After the *Katà Khristianon*, Porphyry wrote the *Perì Apokhês Empsykhon*, a neo-platonic and neo-pythagoric treatise on animal life, vegetarianism, justice and peace. Borrowing from Plotinus, Porphyry maintains that we are all intellectual entities tied to the sensible because of two forces: our incapacity to remain endlessly bound to the intelligible, and a gravitational force towards the "lower world". The most efficient ways to "go back" in the realm of intelligible are justice and vegetarianism. Porphyry fears the destruction of truth and justice, but – unlike >**Plato** – does not speculate only abstractly on the issue: he wants *lógoi* and *érga*, knowledge and action.

His idea of vegetarianism implies a radical change, which goes in contrast with the customs of the *polis*, the ritual slaughtering and sacrifices of religion. Like Plato and Pythagoras, Porphyry considers violence on animals as an "appetizer" for war.

The first instruments used for killing non-human animals are exactly the same used in the first conflicts among humans. Hunting and war are inevitably bound, both metaphors of fraud and falsity, both result of an original violation: not really eating the "apple", but eating meat (Taylor 1823: 126).

Among the arguments used by Porphyry to fight the "enemies" of animals, at least three fall under the domain of **>ethological zoosemiotics**:

1) Animals do think and communicate. In contrast to the stoics and peripatetics, Porphyry maintains that we can find the *logos*, the discourse, among non-human animals, and that this discourse can also reach perfection.
2) Animals also have the "inner discourse": the general organization of their organism is similar to the human one, e.g. they suffer the same pathologies. Animals are not less sensible than humans. To maintain that a different physical constitution corresponds to the absence of reason and sensibility, is like saying that gods are not sensible either, because their physical constitution is also different from the human one. The difference between humans and other animals is (in a proto-darwinian fashion) a matter of more/less, rather than presence/absence (Taylor 1823: 114).
3) Animals are intelligent and rational. Porphyry mentions the great amount of information collected by ancient philosophers on the topic. To be able to take care of one's own interests is a first important sign of intelligence ("each animal knows where it is weak and where it is strong, and it protects the former and makes use of the latter, as the leopard uses its teeth, the horse its hooves and the bull its horns, the cock its spur and the scorpion its sting"). Ratio, to Porphyry, does not originate from learning, nor from memory, and that applies to all beings, including gods. The sole fact that we cannot see the world through their own senses and figure out their own way of reasoning, is not a good excuse to state that ratio is missing. Moreover, non-human animals understand our **>language** to many extents, and perceive the diverse signs.
4) Both humans and non-humans are part of the same ethical system. Both search for and have a sense of justice ("Who does not know how animals that live in groups observe justice towards each other?"), both are victims of cruelty and fights. Violence is, in both cases, a sign of starving and desperation. Moreover, they are reciprocally necessary, and that is when humans break the balance: in exploiting and killing other animals that are not necessary to their surviving, human beings show off a superiority that reveals their evil nature. This unnecessary violence can be avoided, and the first important step, concludes Porphyry, is vegetarianism (Taylor 1823: 65).

Portmann, Adolf (1897–1982)

Swiss zoologist. Professor of zoology in Basel University, Portmann was a specialist in marine biology and comparative morphology of vertebrates. He is a figure of some importance in the zoosemiotic context because of his interdisciplinary and philosophical approach to animals studies (he would often promote/label his works as “Philosophical zoology”), and for often dealing with topics of clear zoosemiotic interest, such as camouflage (>**Mimicry**) (a.o., 1959), >**sociality** (a.o., 1961), and physical appearance (a.o., 1952). Particularly in the latter topic (and in the entire field of morphology), Portmann proposed his most innovative theories, postulating a continuity between the animal’s external appearance and its inner status, in open contrast with physiology and genetics.

Pragmatics

In Charles >**Morris**’ original formulation (1937: 64), pragmatics is that branch of semiotics that investigates the relation between signs and sign users, dealing with “the origin, uses, and effects of signs within the behavior in which they occur”. A pragmatic investigation, thus, deals with the biological aspects of sign functioning, including the processes and the determinants of propagation of the signs: “Zoopragmatics may be said to deal with the origin of signs in the source, or sender, the propagation of signs through a medium, or channel, and the effect of signs on the destination, or receiver” (>**Sebeok** 1990: 43).

Principles of Picture-Making

While studying drawing and painting activities in captive chimpanzees (>**Pictorial signs**), Desmond >**Morris** (1963) noted that the various drawings presented a number of characteristics in common, and seemed to obey to a limited number of general rules (related to motivation, activation, structure of the elaborate, etc.). Such rules, which he called **principles of picture-making**, were altogether six:

1. *Self-rewarding activation.* In order to start drawing, there is – in chimpanzees as in humans – a sort of input, an intrinsic motivation that does not derive from any primary need. It is the prospect of drawing itself that encourages and rewards the subject (Morris 1963: 158–159)
2. *Compositional control.* To Morris, all that is concerned with filling spaces, respecting boundaries, creating balances within the drawing, whether they are rhythmical, chromatic or something else, is part of compositional control. In other words, drawing constitutes a game of rules, that the subject decides to play by creating, and then respecting, an articulated geography of spaces, shapes and colours (Morris 1963: 161).
3. *Calligraphic differentiation.* Style and pictorial skills are not fixed, but rather follow a gradual and constant evolution (Morris 1963: 162).
4. *Thematic variation.* There are several phases in the approach to drawing, which can be interestingly classified. Firstly, the chimpanzee experiences a phase of exploration of the sheet, usually tracing apparently random signs, as if he wanted to achieve familiarity with the sheet (in sports jargon, that would be called warming-up). At some point, a certain graphical pattern (e.g., a roundish figure)

attracts the artist's attention, and s/he starts repeating it over and over, with apparent enjoyment. Little by little, the circular pattern evinces small variations, which constantly increase, until the roundish figure is totally replaced by new patterns. Such a rule may not be simply individual, but, rather, may follow particular inclinations within a community (Morris 1963: 164).

5. *Optimum heterogeneity*. Like the previous principle, this one basically answers a cognitive question: is there a concept of "the artwork is over" in chimpanzees? In other words, when the artist stops drawing, is it because s/he is bored, hungry, etc., or because s/he really means to stop at that point? Optimum heterogeneity is the name given to this ideal sensation that the work is finished. A possible interpretation is that the artist has a certain subject in mind and, consequently, s/he is portraying something specific. Although in most cases chimpanzees' drawings look like an abstract form of expression, in few cases, especially when s/he is able to interact linguistically, the chimpanzee may provide scholars with rather accurate indications, if not actual titles, about his/her own work.
6. *Universal imagery*. Some elements in (not only visual) art imagery are possibly universal. Not necessarily complete and complex figures, but simple graphical patterns may be drawn by different subjects, without any possibility of reciprocal influence. A typical case is the so-called "fan pattern", which emerged in a significant percentage of the drawings analyzed. According to Morris, there are three factors determining the preference for certain patterns:

   - *Muscular factor*. Some movements of the arm are more pleasant and easy-to-perform than are others, thus more recurrent.
   - *Optical factor*. Some forms and proportions are more "acceptable" by the visual apparatus than are others, thus one tends to prefer them.
   - *Psychological factor*. Each individual is equipped with a visual imagery, made of partly cultural and partly universal elements. In growing up, one may occasionally avoid obeying muscular and optical constraints, while the recurrence of certain archetypes does not seem to be abandoned.

Prodi, Giorgio (1928–1987)

Italian oncologist, writer and intellectual. He pioneered >**biosemiotics** by being the first one to label the study of biological codes under the name "Nature Semiotics". He then proposed a number of reflections that went hand in hand with >**Sebeok**'s development of the field and with Jakob von >**Uexküll**'s theory of Umwelt.

Prodi is also discussed in Chapter 1, Section 1.1.4.

Proprioceptive

>**Semiosis** is proprioceptive when sender and receiver of a text are the same subject, that is, when the animal is sending messages for its own use and benefit. A typical

case of proprioceptive semiosis is the >**echolocation**, by means of which animals like bats, cetaceans and some birds can receive information about the surrounding environment.

It has also been suggested that all forms of semiosis are in a way proprioceptive, in that there is always a degree of self-semiosis (interpretation, signification, etc.) in an organism that interacts with the environment. In this Companion, however, in order not to generalize (thus, perhaps, banalize) the concept, a choice has been made to stick to the original, narrower, definition.

Proprioceptive semiosis is also discussed in Chapter 2, Section 2.1.

Pure Zoosemiotics

The portion of zoosemiotic research that pertains to the elaboration and establishment of theoretical models of zoosemiosis.

Referential

One of the six >**Functions of communication** postulated by Roman Jakobson (1963). A message is referential when its focus is the >**context**.

Regan, Tom (1938)

American philosopher. Professor emeritus of philosophy at North Carolina State University until his retirement in 2001. Along with Peter >**Singer**, he must be considered the most important contemporary animal rights philosopher, particularly after Regan 1983, which gave him international notoriety. Unlike Singer, who roots his philosophy in the tradition of utilitarianism, In The Case for Animal Rights, Regan argues that non-human animals bear moral rights, Regan's philosophy refers primarily to >**Kant**'s heritage, although Regan specifically disagrees with Kant's assumption that respect is due only to "rational", therefore human, beings: an inherent value, and therefore legal rights, Regan argues, are ascribed also to infants and mentally impaired subjects. Regan's main point, indeed, is that "life", and the care that each subject has for it, not "rationality", is the key-word for the attribution of an inherent value to a subject. Each individual is a "subject-of-a-life". Such value includes the right to be treated with respect and not to be harmed.

Like Singer and others, Regan's importance for >**zoosemiotics** lies first and foremost in his contribution to animal >**ethics**, a subject whose inclusion in the zoosemiotic paradigm is suggested in this companion (seeChapter 5). However, exactly like his colleagues, his crucial role in animal liberation philosophy also has a bearing on the general methodology of animal studies (see Chapter 3 of this companion).

Representation

In general semiotics, Representation is the process through which a message is presented or depicted in some way. For the purposes of this companion (within the framework of a tripartite definition of >**semiosis**, as discussed in Chapter 1, Section 1.1), this property of representation comes to designate the virtual independence of sign processes from an actual receiver of the message. Representation occurs independently from a specifically identifiable destination: animals codify sense *whether or not* this sense will reach a specific addressee (which can be only virtual, or ideal, or may also not exist in the first place). In representation, therefore, the meaning is produced first and foremost by the sender of the message.

Rhetoric

See **Biorhetorics**

Ritualized Fight

See **Ritualization**

Ritualization

A crucial notion for both ethological and zoosemiotic interpretations of animal behavior. In >**ethology**, ritualization is the process of alteration of a behavioral pattern, in terms of intensity, recurrence, articulation and dynamics, in a way that increases and/or varies its effectiveness as a sign to other specimens. In >**zoosemiotics**, with reference to the traditional notion of >**semiotic threshold** (Eco 1976), ritualization scan be defined as the transition from a non-semiotic to a semiotic behavioral pattern. This may occur in both incidental circumstances that become intentional through selection, or in the use of a specific pattern (part of a more general one), originally non-communicative (e.g., simply physiological) for communicative functions (normally functioning as a synecdoche for the general pattern, >**Biorhetorics**). An example is the transformation of aggressive patterns into "warnings" (a "modern" dog grinds his/her teeth not for fighting, but to ritually warn his/her opponent: before dogs learned this rhetoric use of the sign, the act of grinding was a simple physiological "accompaniment" to the actual aggression, and therefore had no semiotic function above the threshold), or of functional patterns into socializing, >**phatic** signs (e.g., social grooming).

>**Eibl-Eibelsfeldt** 1970 (100–1) discusses ritualization as a process that produces the following changes:

1) Function. After ritualization, a pattern previously serving a certain function in the animal >**ethogram** comes to serve another one.
2) Motivation. The ritualized pattern can emancipate from its original motivation, and be now animated by a different one.
3) Redundance and exaggeration. Ritualized signs are generally more redundant and exaggerated than non-ritualized ones.
4) Threshold values. The ritualized pattern is generally more easily released than the non-ritualized one.
5) Postures. In the ritualization process, several bodily movements tend to be framed into still or quasi-still postures (exactly as in the case of the dog grinding his/her teeth, who "freezes" his/her aggression into a simple warning)

6) Orientation. Elements in space orientation may be changed during the ritualization process.
7) Intensity. The intensity (and its variations) of a sign may become more stereotyped (and therefore more constant) in a ritualized display.
8) Complexity. A ritualized sign reduces the element of complexity of the non-ritualized sign it refers to.
9) Body structure. Selection may favour variations in the body structure for the ritualized processes (typically, through the addition of ornamental elements in an animal's body).

In its modern, ethological configuration, the notion first appeared in Huxley 1923. In this companion, the concept is briefly discussed in the context of **>zoosemiotic universals** (Chapter 2, Section 2.3.2).

Rorarius, Hieronymus (1485–1556)

Italian philosopher and *Apostolic nuncio* at the court of Ferdinand of Hungary. In a rather peculiar pamphlet called *Quod animalia bruta saepe ratione utantur melius homine* (1548), he maintained that animals make better use of reason than humans do. The book, which failed to have any remarkable impact in his times, was rediscovered and republished in 1648 by the French intellectual Gabriel Naudé, in the particularly propitious period of heated debates around **>Descartes**' mechanism, and since then several reprints of the book were published. In the 1728 reprint (for the German publisher Weigand), editor Giorgio Enrico Ribovio added an extensive "Dissertation historical-philosophical on animal soul", making Rorarius' essay almost a forerunner of the Enlightenment. Additional exposure was given by Pierre **>Bayle**, in his *Dictionnaire historique et critique*, where Rorarius' arguments are used as pretext to discuss the nature of the soul.

Rothenberg, David (1962)

American philosopher and musician, with a strong interest in **>zoomusicology** and **>interspecific >semiosis**. Currently Professor of philosophy and music at the New Jersey Institute of Technology. He studied several cases of interspecific music, focusing particularly on cetaceans (a.o., Rothenberg 2008) and birds (a.o., Rothenberg 2005). His approach combines philosophical and musicological elements, with a particular emphasis on the musician's point of view. He currently works on the relationship between evolution and beauty (**>aesthetics**).

Savage-Rumbaugh, SUE

See **Kanzi**

Sayers, William (1935)

Canadian historian and etymologist, currently working at Department of Comparative Literature and Graduate Program in Medieval Studies, Cornell University. Sayers writes on medieval western European language and literatures, in particular Old Norse, Old Irish, and Anglo-Norman French. A current interest is

English etymology, and lexicographical approaches to technical language and language in the popular register: in this context, he has discussed, for a special issue of *Sign System Studies* on **>zoosemiotics** (Sayers 2009), the issue of Anglo-Norman French and Middle English nomenclature for animal names and vocalizations. This is important also for the fact that issues of this kind are rarely tackled within an explicit zoosemiotic context, and in that sense it contributes significantly to the development of **>Anthropological zoosemiotics**.

Schopenauer, Arthur (1788–1861)

German philosopher. Primarily known for his Philosophy of the Will, Schopenauer found several applications of his thought to questions pertaining non-human animals, particularly **>ethics**. All animals, human and non-human, are to Schopenauer phenomenal manifestations of Will, and therefore embody vitality, strength, energy, and desire. Their inner experience is directly connected to their external essence and behavior: these elements make human beings and other animals indistinguishable from an ethical point of view: they all recognize themselves in each other, they all suffer. Notable works of zoosemiotic interest are *Über das Fundament der Moral* (1840, where an articulated criticism to Kant's conception of the human being as final goal also appears), and *Parerga und Paralipomena* (1851, where an extensive discussion on animal ethics appears in the first chapter).

Schopenauer was also a supporter of the Society for the Prevention of Cruelty to Animals, in London, and the Animals' Friends Society, in Philadelphia. As it is done in this Companion, Schopenhauer, too, maintained that the use of the pronoun "it" in reference to animals was inappropriate, in that it would reinforce their consideration in a cartesian sense.

Sebeok, Thomas Albert (1920–2001)

Hungarian-American semiotician. Quite simply, the most important figure in **>zoosemiotics**. His contribution to the field is incalculable, and it includes the following actions/innovations:

1) Giving birth to zoosemiotics, as a concept and as a term, in 1963;
2) Giving credibility to the field, by defining its theoretical foundations and by creating an important following among semioticians and non-semioticians;
3) Setting the zoosemiotic agenda, for nearly all of its main topics, either by discussing them extensively or by pioneering them and opening doors for other scholars;
4) Leading the field throughout nearly 40 years of its history, until his death; and finally
5) Producing an outstanding body of zoosemiotic literature, that remains in many cases the point of reference for a given topic.

Born in Budapest, and American citizen since 1944, Sebeok starts his carreer in semiotics at Chicago University, by studying linguistics and then anthropology with Charles >**Morris**, his first major influence. Along with Morris, Roman Jakobson, Heini >**Hediger** and Jakob von >**Uexküll** are decisive figures in the formation of Sebeok's "biological way of thinking". At the time of his PhD in Oriental Languages and Civilizations (1945), he joins Indiana University, in Bloomington, an institution to which he remains "faithful" for the rest of his life. During a period of study at the Stanford Center for Advanced Study in the Behavioral Sciences in 1960, Sebeok has the opportunity of deepening his knowledge on biological sciences (ethology and animal >**communication** in particular), which convinces him to seriously explore the potentials of a "biological way" to semiotics (an instance already advanced by >**Peirce**, but only at a philosophical level). The conditions for the birth of zoosemiotics are set, and few years later the chance is taken. For most of the 1960s and 1970s, Sebeok's efforts are directed towards the promotion and the development of zoosemiotics, first as an umbrella-term gathering different approaches on animal communication, then as a field on its own, with an autonomous paradigm and a focus on animal >**semiosis**. He provides from scratch, or adapts from other scholars, most systematics of the discipline, and proposes concrete research on a multitude of topics, including >**language, >interspecific semiosis**, >**cognition**, >**aesthetics**, human-animal relationship (>**Anthropological zoosemiotics**).

From the mid-1980s onwards, Sebeok developes a more global and philosophical approach to semiotics (see 2001a), and – after extending the discipline to the non-human world – he opens it up to the non-animal too, giving an institutional birth to >**biosemiotics**. He witnesses the end of the twentieth century ("Sebeok's century", as John >**Deely** once called it) as the most charismatic living figure in semiotics, after Umberto Eco. He passes away in 2001, leaving an impressive body of work and a very strong legacy, not only in zoosemiotics.

Self-awareness

The explicit acknowledgment of one's own existence as individual, that is, separated from other subjects, with an individual mental activity (including, possibly, the acknowledgment that other subjects possess self-awareness too). It may be considered a slightly "lower" degree of >**self-consciousness**, which additionally implies the development of a sense of identity. For practical reasons, but also because in several studies the two concepts are treated as synoniyms, in this Companion a discussion of the two notions is found in the entry "Self-consciousness".

Self-consciousness

The reflective mental activity thanks to which a subject becomes aware of him/herself and of his/her identity, and from which a process of investigation of the self may be activated. A long-debated issue in the whole history of thought, self-consciousness was a problem discussed by all the great thinkers of Western philosophy, including, a.o., Socrates, >**Plato**, >**Aristotle**, >**Aquinas**, >**Descartes**, >**Locke**, Spinoza and >**Kant**.

Although often anticipated, in a speculative sense, by some of the above-mentioned, and other, philosophers, the systematic study of self-consciousness non-human animals started with the seminal Gallup 1970 and the famous "Mirror Test". As it was already known that chimpanzees could use mirrors to inspect their own body, Gallup intended to empirically determine whether this form of awareness corresponded to a true recognition of their identity. Gallup would anesthetize chimpanzees that were already familiar with mirrors and, during their sleep, he would mark their heads with a colorful dye. Once awake, some of the marked apes would be put in front of the mirror, and, indeed, they would be observed touching their foreheads in correspondance of the mark, significantly more often than those who would not be allowed to the mirror. Gallup repeated the test with other apes, and also with some monkeys, however only Great Apes, and particularly orangutans, besides chimpanzees, showed remarkable results. More recently, other species that successfully "passed" the mirror test were bottlenose dolphins (Reiss and Marino 2001), elephants (Plotnik – DeWaal – Reiss 2006) and magpies (Prior – Schwarz – Güntürkün 2008).

Semantics

In Charles **>Morris**' original formulation (1937: 64), semantics is that branch of semiotics that investigates the relation between signs and signified objects, dealing with "the signification of signs in all modes of signifying". An investigation of the semantic type thus focuses on the meaning of a given message, or combination of messages. Zoosemantics "is devoted to the signification of signs, and must take account of the context referred to by the source and apprehensible by the destination; this is the least well understood dimension of animal communication studies" (**>Sebeok** 1990: 43).

Semiosis

In general semiotics, semiosis is any sign action or sign process, or, following Charles **>Morris**, the process in which something is a sign to some organism. In this companion (Chapter 1, Section 1.1), semiosis is defined as the primary focus of **>zoosemiotics** (here, indeed, defined as "the study of *semiosis* within and across animal species"). Such definition is also useful to avoid confusion between semiosis and **>communication** (which is a special, therefore smaller, case of semiosis). Constituting parts of semiosis are at least the following:

a) **>Signification**, occurring when the receiver is the only subject taking part in the semiosis, and a true sender is missing;
b) **>Representation**, occurring when the sender is the only semiosic subject; and
c) **>Communication**, occurring when sender and receiver take both part in the semiosic phenomenon, and therefore the above-mentioned "sense" (or text) is exchanged, understood or misunderstood.

Semiosphere

The globality of environments where sign processes operate and interact in each respective >**Umwelt**. Developed by Yuri Lotman in 1984 (for a thorough description, see Lotman 1990), the notion was widely succesful in nearly all fields of semiotics. In >**biosemiotics**, because of a basic assumption that semiosis and life overlap, it has been suggested (in different articulations and by different authors) that the concept of "semiosphere" in fact corresponds directly to that of >**biosphere**. For a review of these suggestions, see Kotov 2002.

Semiotic Threshold

A Multifaceted notion, originally formulated by Umberto Eco (e.g., 1976: 16–28) to separate research topics that fall under the domain of semiotics from topics that do not. To Eco, the boundary roughly corresponds to the difference between a direct action and an action mediated by signs. Most indexes (symptoms, physiological stimuli, physical information...) are below the threshold, while forms of interaction that imply the presence of codes and conventions, and require interpretation, are above. In subsequent, commonsensical, interpretations of the notion, the threshold came to separate natural sign systems from cultural/artificial ones, unintentional from intentional interactions (>**Intentionality**), or – in hardcore semiological formulations – non-human from human semiosis.

Since then, the idea itself of threshold has always been very charming and controversial within semiotic discussions, there have been many systematic attempts to redefine it, by establishing lower or higher boundaries. Among the formulations of interest for >**zoosemiotics**, worthy of mention are the biosemiotic idea of the threshold (based on the Peircean holistic view of semiotics, and therefore a practical negation of the concept altogether); John >**Deely**'s notion of the "Semiotic animal" (2005, where a threshold between "semiosic" and "semiotic" is based on >**distant space-time semiosis**); Susan Petrilli's concept of >**metasemiosis** (1998, metasemiosis itself being the line of separation between human and non-human semiosis); and, finally, >**Cimatti** 1998, which, instead of changing, accepts Eco's original formulation of the threshold as a phenomenon marked by interpretation and codes, and simply works in the direction of proving which part of non-human semiosis is characterized by those elements as well.

Signification

In general semiotics, Signification is the process by which signs and meanings are produced. For the purposes of this companion (within the framework of a tripartite definition of >**semiosis**, as discussed in Chapter 1, Section 1.1), this main property of signification comes to designate the virtual independence of sign processes from an actual sender of the message. Signification occurs independently from a specifically identifiable source: animals make sense out of each other and out of their environment *whether or not* this sense is explicitly codified and conveyed. The act of producing meaning, therefore, is first and foremost performed by the receiver of the message.

Significational/Representational Anthrozoosemiotics

See **Anthropological zoosemiotics**

Singer, Peter (1946)

Australian philosopher, currently Professor of Bioethics at Princeton University. Together with Tom Regan, the most important figure in animal rights philosophy. Similarly to Regan himself and to >**Bentham**, who he follows to some extent, Singer's work is not directly relevant in terms of pure >**zoosemiotics** (not, of course, unless one includes the reflections proposed in the Chapter 5 to this Companion), yet his impact and influence in animal ethics is so poignant that he cannot be ignored by any animal-related fields of inquiry.

Singer's groundbreaking work is *Animal Liberation* (1975), arguably the founding paradigm of the modern animal rights movement. As a utilitarianist philosopher, Singer is not fully comfortable with the idea itself of "rights", which he prefers to replace with the notion of "interest": the interest of non-human animals must be taken into consideration because of their ability to suffer and feel pain. The "greatest good of the greatest number" – argues Singer – is the only valuable parameter for measuring ethics: in that sense, there is no argument in principle not to apply this line of reasoning to non-human animals.

Singer, together with Richard Ryder (author in the same year of the book *Victims of Science*) is also responsible for the coinage and popularization of the term (and concept of) >**speciesism**, and – together with Paola Cavalieri – founded the Great Ape Project, promoting, among other things, the ethical and juridical status of >**personhood** in Great Apes.

Singing

See **Zoomusicology**

Smith, William John

American ethologist, with a specific interest in animal >**communication**. Currently affiliated to University of Pennsylvania, Smith developed a personal ethological paradigm that makes his research very zoosemiotic-friendly, particularly in relation to his interest in structure-function and message-meaning relations and distinctions in animal signs (in a way rather similar to Peter >**Marler**). In general, Smith's position in defining communication is critical towards the ordinary understanding of the concept: "preconceived categories of behavior must be questioned continuously and modified or replaced as necessary, and a considerable range of events must be studied" (Smith 1991: 214). In his most important book so far (1977), Smith argues that animal communication should be classified according to seven features:

1) Distinctive displays within sign repertoires (which for Smith, vary from 15 to 45);
2) Subsequent division of such displays into specific messages;
3) The critical role of context in the meaning articulation of each message;
4) Message-meaning association within the following 12 contexts: identification, probability, general set, locomotion, attack, escape, non-agonistic subset, association, bond-limited subset, play, copulation, frustration;
5) Relation between messages and current events;
6) Relation between displays and social interaction;
7) Role of the motivation and the emotional state within communication.

Controversial in many respects, Smith's theories are anyway pretty close to >**zoosemiotics**, particularly in the great prominence given to choice and interpretation in the receiver's role.

Sociality

"The basic assumption of >**zoosemiotics** is that, in the last analysis, all animals are social beings, each species with a characteristic set of communication problems to solve." (Sebeok 1963: 465) This sole quotation should be enough to give the notion of sociality the role of quasi-synonym to >**semiosis**. Sociality, not to be confused with socialization (which is a concrete action), is the condition or quality of being social (tendency to intersubjective interaction and association, to form groups, to establish social roles, to mentally represent other entities, etc.). Sociality is at the basis of nearly all processes and activities related to animal semiosis, up to the point of becoming indistinguishable from the latter. The two phenomena, in particular, are in a mutual causal relation, that allows reciprocal increase of specialization and complexity. Semiosis, for instance, may increase the level of an animal's sociality, by taking a >**cognitive** role and therefore producing and exchanging more complex forms of behavior (conveyed through an increasingly sophisticated information level). Such exchange is possible when the subjects involved are able to create >**mental represention**s of the environment and recognize the entities within it (including other subjects). This process, itself a result of the semiosis-sociality alliance, is in turn possible thanks to another effect of this combination: the sharing of a similar form of mental categorization (i.e., a >**code**).

Sociobiology

A relatively recent branch of >**ethology**, sociobiology studies the structure and the dynamics of social behavior in animals within a strictly Darwinian framework (for instance, emphasizing the evolutionary advantages produced by a given pattern). Sociobiology, in other words, is the name given to an ethologist's inquiry whose focus implies the interaction between two or more animals, with the essential difference that, unlike ethology, which mostly investigates the mechanisms that maintain social systems, sociobiology concentrates on the ultimate factors that determine social behavior. Methodologically, it largely relies on notions like >**Tinbergen**'s >**Four questions**.

The main goal, as an evolutionary science, is that of proving that social behavior is also the result of natural selection, even in those cases (like aggression) where it seems clear that context and stimuli play a prominent role. For this reason, particularly when it comes to its application to human behavior, sociobiology has been the subject of several controversies.

Wilson 1975 is generally regarded as the landmark in the development of the field, although it must be reminded that proto-sociobiological approaches to animal behavior had existed since the end of nineteenth century already (e.g., in Kropotkin 1902), while the term itself "sociobiology" was allegedly circulating since the 1940s in biological and psychological environments.

Species

Technically, a species is the last taxonomic rank before the "individual" (with the exception of those cases where a "subspecies" is identified). Theoretically, it is one of the most controversial notions. As nearly all cases in taxonomy, the concept of "species" has to face the basic problem of being a digital/discreet category designated to represent an analogical/blurring body of entities. There is no point, in space or in time, in which, say, a nightingale is *neatly* and *unmistakably* a *Luscinia luscinia* and not a *Luscinia megarhynchos*. The sole fact that the boundaries across species are constantly (a) redifined, (b) updated and (c) contested confirms the high level of problematicness of the topic. (for a practical illustration, the reader may check the taxonomic "histories" of species that are particularly complicated to classify, as those of the genus Haplochromis. In this case, no less than currently-acknowledged 41 genera, for a total of 354 species, all part of the Haplochromini tribe, were at some point considered a single genus "Haplochromis", before researchers would find significantly distinctive features across them). Not to mention the presence of the so-called "ring species", that is, species that interbreed with closely related groups creating constant hybridizations (as in the case of the genus *Larus*).

Since Linnaeus' first (and, partly, still valid) attempts to systematize taxonomic classifications of living organisms, the definition and conceptualization of "species" has undergone several revisions and discussions (including a number of comments by >**Darwin** himself) until the publication of Mayr 1942, which – although far from *unifying* the diverse theoretical approaches to taxonomy – established a dominant paradigm, and a general agreement in defining species as "groups of actually or potentially interbreeding natural populations, which are reproductively isolated from other such groups" (Mayr 1942: 120). However, already at this stage, several contradictions appear (e.g., how about those species who reproduce asexually? What about the many cases of hybridization? What about the "ring species" phenomenon?), and in general it became soon clear that the notion is unclassifiable under the Aristotelian logic of the "necessary and sufficient features" (as in the notion of "triangle", where three corners are *all* and *only* needed), but rather falls in that group of problematic semantic fields (like Wittgenstein's example of "Game") not definable in absolute and instead requiring a contextual categorization.

In that sense a species can at least be defined in terms of the following types:

(1) *Typological*, when they display fairly fixed properties (in the Linnaenian sense);
(2) *Morphological*, when indeed a specific separation is possible on the basis of morphological characteristics.
(3) *Biological*, an all those cases when Mayr's definition is applicable.
(4) *Reproductive*, when the reproduction generate offspring that is in turn fertile.
(5) *Recognition-based*, when it is based on shared reproductive systems (possibly, in the semiotic sense, too).
(6) *Mate-recognition-based*, when a group of organisms appear able to recognize each other as potential mates (again, this definition is interpretable in a rather direct semiotic sense).
(7) *Evolutionary*, when a group of organisms shares an ancestor, therefore maintaining a lineage's integrity and distinctiveness.
(8) *Ecological*, when a group of organisms adapt to (and create) a particular niche.
(9) *Genetic*, when it is based on similarity of DNA.

The same uncertainty in definitions and boundaries occurs when strictly semiotic problems are at stake. The identification of such semiosic categories like **>intraspecific**, **>interspecific** and others, is subject to identical controversies. In this companion, a choice was made to maintain the use of "species" as basic terminological point of reference, but it must be clear that such choice is more a commonsensical and solomonic strategy, rather than a theoretical stand. When possible, the term has been replaced with words like "community" or "group".

Speciesism

Expression coined in 1975 by Richard Ryder and Peter **>Singer** (in two different works) in order "to describe the widespread discrimination that is practised by man against other species ... Speciesism and racism both overlook or underestimate the similarities between the discriminator and those discriminated against". As such, it has both an ethical and methodological impact on animal-related studies, and it is closely related to the notion of **>anthropocentrism**.

Speciesism is further discussed in Chapter 1, Section 1.2.3.

Species-Specific

Any form of **semiosis** that occurs typically and exclusively in one **>species**, up to the point of tyipifying it, in a certain respect (as in the case of **>bee dance** for the *Apis mellifera*). The term should not to be confused with **>intraspecific semiosis**, which occurs *within*, but not exclusively in, one species.

Symbiosis

Umbrella-term that describes various phenomena, like mutualism, commensalism, amensalism, or parasitism, which share the common denominator of two dissimilar organisms sharing a number of complementary and reciprocally useful interests. The concept is particularly interesting for **>zoosemiotics**, as it implies the establishment of an **>interspecific >code** between two species that may be biologically very different (like in the case of "cleaning symbiosis" between so-called cleaning and client organisms). The interaction between these two species implies at least semiosic formulations that linguistically correspond to "I'm not an enemy", "I want to do something useful for you", and "this is what I am about to do". And, indeed, as all semiotic actions, symbiosis has its fair share of cheaters as well (as those cleaning organisms that, instead of eating their client's ectoparasites, feed mainly on its tissues.

Symbol

In Charles S. **>Peirce**'s theory of signs, a Symbol is a sign that relates to the object by means of convention, habit or social rules. A symbolic sign is thus an arbitrary representation of the object, of which – in principle – it may bear no whatsoever resemblance (although many are the cases where such resemblance can be traced at some point of the history or the articulation of the sign, a typical case being onomatopoeias).

A great deal of skepticism, within the semiotic environment, surrounds the notion of symbolic **>semiosis**, as applied to non-human animals. As a matter of fact, this seems to be one of the points of disagreement between (some) semioticians and ethologists. While the latter, especially those belonging to the latest generations, offer an increasing body of evidence (empirical and/or theoretical) in favor of this application, most semioticians base their own definition of humanity, in the semiotic sense, upon the label of "symbolic animal" (**>Symbol**). Such skepticism applies both to the alleged capacity of non-human animals to use symbols belonging to a species-specific human **>language**, and to the possibility that forms of symbolic semiosis occur spontaneously *within* the natural interactions of a given species.

However, even excluding the ever-increasing variety of examples provided by ethologists, and focusing exclusively on instances of purely semiotic interest, it becomes rather adventurous to classify certain instances of zoosemiosis as something else than symbolic. The condition, quite simply, is to take Peirce's definition of "symbol" literally, without – perhaps unnecessarily – narrowing the concept. And, despite the fact that Peirce himself essayed to define the term on dozens of occasions, it is always extremely clear that the necessary requirement for a symbol to be recognized is no more and no less than the existance of a convention, and its main characteristic is arbitrariness. A sign which bears no resemblances or physical connections with an object, and yet represents it, via a more or less explicit agreement between two or more subjects, *must be* considered a symbol, not least because it cannot be considered either as an icon or an index.

Once that has been accepted (and not so many semioticians seem to be eager to do it), the instances of symbolic semiosis in non human animals suddenly become abundant in number and variety. Part of the **>bee-dance** sign repertoire is arbitrary and conventional (see also Sebeok 1990: 43, for a specific interpretation in this respect), so are many social rituals in different species, including mating and/or **>aesthetic** ones (the process itself of **>ritualization** of a pattern implies the transformation of the specific function of that pattern into a symbolic one). Symbolic is the nature itself of a **>dialect**, instances of which are found in numerous species. In Chapter 2, Section 2.2.2, a specific discussion on these examples is offered.

To this, one shall also add the several experimental contexts in which non human animals were taught to understand, use and re-elaborate symbols created in human language. The cases of **>Washoe**, **>Koko**, **>Kanzi**, **>Chantek** and other great apes reveal encouraging success in that direction.

Syntactics

In Charles **>Morris**' original formulation (1937: 64), syntactics is that branch of semiotics that investigates the relation between signs and other signs, dealing with "the ways in which signs are combined". A syntactic investigation concerns thus the sign repertoires of living systems in themselves, their structure, their spatial-temporal features, the way they form meaningful combinations. In **>Sebeok**'s words, zoosyntactics "deals with combinations of signs abstracted from their specific signification or their ecological setting" (Sebeok 1990: 43).

Syntax

In general, the term refers to the principles (or study of them) that regulate the structure of a sentence structure in a natural or artificial **>language**. It is generally considered a distinctive trait of human language, although a number of ethological studies have provided evidence of the existence of both *Phonological Syntax* (a combination of sounds that, taken alone, do not necessarily have a specific meaning), and *Lexical Syntax* (a combination of sounds that are also meaningful as single units) in non-human animals (Robinson 1984, Mitani-Marler 1989, Arnold-Zuberbühler 2006). This issue is further developed in Chapter 2, Section 2.2.2.

In his theory on **>zoosemiotic universals**, Felice Cimatti (1998: 179–190) formulates the hypothesis that a syntactic attitude (which he calls "Syntax of perception") exists in principle in animal minds. Animals perceive the environment as gestalts, through a figure-background scheme. In the data available, animals pay attention mainly to those that are pertinent to their experience and **>Umwelt**. (more details in Chapter 2, Section 2.3.2.

Tactile Channel

In this glossary, the umbrella term "Tactile Channel" is used to identify all the sensory modes that pertain to a physical contact, that is the **>thermic**, the **>electric** and indeed the **>tactile**. The messages transmitted through these channels all occur by direct body contact or by vibrations produced in the close vicinity of the receiver. Of the three, the human animal possesses the tactile, and – possibly – the thermic channel (in the sense that a certain degree of information is conveyed also through body temperature).

The main advantage of this channel is variability in time and intensity. The main disadvantage is the total absence of distance transmission (as physical contact between sender and receiver is compulsory). One may think of human bodily contact as an example: the degree of variability is extremely high, and people can communicate a whole set of messages through a simple touch (caresses, slaps, scratchings, contacts of sexual type, etc.). At the same time, unless sender and receiver are at touching distance, no message can be transmitted. The case of thermic communication, widely documented in ecosystem, is still not empirically studied among animals (although it appears obvious that bodily temperature is a carrier of many different messages), while for the electric one, there are several examples among fish families, like the *Mormyridae*, the *Gymnotidae*, and, famously, the *Torpedinidae*. Through the electric signals, besides predatory purposes, these fishes can exchange a very high variety of messages, including detection of objects, social status, conflicts, courtship and others. Species like the Atlantic torpedo, *Torpedo nobiliana*, which is the biggest Torpedo still in circulation (it can reach up to 90 kg of weight), is capable of delivering a 220v electric shock.

Taylor, Hollis (1951)

Australian-American musicologist, with a strong interest on **>zoomusicology** and **>interspecific >semiosis**. In her PhD at University of Western Sydney, Taylor specialized in the singing activity of the Australian Pied Butcherbird *Cracticus nigrogularis*. She developed a template that combines biology, **>ethological zoosemiotics**, avian neuroscience and current state-of-the-art sonographic analysis with a wide range of both historic and novel musicological techniques. A paradigmatic aspect of her research is the assumption (and application) of a musician's point of view in analyzing her case-studies.

Tembrock, Günther (1918)

German ethologist. Professor at Humboldt University in Berlin, he there founded the first German research facility for **>ethology**, in 1948 (in times when the field was still known as *Animal Psychology*), and the European largest Animal Sound Archive (*Tierstimmenarchiv*) in 1951 (the archive currently consists of some 120,000 recordings). His contribution to ethology is enormous, and his interests intimately keen to >**zoosemiotics**, most notably in the study of acoustic >**communication**. In this area of inquiry, Tembrock provided an extensive body of research, including theoretical and methodological innovations. The term and concept of >**Biocommunication**, developed in semiotics by Günther >**Witzany**, was already postulated in Tembrock 1971, while his promotion of the field of >**Bioacoustics** is also notable).

Although not a semiotician, Tembrock has always been sympathetic with the zoosemiotic program, regularly contributing to semiotic publications with an explicit semiotic approach (as in Tembrock 1993).

Terrace, Herbert

See **Nim Chimpsky**

Thermal Channel

See **Tactile Channel**.

Thermic Channel

See **Tactile Channel**.

Theophrastus (ca. 370–287 b.c.)

Greek philosopher. A follower of >**Aristotle**'s, Theophrastus shared with the great stagirite the poliedric interest for many fields of inquiry, including biology and >**ethics**, which make him a relevant figure for >**zoosemiotics** as well. He wrote extensively about human-animal biological continuity, human rationality (as opposed to non-human animal irrationality), and the ethical implication of physical cruelty on non-human animals. The vast majority of his works, surviving only in fragmentary form, or completely lost, are reported by Diogenes Laërtius' *Lives and Opinions of Eminent Philosophers*.

Things

See **Things-Objects**

Things-Objects

Notion introduced by John >**Deely** (e.g., in 2001) postulating that any organism establishes two types of relation with the environment: one independently and one dependently of the (sensorial, semiotic, etc.) experience:

> [...] experience, for any organism, does not simply consist of anything that is 'there' prior to and independently of the experience, but only of 'what is there' within and dependently

> upon the experience. So that however many or few relations within the experience may also obtain independently of the experience, these relationships have meaning only insofar as and as they are incorporated with that larger network of relations, which constitutes perception in contrast to (while inclusive of) sensation, upon whose pattern the appearance of objects as such depends. And this larger network involves relations which would not obtain but for the biological constitution of the perceiving organism acting as interpretant even of what is given in sensation along with, indeed, within, the perception of objects as objects (Deely 2001: 128–9)

A "thing" is thus a "notion of what is what it is regardless of whether it be known or not" (Deely 2001: 129), while an "object" requires "a relation to a knower, in and through which relation the object as apprehended exists as terminus" (Deely 2001: 129).

The implications of this distinction are that an object *may* or *may not* be a thing, and a thing *may* or *may not* be an object, but as Deely points our this is hardly playing with words:

> [...] to say that a thing may or may not be an object is merely to say that any given element in the order of what exists independently of finite knowledge ('things') may or may not be known, whereas the inverse saying that an object may or may not be a thing is to say that what is not known is not an object, or, equivalently, to say that whatever is known is an object. And since whatever exists as an object does so only within that network of relations (what Sebeok characterized as 'a semiotic web' and Uexküll called an 'Umwelt') indifferently from nature and from mind (yet according to a mixture or pattern wherein those relations within and by cognition itself tend to predominate in the presenting of an object as this or that), we see at once that 'what an Umwelt is' amounts to a species-specific objective world, with elements of the physical environment made part of a larger, 'meaningful' whole or 'lifeworld' wherein the individual members of a given species live and move and have their being as members of that species rather than some other. (Deely 2001: 129)

The notion of things and objects is closely connected to that of >**Umwelt**, as it takes into account the connection between objective and subjective environment in an organism's >**semiosis**.

Tinbergen, Nikolas (1907–1988)

Dutch ethologist and ornithologist, and Nobel Prize winner in 1973 together with Karl von >**Frisch** and Konrad >**Lorenz**. One of the most important figures in modern >**ethology**, Tinbergen's interest covered a wide range of topics of interest for >**zoosemiotics**, including studies on the notion of >**instinct** (a.o., 1951) and social behavior (>**Sociality**) (a.o., 1953 and 1974). His contribution extends to the very foundations of methodology in ethological sciences, often influencing other fields of inquiry like >**sociobiology** and >**social sciences**, as in the well-known case of the so-called >**Four Questions** (Tinbergen 1963)

Tools

The use of tools in an animal >**species** is usually defined as the capacity of using material elements from the environment, in order to reach an otherwise unattainable goal. Such a definition includes both goals that could *not* be reached altogether, and goals that could be reached only *partially* (smaller in quantity, shorter in distance, etc.): in that case the tool acts as an *upgrade/extension* of the body. In Beck 1980, a seminal study on the subject, it is suggested that a tool, in order to be defined so, has to be an object clearly detached from the body of its user: it may be animated or not, and it may be a *product* of the user's body or not, but it cannot be a *part* of its body. In Parker-Gibson 1977 there is an important distinction between actual and *proto*-tools, the latter consisting in those instances where tool and goal coincide (as in the case of those birds, like the vulture *Neophron percnopterus*, crashing a nut against a rock). Tools, for Parker and Gibson, can also be context-specific and intelligent. Intelligent tool-use involved trial-and-error processes and "accomoda-tion to the specific situation and exploring and manipulating physical causality in a generalized manner" (Parker-Gibson 1977: 628).

Baber (2003: 21–24) suggests that the relevant parameters in tool-use are species, object, function and activity. Insects, birds and mammals are the tool-makers and users most witnessed and studied. Apart from the Egyptian vulture discussed above, recurrent examples of tool-usage include:

1) the weaver ants *Oecophylla smaragdina*, "sewing" leaves with a type of silk secreted by a gland (see, a.o., Hölldober–Wilson 1990);
2) the Galapagos finches *Cactospiza pallida*, inspecting trees' cavities with twigs and cactus spines (see, a.o., Tebbich 2002);
3) the sea otters *Enhydra lutris*, who notoriously dive to the bottom of the sea to collect a mollusk and a rock, then, swimming on their backs, place the rock on their abdomens and repeatedly hit the mollusk against it in order to break open the shell (see, a.o., VanBlaricom 2001). Otters often keep the same stone throughout their entire life, placing it inside their armpit for later use;
4) the digger wasps *Ammophila*, using fragments of stones and twigs in order to secure the entrance of the cavities where their eggs are laid (see, a.o., Brockmann 1985, although the first observations date back as early as 1892, thanks to Samuel W. Williston).

However, it is certainly within the family of Primates (not only great apes, but also such species as the widely-documented capuchin monkeys *Cebus apella*), that scholars found the most impressive examples of tool-making and use, both in natural (a.o., >**Goodall** 1964 and 1986, Beck 1980, and Visalberghi 1993) and captive (a.o., Tomasello-Call 1997) contexts.

Within >**zoosemiotics**, it was >**Sebeok** (1981: 239–49) who devoted partic-ular attention to this phenomenon, classifying tool-use as part of the group of >**Architectural signs**. Numerous examples of fabrication and use of tools are pro-vided there, according to different degrees of (humanly perceived) complexity: (a) the body itself (or a part of it) as a utensil; (b) ready-made, nature-provided tools; and (c) *ad hoc*-fabricated tools. In this case, thus, the body itself *is* classi-fied as tool. Sebeok also wonders whether tool-use may be ascribed an aesthetic quality (> Aesthetics), besides the obvious functional/practical one (Sebeok 1981:

245–6). Examples of an aesthetic use of tools are not rare, mostly within what Juhani Pallasmaa (1995: 55) calls the "communicative function" of architecture, and particularly within the courting-related strategies.

In Chapter 2, Section 2.4 and in the entry >**Pergolate**, the behavior of the male Satin Bowerbird is extensively reviewed, but there are more instances, like the case of the males of the fly *Hilara sartor*. The courting ritual is performed by several of them around one single female. Each male carries a sort of big balloon produced by an anal secretion, apparently serving the purposes of a "gift" (see, a.o., Kessel 1955). The pattern probably developed from a more "useful" one (>**Ritualization**), which is still displayed by some species related to the *Hilara sartor*: gifting the female with an actual prey (i.e., food, and not a pleonastic anal secretion). In a way, it is like presenting candies instead of flowers.

Certain species of spiders, too, display "food-offering" patterns. The remarkable novelty, in this case, is that they package their gifts. The prey, already killed, is packaged with silk then offered to the female (Akimushkin 1988: 108–9).

Tool-Making

See **Tools**

Tool-Use

See **Tools**

Training

The procedure of teaching one or more animals specific responses to specific stimuli, generally by using the methodologies of >**behaviorism**, based on reinforcement and/or punishment. Training is generally practiced in contexts like companionship, entertainment, labour and others.

Training is one of the key-concepts in Communicational >**Anthropological zoosemiotics**, and has been theorized by several semioticians or quasi-semioticians, including Gregory >**Bateson** (a.o., 1972), Thomas >**Sebeok** (a.o., 1990) and – most notably – Heini >**Hediger** (a.o., 1968). According to the latter (1968: 120), there are two types of training: (a) apprentissage, i.e. laboratory training; (b) dressage, or circus-like training. The difference consists basically in the typology of relation established between trainer and trainee: almost non-existent in the first case, quite intensified in the second. Sebeok reinforced the concept by emphasizing that "the two procedures are distinguishable in at least two respects: the semiotic character of the sign which initiates the requisite action (unmarked vs. marked), and the degree of emotional intensity coupling the interactants (minimal vs. maximal)" (Sebeok 1990: 45). The question is discussed at length in Chapter 3, Section 3.2.1.

In Bateson 1972, a particular accent is put on the relation between training and the animal's intelligence, particularly on how the former is (erroneously, for

Bateson) used as a criterion for testing the latter. This aspect is dealt with in Chapter 3, Section 3.3.5.

Trans-Specific

Form of >**semiosis** that occur in different >**species**, not to be confused with >**inter-specific** semiosis, which occurs *between* different species. Trans-specific is a sign, a >**code**, or else, that appears *independently*, but in the same, homologous (> **homology**), configuration in more than one species, but that does not *necessarily* serve the purpose of establishing an interaction between these species.

Trophallaxis

The practice of exchanging liquid food by means of regurgitation within the members of a community of social insects, or between these and the so-called *guests*. The exchange may occur mouth-to-mouth (oral trophallaxis) or anus-to-mouth (anal trophallaxis). It serves an important communicative function of enhancing a group's cohesion, and in that sense it qualifies as a form of >**phatic** >**semiosis**.

Turovski, Aleksei (1946)

Parasitologist, zoologist and zoosemiotician at Tallinn Zoo, Estonia, Aleksei Turovski is one of the leading figures in today's >**zoosemiotics**, with a strong emphasis on empirical and field-work. A specialist in >**comfort** behavior, Turovski has published several articles where the topic is given an openly semiotic interpretation (like 2000 – where he also comments on the figure of Heini >**Hediger**; 2001 and 2002). He is also very committed in popularizing zoological topics, especially through Radio programs, an effort for which he was granted the *Guardian of Estonian Life Science* prize in 2007.

Uexküll, Jakob Johann Von (1864–1944)

German-Estonian biologist, possibly the most decisive figure in >**zoosemiotics** after >**Sebeok**. Although not a semiotician (the closest he ever got to it was being in contact with Ernst >**Cassirer**), Uexküll and his work have been promptly revisited as major influences for the establishment of both >**biosemiotics** and zoosemiotics, mostly after publications by his son Thure and Thomas Sebeok (both extensively analyzing his work from a semiotic perspective). He himself was defined as a cryptosemiotician, i.e., a semiotician who was not aware of being such. Leaving aside the dubious validity of such labels, it is undeniable that Uexküll's impact on semiotics has possibly been bigger (and in general, on human sciences) than on biology itself, although he was certainly credited by ethologists as an important influence:

> Great influence was exerted upon the development of ethology by J.v.Uexküll (1921), who conducted experiments to investigate the interrelations between organisms and their

> environment. He showed that an animal can perceive only a limited part of its potential environment with its sense organs. Some of these perceived characteristics of the environment serve as specific cues. According to Uexüll only those objects serve as cues which are of siginificance in the life of an animalm, thereby becoming the bearers of meaning for the subject. [...] K.Lorenz was the first to appreciate the full significance of these discoveries. (Eibl-Eibesfeldt 1970: 6, 7)

However instrumental the role of Uexküll's in the development of >**ethology**, that still is little compared to the huge debt owed to him by any biosemiotician or zoosemiotician. Other illustrous scholars who acknowledged Uexküll's influence on their work were Martin Heidegger, Gilles Deleuze and Maurice Merleau-Ponty.

Most of Uexküll's fame revolves around the revolutionary notion of >**Umwelt**, which is here discussed at length in a specific entry and in Chapter, Section 1.1.4. Recently, however, another portion of his work became part of a biosemiotic project of anti-darwinian revisionism:

> [...] the redeployment operation had to face the fact that Uexküll's philosophy of life was, to put it lightly, somewhat dated. He was not only a staunch anti-Darwinist, but also a feeble evolutionist (verging on creationism), had strong sympathies for Hans Driesch's vitalism, and often resorted to musical metaphors to explain the perfection of the living world (Barbieri 2006: 104–5).

For more on this issue, see the entry Karl Ernst von >**Baer.**

Umwelt

One of the most important notions within >**zoosemiotics**, "Umwelt" means in German "Environment", but in Jakob von >**Uexküll**'s formulation came to designate a much more complex notion, usually referred to as "subjective universe" (Sharov 2001: 211), or also "semiotic world of organisms" (Kull 1998: 304). The starting assumption is that the environment inhabited by an organism is not merely the actual environmental niche, but is a larger not purely physical "environment", of which the niche is just a part, that is perceivable and meaningful in its entirety only from the perspective of that particular organism. Uexküll (1982: 29–30) exemplifies this concept by describing the completely different meanings that a flower may have to a young girl (an element of decoration), an ant (a path for reaching food), a cicada-larva (construction material), and a cow (fodder). In John >**Deely**'s words, the same *thing* becomes four different *objects* (>**Things-Objects**).

An Umwelt is the result of a *Merkwelt*, i.e., the specific perceptive field of a given organism, and a *Wirkwelt*, i.e., the field of actual interaction between the organism and the environment (the latter being also the organism itself, as the example of >**proprioceptive semiosis** paradigmatically explains. Perceptual and operational elements come to establish the specific Umwelt of the given organism, which is exclusive for each >**species**, in particular, but also – broadening the concept – for each community, individual, class, family and so forth. The Merkwelt and the Wirkwelt are constantly in action, as the organism (also called a "structure", or a "receiver of meaning") affects and is affected by the environment (the "counter-structure", or "carrier of meaning"). In that sense, the process is described as an "Umwelt circle".

Probably because of its original meaning in the German language, but also because of the complexity of the concept, the Uexküllian "Umwelt" has often been confused with concepts like "environmental niche", "habitat", or, indeed, "environment". It is evident, though, that Umwelt does not propose a tangible category, but rather an array of subjective and perceptive elements:

> We see then how different and richer is the concept of Umwelt than the subalternate concept of 'environmental niche'. The concept of environmental niche simply identifies that part of the environment as physical upon which a given biological form mainly depends in deriving the physical aspects of its sustenance. The concept of Umwelt, by contrast, shows us how a given 'environmental niche' is merely the physical part of a larger, objective, not purely physical whole which is, as it were, fully comprehensible only from the perspective of the particular lifeform whose world it is, whose 'environment' is meaningful in the specific way that it is thanks only to an irreducible combination of relations many of which have no being apart from the lifeworld and all of which contribute to the contrast between the physical environment as neutral or common respecting all organisms, on the one hand, and parts of that same physical environment interpreted and incorporated within a meaningful sphere of existence shared by all the members of a species, on the other hand. Only things which are objects make up part of these species-specific worlds, but within these worlds are many objects which also are not things apart from the worlds. (Deely 2001: 129–130)

As already mentioned, the Umwelt theory is crucial for the whole paradigm of zoosemiotics, but in fact it is a central concept in the entire biosemiotic area. For this reason, there is a very rich literature on the subject, and different theoretical applications, including the above-mentioned Kull 1998, Sharov 2001, Deely 2001, but also Kull-Torop 2003, Martinelli 2006e, Turovski 2002, Uexküll (Thure von) 1987, and others.

The concept of Umwelt is further discussed in Chapter 1, Section 1.1.4.

Vilkka, Leena (1964)

Finnish philosopher. Currently associate professor of philosophy at the University of Helsinki and Hyvinkää city council member, she is one of the leading European animal right philosophers. Similarly to Tom >**Regan** and Peter >**Singer**, her theories are of remarkable relevance for >**zoosemiotics** mostly when it comes to general ethical and methodological discussion. Of particular interest is Vilkka's 1997 definition of >**biocentrism** that is adopted in this companion, and in other semiotic uses of the concept (e.g., Martinelli 2006a).

Visual Channel

The sensory mode (>**channels**) connected with the production, emission and reception of visual signs. From a >**cognitive** point of view, it perhaps represents the main perceptive vehicle for the human >**species**, even though it is hardly the best-specialized. Visual signs can be divided into extrinsic and intrinsic. An extrinsic sign is produced in the organism's environment (tracks, nests, traces, etc.), while an intrinsic sign is part of the organism's body or behavior (orientation towards the receiver, body shape, colour, movement patterns, etc.). Several forms of animal nonverbal >**communication** are to be considered intrinsic. The main advantages of visual >**semiosis** are the specialized orientation in space, and the great (virtually endless) quantity of information that may be conveyed. The biggest limitation is the need for light (which, contrary to common belief, is a handicap also for nocturnal species, whose perceptive qualities are anyway limited in comparison with daylight conditions).

Washoe (1965–2007)

Named after the Washoe County, Nevada, when she was first raised, Washoe was a female chimpanzee of the species *Pan troglodytes*, and the main subject of one of the oldest >**interspecific communication experiments**, started in 1967, and performed in different sites and US countries. The program was designed and initially carried out by Allen and Beatrice Gardner, then taken over by their former student and current project leader Roger >**Fouts**. Over the years, the project accumulated more chimpanzees, until a whole colony was formed, that allows the program to continue, in spite of Washoe's death at age 42.

The Washoe-project was a true watershed between the past and the present of interspecific communication experiments. It was the first one to be generally acknowledged as a success, it set several methodological indications eventually followed in other programs, and it inaugurated a series of experiments based on linguistic sign systems alternative to speech. Washoe was trained with an *ad hoc* version of the >**American Sign Language** (ASL). The Gardners wanted to prove that (1) chimpanzees are able to learn a human >**language**, and (2) real >**interspecific** >**communication** between humans and other animals is possible. The >**training** was organized so that Washoe could imitate her trainers' gestures, and – at the same time – through direct manipulation of her arms. The training environment was set in a cosy, relaxing, non-laboratory-like atmosphere, an aspect which the Gardners and later Fouts considered very important in order for Washoe to be trained properly (Gardner-Gardner 1969: 666)

The first signs taught were the *passepartout* "more" and "come here", the latter being not only a request for a trainer to approach her, but also for objects and – associated with other signs – for certain actions (e.g., "come here tickle"). After the initial training stage, Washoe showed increasing curiosity in learning signs. Some were very close to the natural >**intraspecific** gestures by which a chimpanzee interacts with his/her fellows. The sign for "come here", for instance, was homologous (>**analogies/homologies**) to the "grooming" sign. Instead of considering that a problem, the Gardners encouraged such mixtures, in order to facilitate Washoe's work, claiming that the same process occurs in human language, when abstract signifiers often carry characteristics of the signifieds (as in onomatopoeic words). Plus, their resemblance to nature sounds do not make these words *less words* than – say – "wardrobe", or "acknowledgment".

The results of the project were remarkable: Washoe spontaneously used the signs, even to name things, rather than just making requests (differently from usual reports by semioticians, who often maintain that all these animals were capable of doing was requesting food), then started to associate the ASL signifiers to more complex meanings – for instance, the words *flower* or *car* were used not only for real flowers and real cars, but also for pictures portraying those items (Gardner-Gardner 1969: 667). Signs were no longer strictly related to the context in which they had first been produced: the sign "open" was used not only for doors, but also for refrigerators, windows, taps: Washoe had turned those signs into *cognitive entities* (Gardner-Gardner 1969: 670). Furthermore, Washoe learned that signs had purely abstract meanings, and were not necessarily connected with perceptive resemblances: in the beginning the sign for "flower" represented all smelling objects, and started to designate a flower only when Washoe was taught the sign "perfume".

Later, Washoe showed interest in adjectives, and compiled metaphoric object-attribute relations (e.g., the term "dirty" acquired a moral connotation: Washoe would use the expression "Roger dirty" every time she wanted to insult her trainer for not having granted one of her requests). In formulating sentences of the object-attribute or object-action type, Washoe had her personal rule on how to establish syntactic orders: with no exception, the perceptively most relevant or most dynamic entities were signed for first. So, it was "Roger good" and not "Good Roger"; "Bottle red" and not "Red bottle"; "Look there" and not "There look", etc. This "gestaltic" process appears in most interspecific communication experiments with other animals, and seems to reflect a basic *perceptive* syntax.

Washoe was also given ASL-trained company (including equally-talented chimpanzees like Moja, Tatu and Ally, who used ASL quite regularly in **>intraspecific** communication, as well – possibly suggesting that the intraspecific use of ASL was meant more like a game for the chimps, or like a rehearsal of the idiom that is necessary in order to communicate with the humans), and was also allowed to form a family, with natural and adopted offspring. Her adopted infant, Loulis, by Fouts' deliberate decision, was not taught ASL in the first 5 years of his life, but nonetheless acquired more than 70 signs by watching the other chimps (reflecting the manner in which human children acquire language). Later in time, Washoe taught ASL to her natural infant Sequoyah by her own initiative.

Along with **>Koko**, **>Kanzi** and **>Chantek**, Washoe acquired over the years a wide popularity, almost becoming the "speaking ape" by definition. In 1981, the no-profit organization "Friends of Washoe" was founded in order to promote the welfare of chimpanzees. At the time of her death, thousands of condolence messages were sent to the website of the organization (www.friendsofwashoe.org), prompting the opening of a specific tribute page (www.friendsofwashoe.org/washoe). Washoe's story is also described in several programs and "popularizing science" literature, most notably the very informative Fouts 1999.

Wenner, Adrian (1928)

American biologist. Wenner is very well known, besides his scientific merits, for raising a long-lasting controversy around Karl von >**Frisch**'s theory on >**bee dance**, by maintaining that bees actually locate foods exclusively through the >**olfactory channel**. Officially the "loser" in the challenge (in 1973 von Frisch was awarded a Nobel Prize, and Wenner withdrew from bee research), Wenner recently had some compensation in the findings of a study conducted in 2005 by the Rothamsted Research (consisting in bees-tracking with radars), which confirmed that odor is *also* a guide for the bees.

Besides this episode, Wenner has been active also in the specific zoosemiotic department, by contributing to one of the classics in the field (Sebeok-Ramsay 1969) with an important overview of animal >**communication** studies. Wenner 1969 is, first and foremost, a plea against superficial discussion on this complex topic, and a warning about the many risks related. It reminds scholars (as this companion also does in Chapter 1, Section 1.2.3) that the expression "animal communication" itself is in principle superficial, as any discussion on communication should always focus on *that* specific organism engaging in *that* specific type of communication, instead of using an expression that put in the same cauldron flies and chimpanzees, alarm calls and play-bows.

Wenner thus proposes a classification of communication typologies that, among other things, include an intra-individual, inter-individual and animate-environmental level of signalling, and specific elements/parameters such as the variability of the signals, the redundancy within/between signals, the ontogenetic development and role of learning, and others.

Witzany, Günther (1953)

Austrian philosopher and semiotician. Mostly focusing on >**biosemiotics**, Witzany's work is often of specific zoosemiotic interest, particularly for his investigation on Coral >**communication** (Witzany 2007: 85–118), and for elaborating the concept of *Mitwelt*, in relation to "all rule-governed sign-mediated interactions of organisms" (Witzany 2007: 207). He is responsible for a systematic theorization and application of the concept of >**Biocommunication** in semiotic contexts (Witzany 2007).

Zooanthropology

See **Anthrozoology** and **Marchesini**, Roberto.

Zoocentrism

See **Biocentrism**

Zoomorphism

According to Webster's Dictionary, Zoomorphism stands for: (1) the transformation of men into beasts; (2) the quality of representing or using animal forms; as, zoomorphism in ornament; (3) the representation of God, or of gods, in the form, or with the attributes, of the lower animals. In biological sciences, zoomorphism, similarly to **>anthropomorphism** (to which it is strictly related) is regarded suspiciously by scholars, and normally labelled as a methodological bias. The dictionary of **>ethology**, edited by one of the most authoritative Italian ethologists, Prof. Danilo Mainardi, defines the concept in the following way:

> Zoomorphism is an attitude that is less spontaneous, more intellectual, and opposed to anthropomorphism. It consists of considering the human being as an animal like all others, giving more emphasis to the characteristics it shares with other species, and very little emphasis, if any, on human species-specific features. A well-known example of zoomorphism is Desmond Morris's The Naked Ape, a best-selling essay in which our species is described without any stress on its capacity to produce and transmit culture. (translated from Mainardi 1992: 48).

Unlike anthropomorphism, zoomorphism is thus not considered a spontaneous, but a culturally-shaped phenomenon, which implicitly admits that the instinctive human interpretation of the surrounding environment is centripetal rather than centrifugal. Curiously, the primary difference between humans and animals indicated by Mainardi is the presence of **>culture** in the former. It is a curious line of argumentation because it was Mainardi himself who promoted the existence of culture in non-human animals, dedicating a whole essay to the issue (see Mainardi 1975). In addition, in the dictionary of ethology itself, the concept of culture is illustrated as something that is in fact present in most animals: "The capacity to produce and transmit culture is present in some insect, in some fish, in few reptiles, and in numerous birds and mammals" (translated from Mainardi 1992: 227).

As anthropomorphism, *in primis*, but also zoomorphism, are perceived by the scientific community as serious mistakes, unable to provide ethological knowledge with any useful contribution, most zoosemiotic theses are highly problematic, since, depending on the situation, they can be accused of being affected by either the former or the latter attitude. To assert the existence of, say, **>language** in honey bees indeed can be (a) anthropomorphic, when one considers language as an exclusively human characteristic; and (b) zoomorphic, when one interprets such assertion as an attempt to reduce language to a sub-human level. All in all, however, zoomorphism is considered a less "dangerous" attitude than anthropomorphism (at least, a less practiced one), but still a serious "disease" in scientific research. A statement like "human beings are animals" is certainly true and indisputable, but for whatever reason some of its implications are not acceptable to most researchers (**>Animal**).

The risk, it is claimed, is to simplify, or reduce all that natural evolution has granted to humans and made them different and, most of all, "unique" in the animal kingdom.

In actual fact, following the theses defended in this companion, the concept itself of "zoomorphism" seems rather unnecessary, as based on a *petitio principii*, that – more than anything – reveals yet another manifestation of qualitative binary **>anthropocentrism**. That is, it puts human species into an autonomous and independent niche, while all other animals have to share the same big pot. What seems inaccurate here is that a self-styled evolutionary approach (as ethology, and other animal sciences, at this historical point, generally pursue) should not forget the basic assumption of natural evolution, i.e., the gradual, layered organization of living species and, most of all, the commonalities between humans and other animals. Secondly, and more specifically for the semiotic context, one should not forget the implications of the **>Umwelt** theory. A dog offering a dirty and disgusting (to humans) bone to his/her owner, is said to act zoomorphically. However, the point is that the dog is not acting zoomorphically, but rather "cinomorphically"; i.e., s/he observes and evaluates the world through the eyes of a dog, not of a generic animal. Not only would a *Homo sapiens* not know what to do with a bone, but neither would a lizard, an ant, a cow, or a jellyfish. At the same time, there is possibly nothing wrong in dealing with humans as animals, as long as observations concern the human animal. However exaggerated, the question of anthropomorphism at least makes sense, but a word like zoomorphism should probably not even exist: can anyone accuse an animal (including the human one) of looking/behaving like an animal? And if so, which animal are we talking about? Which **>species**? Apparently, there is quite a difference between an ant and a cow. To legitimate zoomorphism is then, possibly, the real mistake, i.e., to admit one's own approximate, binary, and anthropocentric interpretation of the world.

Zoomusicology

Branch of both musicology and semiotics that studies the "aesthetic use of sound communication among animals" (Martinelli 2002: 7), or – in other words – what **>Sebeok** had called "Musical signs" (1981: 210–259). Its area of investigation includes all those sound manifestations that zoologists and ethologists, partly as metaphor, and partly for suggesting a homological correspondance, have described with musical terminology (songs, duets, antiphonal singing, etc.) The idea of zoomusicology, in the modern sense of the term, originated with composer and musicologist François B. Mâche (1992), although the idea itself of animal music dates back to the dawn of human thought (it is already mentioned by the likes of Democritus and **>Aristotle**). Similarly to **>zoosemiotics** with **>semiosis**, zoomusicology approaches "non-human animals" from the direction of human sciences, and music from the direction of biological sciences. Its basic innovation is the assertion that music is not an exclusively human phenomenon, but rather a zoologically-based one. Hence, to adopt the zoomusicological paradigm means to question present definitions of music, starting from its anthropocentric connotations.

The use of the expression "sound communication" in Martinelli 2002 explicitly takes zoomusicology into the domain of semiotic analyses of the musical phenomenon. Musicologists for a long time (and to some extent still) have refused to consider music as an exteroceptive phenomenon, regarding it purely and exclusively as an aesthetic expression 'closed in' on itself. Referring to music as "communication" removes that *closure* and brings back musical processes into the territory they always occupied before the more recent "art for art's sake" interpretations.

In classifying its analytical domain, zoomusicology adopts the same basic and well formulated tripartition proposed by ethnomusicology (the closest field to zoomusicology, mostly for the common fate of having to deal with musical *otherness*, and with western- or anthropocentric definitions of music):

1) *Structures*. The musical traits in themselves. Analysis of this level implies a large use of sound material, such as recordings and spectrograms, and aims to define the organisation of sounds in the species observed; e.g., range of sounds covered, recurrent intervals, timbres, scales, and so on. Mâche 1992 proposes exactly this typology of research;
2) *Processes*. Acts and behavioral patterns related to the structures, in the fields of emission and reception. This is the realm of the para-musical, and it includes the whole cultural dimension of making music, with its rituals (like the almost inseparable relation between musical and >**kinaesthetic signs**), social rules and so on. In zoomusicology, this level constitutes the best-known part of the research, with many of its aspects having already been investigated by ethology. The same type of analysis is also found in Sebeok 1981;
3) *Experience*. This is the level that musicologists like Gino Stefani (1998), François Delalande (1991), and Philip Tagg (1987) have proposed in their discussions about musical universals. Since musical experience may be considered a general experience that takes place between a subject and an object (musical event), they advanced the idea that a universal feature in each experience is the restatement of particular conducts and competences. This view is used for zoomusicological purposes with similar presuppositions. If the first level (structures) was that of the objective, and the second (processes) of the cultural, the investigation of music as an experience lived by an individual is surely the level of subjectivity, although it is clear that many of these experiences follow general rules.

Zoomusicology is an increasingly adopted paradigm, and, by now, an established branch of musicology (with publications, seminars and other events – see Fig. 4.6). Contaminations between zoomusicology and zoosemiotics, besides Martinelli, appear also in the works of musicologists/musicians like Hollis >**Taylor**, Emily Doolittle and David >**Rothenberg**.

Fig. 4.6 Nightingales were the subject of the first international symposium in zoomusicology (in Jäärvenpää, Finland, 2008)

Zoosemiotic Canon

During a biosemiotic seminar in 2004, in Estonia (see Martinelli 2006a: 276–288, for a lengthier account), the issue of the anti-semioticness of the >**Morgan's Canon** was raised, and a proposition of its exact contrary (i.e., "in no case should actions or behaviors be interpreted as the result of an inferior psychic faculty, when it is possible to interpret them as a result of a superior faculty") was named by biosemiotician Kalevi >**Kull** *Martinelli's Canon*. The objective of that notion was to formulate the main semiotic contradictions (and therefore incompatibility) of Morgan's Canon. For obvious reasons of conflict of interests, in this Companion the notion was renamed "Zoosemiotic canon" (a much more detailed discussion is available in Chapter 3, Section 3.3.5):

1. In talking about the "possibility" of interpreting animal behavior in a given way, Morgan does not seem to make clear a distinction between analogous and homologous behavioral patterns (>**Analogies-Homologies**): does the animal *seem*, or is it *really* behaving in a given way?
2. Also, the concept of "superior" and "inferior" psychic faculty is not made clear, but it hints a reference to a greater or lesser similarity to the human psychic faculties. In that sense, it results in a biased case of >**anthropocentrism**.
3. Morgan seems to promote simplicity by "chaining" observation in such a way that it gives the least possible disturbance. In that sense, the risk of "banalizing" the interpretation is greater than the risk of making it unnecessarily complicated.

4. Further important remarks on the inefficiency of Morgan's Canon are discussed in the entry >**Critical Anthropomorphism**.

In the light these points, a fair formulation of an alternative semiotic-inclined canon (as Martinelli's Canon was formulated in Martinelli 2006a: 286–8) should take into account at least the following:

1) Unlike Morgan's, a zoosemiotic canon (ZC) shall *approach the problem of inferior and superior psychic faculties in terms of abductive* (>**Abduction**), *not rhetorical, possibility*.
2) A ZC shall *aim at economic and ergonomic research.*
3) A ZC shall *reject the criteria for establishing differences between psychic faculties in the way Morgan's Canon does.*
4) As a semiotic-centred formulation, a ZC shall *treat any behavioral phenomenon as complex and multi-layered and rejects reductive one-sided interpretations.*
5) A ZC shall *fully support the notion of* critical anthropomorphism, *and therefore rejects reductive hypotheses also on an ethical basis.*
6) On a more strictly scientific level, a ZC shall *promote methodological empathy and*, more generally, *support a more balanced and less radical use of* >**anthropomorphism**.

Zoosemiotic Universals

Behavioral patterns, of semiotic type, shared across different animal species. The expression is borrowed from ethnological studies, where the characteristics common to every or most human communities are called "universals", or "universalia". The search for universals is methodologically based on the principles of continuity >**mind-body** and on the criterion of homology (>**Analogies-Homologies**) between non-human and human >**cognition**.

The universals indicated in this companion include interdisciplinary proposals from Cimatti 1998 (>**Sociality**, Arbitrariness of categorizations, >**Mind**, Principle of complexity, >**Ritualization**, Self- and Hetero-perception, >**Syntax** of perception); San Martin and Pintos 2001 (the phenomenological theses on >**Animal Ontology**); and Partan and Marler 1999 (>**Multimodality**).

The issue of zoosemiotic universals is discussed at length in Chapter 2, Section 2.3.2.

Zoosemiotics

As this entire book, and particularlyChapter 1, is concerned with the definition of the concept, the methodologies, the theories and the applications of zoosemiotics, this entry was conceived for allowing a short selection of definitions, taken from diverse, semiotic and extra-semiotic, sources. In chronological order:

"The term zoosemiotics – constructed in an exchange between Rulon Wells and me – is proposed for the discipline, within which the science of signs intersects with ethology, devoted to the scientific study of signalling behavior in and across animal species. The basic assumption of zoosemiotics is that, in the last analysis, all animals are social beings, each species with a characteristic set of communication problems to solve." (Sebeok 1963: 465)

"*Zoosemiotics* is a term coined in 1963 to delimit that segment of the field which focuses on messages given off and received by animals, including important components of human nonverbal communication, but excluding man's language and his secondary, language-derived semiotic systems, such as sign language or Morse code." (Sebeok 1990: 105)

"Zoösemiotics [is] the study of sign processes occurring within and between species of animals". (Deely 1990: 98)

"Zoosemiotics, the study of the semiotic beahaviour of animals, is a transdisciplinary field of research. Situated between biology and anthropology, it investigates a domain located between nature and culture." (Nöth 1990: 147)

"Zoosemiotics deals with the rules of animal communication by using the theory of information (e.g. mathematic analysis of signals) and the theory of communication. Situated between traditional ethology and sociobiology, it deals with topics of particular interest: (1) the nature of communicative channels (visual, tactile, electric ...) in relation with the environment; (2) the meaning of a message in relation with the contest it is emitted; (3) the ability of social species to construct symbolic languages. The latter shows similarity between zoosemiotics and cognitive ethology." (Malacarne, in Mainardi 1992: 817–8)

"Study of the use of signs and signals by animals for the purpose of communication and information transfer. See Biocommunication." (Scott 1995: 1285)

"Zoosemiotics can be defined today as the study of semiosis within and across animal species." (Martinelli 2006a: 28)

"Zoosemiotics is the name for the study of animal semiosis, communication and representation. [It] stems from the semiotic tradition that does not limit sign processes to human species. Such an approach is developed most clearly in the pragmatic semiotics of Charles S. Peirce and Charles Morris. Other main sources of the zoosemiotic paradigm established by Sebeok include Jakob von Uexküll's Umwelt theory that describes meanings in animals' subjective worlds, the communication semiotics of Roman Jakobson and Karl Bühler, as well as ehotological studies by Konrad Lorenz, Karl von Frisch and others." (Maran, in Cobley 2009: 357)

"The study of the sounds and signals used in animal communication, as song in birds or tail-wagging in dogs." (www.dictionary.com, retrieved 2009)

"[redirected from *Zoosemiotics*] Animal communication is any behavior on the part of one animal that has an effect on the current or future behavior of another animal. The study of animal communication, sometimes called spencerology (distinguishable from anthroposemiotics, the study of human communication) has played an important part in the methodology of ethology, sociobiology, and the study of animal cognition. Animal communication, and indeed the understanding of the animal world in general, is a rapidly growing field, and even in the twenty-first century so far, many prior understandings related to diverse fields such as personal symbolic name use, animal emotions, animal culture and learning, and even sexual conduct, long thought to be well understood, have been revolutionized." (www.wikipedia.org, retrieved 2009)

Chapter 5
Does Zoosemiotics Have an Ethical Agenda?

5.1 Introducing the Problem

How widely separated is a lavishly presented roast hog served in a fancy restaurant from a sanitized, shrink-wrapped piece of chicken breast bought in a supermarket? They might appear to be aesthetic and cultural opposites, but at the same time they are both commercial presentations of a dead animal as a product. What are, if any, the common denominators between these two (and of course many other) instances of human-other animal relationship? Are the ethical aspects of this relationship a topic of interest for zoosemiotics?

The present section shall attempt to

(a) explore the idea of otherness as applied to the human-other animal relationship,
(b) introduce the concept and distinguish different typologies of anthropocentrism,
(c) classify the anthropocentric attitudes, in order to understand them as a social phenomenon and as the result of a semiotic process,
(d) address the general issue of animal ethics in terms of discourse, and in the light of ethically-minded semiotic theories,
(e) suggest a possible semiotic framework for so-called "biocentrism".

Before proceeding, however, the question that provides the title for this section needs to be answered. It is in fact a question that can be interpreted in two ways, depending whether, in the word "zoosemiotics", one pays more attention to the "zoo-" or to the "-semiotics". One important initial point, indeed, is: does semiotics, in general, have (the right to possess) an ethical agenda? Then, if yes, does it also possess the tools to address ethical issues within the field of non-human animals (the "zoo-" part)?

As for the first question, there has been a certain degree of discussion as to whether semiotics, the way it is constructed as a field of inquiry, has in principle the epistemological prerequisites for including ethics in its theoretical program (for an interesting insight into this discussion, see Bopry 2002: 15–6). Given that the debate itself is quite beyond the scope of this companion, what can be offered, quite simply (and metaphorically), is a typical judicial case based on precedents.

D. Martinelli, *A Critical Companion to Zoosemiotics*, Biosemiotics 5,
DOI 10.1007/978-90-481-9249-6_5, © Springer Science+Business Media B.V. 2010

In several legal systems, precedents are cases that establish a principle that a court then utilizes when deciding subsequent cases that present it with similar issues or facts. So, if it can be proved that semiotics has already approached ethical issues with its own methodological and theoretical tools, it can be affirmed that, in general, it is legitimate to approach other ethical issues. Also even readers who are not deeply familiar with semiotics may recall at least the cases of Roland Barthes' *Mythologies* (1957), and possibly Ferruccio Rossi-Landi's works on language and ideology (like 1972, or the entire enterprise of the journal *Ideologie*, which he himself founded). In addition to these admittedly more general examples, there is an entire (lesser known to non-specialist readers) tradition of specific applications of the semiotic paradigm to ethical issues, up to straight identifications of semiotics *with* ethics (it is the case with Ponzio-Petrilli's project of semioethics, mentioned in Chapter 1 of the present companion). Among these applications, one should at least mention Stefani (1985, 1989) (where musical semiotics is put in connection with Peace studies), Stefani and Guerra (2005) (where several issues in social integration and art-therapy are addressed through a semiotic interface), Habermas (1998) (on communicative ethics), Tarasti (1997) and (2000) (on the general necessity of ethically-minded research in semiotics), and that interesting path on the semiotics of resistance that goes from Eco (1997) to Bankov (2004), then to Tarasti (2005), and finally (on a specific zoosemiotic application) to Martinelli (2008).

Thus, Tarasti (1997) and (2000) both address overall the issue of an ethically-minded semiotics: particularly in 2000 Tarasti makes the claim that a semiotician does not have to fear dealing with ethical issues, as applied to semiotic research. As he himself states about art:

> [W]e cannot exclude from the present investigation the vast domain of ethical problems. One cannot be satisfied, in general, with a study in which a phenomenon is reduced to some of its aspects, but even in a most abstract art work we have to account for the whole weight of the reality that has yielded it and which speaks therein through its own sign systems. The intentions of an author cannot be eliminated as a kind of intentional fallacy. The social context and the "ecoform" of an art work cannot be left without attention as "extrasemiotic conditions". (Tarasti 2000: 87–88)

In what he calls the "discovery of ethics", Tarasti underlines the importance of the subject and its foregrounding and disappearance: "Without the concept of subject there is no ethical choice" (Tarasti 2000: 88). It seems, thus, that a *prescriptive* semiotics, as an implement to a *descriptive* one, does not sound anymore like an *arrogant* semiotics, as it may have done not too long ago, when it was suggested that science should prevent personal opinions from emerging, and – more to the point – that suggesting what people should do is not so scientific as explaining phenomena. The cases mentioned above, and several others which are emerging in the young generation of semioticians, show the scholars' preference for an intellectually honest weakness to a pretended impartiality, which – it is argued – is by principle impossible.

In addition to this, looking at the same issue from another angle, it should not be overlooked that other branches of the human sciences that opened the path of research and discussion around other animals, are making a similar choice of *facing*, rather than avoiding, the ethical consequences of their reflections. One example is phenomenology (mentioned in Chapter 2, in the paragraph on zoosemiotic universals):

> The phenomenological issue of animality thus undoubtedly has consequences for our duty toward non-human animals [...]. [T]oday, a hundred years after the rise of phenomenology, we must try to put one of his most significant discoveries into our practical life: the vindication of subjectivity as the center of the universe simply means the vindication of every subject sense – every subject without exception – be it a human or a non-human animal. Reality, for its transcendental constitution, depends on the intersubjective community of monads, and the latter includes our co-subjects, the other animals, as well. Our being entwined with other animals at a transcendental level, as we have now seen it in phenomenology, also binds us ethically. (San Martin and Pintos 2001: 363).

In a sense, this quotation sumarizes the entire message of this section of the book and the theses proposed, which admittedly remain hazardous suggestions in the context of the zoosemiotic theoretical program. At the end of the day, the authentic spirit of philosophy is the attempt to complete an ideal path that goes from spotting a "problem", finding the right questions that define it, answering those questions and – finally – being able to benefit from the answers in "real" life.

The second part of the title-question refers to the possibility of *specializing* ethical discourse in a circumscribed area of semiotics. That is, is it acceptable that zoosemiotics discusses animal ethics? The point here is not to wonder about the legitimacy of animal ethics in general: it is *obvious*, and confirmed by countless examples in different fields, that animals have by now fully acquired (and deserve constant discussion around) their ethical status, in most cases as "moral patients", but sometimes as "moral agents" too (at least following studies like De Waal 1996). Thus, it is obvious that if anthroposemiotics is entitled to discuss human ethics, then zoosemiotics is within its rights in discussing animal ethics. What must be instead asked is if zoosemiotics, given its young age, and still uncertain paradigm (a condition that, hopefully, publications like the present one will improve), possess the necessary theoretical substance to tackle the ethical task, which is by nature huge, complex and multifaceted. In other words, is zoosemiotics as ready as, say, musical semiotics proved to be (Stefani 1985, 1989, Tarasti 2000, a.o.) for this form of ethical self-analysis? Will zoosemiotics add something relevant to the animal ethics discussion, by means of its specific theoretical resources? This, more than anything, shall be the real challenge launched by this section of the book.

5.2 The (Zoo)Semiotics of Resistance

The aforementioned theoretical path of the semiotics of resistance offers fertile soil to start a discussion. A paragraph of Tarasti (2005: 4–7), entitled "Globalization – new civilization", attempts a list of what he calls a "prophecy on the *condition humaine*," i.e., a series of negative ideological principles and actions that modern society has adopted, and that Tarasti provides as a ground for developing a theoretical and practical attitude of resistance. His charge is as follows: "Do we, as intellectuals, scholars and artists, want to be part of such a world?"

It is not of particular interest, in this context, to adopt fully the paradigm of resistance semiotics, therefore there is no particular need to introduce the theory in general (information can be obtained particularly in Eco 1997, Bankov 2004). It is Tarasti's specific treatment of this theory, and particularly the structural organization of the topics, that serve the present purposes. Is the *condition humaine* also a *condition animale*? Is, in other words, the ideological discourse of modern society described by the Finnish semiotician applicable to the human-other animal relationship? That is: is this relationship itself an *ideological discourse*?

As already said, presenting the issue within this framework is an excellent way to display the various problems in animal ethics where, possibly, zoosemiotics may offer a usable interface for discussion.

5.2.1 Ending the Past

What Tarasti calls an "active loss of memory" (2005: 4) means, in this paragraph's case, the complete, and definitely active, loss of awareness of our biological nature and condition as "animals", in the way described in Chapter 3, while referring to the concept of animal in the semiotic perspective. There is not so much to say here, as the picture should be quite clear already. In all fields of human interaction, we observe a continuous process of detachment of human beings from their biological status as animals. Calling somebody an animal does not normally produce a delighted reaction, despite the fact that we do nothing but make an undeniable, scientific assertion, that we, as part of the human species, are members of a category known as the Animal Kingdom. Yet when using that word, we normally do not mean it in its denotative sense, but rather, as already seen in Chapter 3, in its connotative one. *Animal* denotes a semantic field of which also humans are part, but it also connotes a specific human being endowed with some negative characteristics (being uncivilized, violent, irrational, etc.). Thus, the connotations of words are often more important than their denotations. Generally, we tend to think that it is our perception that influences the words with which we describe a given experience. In fact, the contrary occurs just as often: words influence perception. And words, as applied to non-human animals, are often confused, demagogical, conventionally uttered in good faith, but often used in bad faith. When we say *animals*, we think of every animal species except the human one. Our biological roots are forgotten, and in the process, we efficiently create a cognitive gap, identifying ourselves as different, and hence better. The past, indeed, is ended.

5.2.2 Dominance of One Discourse

In Tarasti's observations, modern society displays a tendency for one particular discourse to become dominant over the others (2005: 5). Tarasti refers here to the economic-technological discourse, but what is interesting here is the dominance of an anthropocentric discourse, which bears the same military-like characteristics that Tarasti emphasizes in his case (not to mention that an anthropocentric discourse bears consequences in the economic-technological one as well). The creation of a dominant discourse has the immediate effect of separating the world (i.e., the perception of reality) into center and periphery. And periphery means otherness. To discuss the idea of *animal* means also to discuss the idea of otherness. Nowadays, non-human animals are perceived as *the others* par excellence, i.e., they are to *Humans* what, in the past – and somewhere in the present – *Women* were to *Men*, *Blacks* were to *Whites*, *Native populations* were to *Settlers*, and so on. Humans basically *think*, *do*, *have*, *are*, etc., what animals *do not*, *have not*, *are not*, and vice versa. Common sayings provide clear illustrations of this mind-set: we cannot "treat anyone like a beast," we condemn "a beastly impulse," we have "beastly weather," etc. Not to mention more specific metaphors such as "chicken-hearted," "a fish out of water," or "to fight like cats and dogs."

5.2.3 Demagogy of Language

A possible addition to Tarasti's list, or better the explicitation of a category that in fact is implied in many of his reflections, could be called "demagogy of language". It refers to a conscious manipulation of words and expressions in order to convey an altered meaning and ultimately lead to a different perception of what these words or expressions are designed to represent. In other words, there is a shift between connotation and denotation, with the former "sold" as the latter. This was already seen with the example of the word "animal", but the problem is more complex. A more specific, but no less significant example comes from the behavior of wolves, which, notoriously, howl. Not only do they howl at the full moon, they also – and most frequently – howl when they are euphoric and excited. This is one of the reasons why, in zoomusicological research, one is entitled to qualify howling as music. In particular, after a successful hunt, i.e., when nobody in the pack was hurt and enough prey was captured, wolves put into action what Jakobson called a phatic use of communication. That is, they celebrate and enhance the group feeling. They all start howling and their euphoria is visible in many ways, from frenetic and vaguely choreographic movements, to proxemics, to displays of affection, etc. Now, the long pre-cognitive ethological tradition (the Lorenzian and Tinbergenian schools, for instance) tended – and still tends – to define this behavior with the expression "mood synchronization," which means that the wolves are happy, that they are all happy for the same reason, and that they display this excitement in a way that is somehow coordinated. But we do not call this *euphoria*, nor do we call it *partying*. The expression "mood synchronization" is a sterile, cold expression, and one that is used in a context which on the contrary should be warm and emotional. Although appearing to be a small detail, in fact this use of language is yet another means of creating distance between humans and other animals. Thus, *we* have fun, *they* simply synchronize their mood.

5.2.4 Metalevel of Activities

Here (2005: 4–5), Tarasti refers to the domination of administration, organization and management of the work, over the work itself. Again, this conception will be adapted to the purposes of the present arguments. Namely, it is the idea of *metalevel* that is here very useful: the idea, in other words, that one is led to deal with a given portion of reality through means that belong to another portion of that reality, one that is only indirectly related. One may call this re-elaboration the *metalevel of discourse*, as created in relation to the socio-cultural representation of the non-human animal.

The visual dimension of this representation is, possibly, the best exemplification of this instance. The animals we see (or do not see) are the result of what we want (or do not want) to see. And, what we do/do not want to *see* is a direct consequence of what we want or do not want to *think*.

Let us imagine an axis, with extremes named "hidden" and "exhibited," and therefore with intermediate points such as "disguised," "visible," "shown," etc. All these are perceptual categories, on which basis we ground our visual relationship with non-human animals. It is not relevant, here, to analyze systematically each category; the aim, rather, is to give an idea of what they represent from an interpretive perspective. The point is that the way we represent (or package, or transform our representation of) an animal is a strong indicator of the entire conception that we have of the animal in question, and of the way this animal is perceived in a given context.

An example, contradictory in itself, therefore twice as interesting, is "meat." The way the animal people will eat is represented illustrates the entire axis mentioned above, and therefore provides several examples of the metalevel of discourse.

On the one hand, the massive commercialization of the product "meat" requires a primarily emotional detachment between form and content. The steak we buy in a supermarket does not even vaguely resemble the animal it used to be. It is merely a slice of reddish matter. In our daily and automatic process of price- weight- and quality-checking, there is simply no place for reflection that that very steak was – not long ago – something that used to breathe and walk. Many children are horrified when they learn *exactly* what a steak is, and many vegetarians declare that their choice was made immediately after making this discovery. A lot of parents complain that they need all their powers of persuasion in order to convince their kids to eat meat, once they discover that the character *Babe* of the eponymous movie appears on their plate together with their French fries. There is a strong cultural conviction, in those parents, that meat *must be* eaten, and this, together with certain, cynical indifference that comes with age, leads them not to consider their children's' complaints and the emotional origin of their aversion.

The butcher's disguising process is thus a success: it creates detachment and indifference. Meat is a product, not a former life. It is not a matter of encouraging ignorance in the customer; it is not like we do not *know* anymore that the hamburger used to be a cow. It is rather that any reflection on that relation, once we are exposed to it, is discouraged. There are extreme cases, as well, such as that of certain tuna fish cans, when, indeed, the customer ends up *missing* the whole relationship, outright (tuna is sometimes mistaken as a non-animal).

On the other hand, there is an equal and contrary visual dimension, i.e., that of ostentation, which can be a mere marketing or even an aesthetic strategy. A marketing-related strategy of ostentation involving meat is that of fish-sellers who exhibit living fishes in order to emphasize the quality and the freshness of their products. And, indeed, how can one find a fish more fresh than a living one. The techniques for keeping the animals alive may range from aquariums packed with lobsters to relatively big bowls where the fishes agonize in the open air, occasionally relieved by a splash of water. The suspicion is that this marketing strategy is feasible only with animals that are morphologically and ethologically distant from human beings: that is, fishes, insects and reptiles. These animals are too small, too silent, or in general too different; their suffering is not dramatic enough, compared to a human's, to elicit our empathy.

Still, in areas that do not specifically involve killing, we would notice that even those animals that are fully able to activate our empathy are not treated very differently. Pet shops are a perfect example. What these shops exhibit in their windows are animals in cages or bowls, looking bored and/or sad, if not pathologically aggressive. And yet that does not shock us: our conscience remains clear as we immediately conclude that (a) after all, those animals are not suffering, since they are well fed and they keep each other company; and (b) nobody is hurting them, and soon someone (perhaps ourselves) will buy and take good care of them.

There are other types of ostentation, also related to aesthetic requirements. Consider those monumental meat dishes, served in certain *haute-cuisine* restaurants, that present the animal in more or less their original form, including head and legs, with a series of decorations (herbs and branches that look like wings, round fruits placed in the mouth or in the eyes, etc.). Such dressing is a symbol of pantagruelic *grandeur*, and perhaps, in its own way, a religious metaphor of the sacrifice to the gods.[1]

What we need to ask ourselves is whether there may be a relation between such diverse forms of perception and the display of the *product* animal. Are there common denominators between the roast hog of the fancy restaurant, and the anonymous plastic-packaged chicken of the supermarket?

Probably yes. Two such relationships have already been noted in Tarasti (2005): first, *symbolic violence*, which one shall directly connect to the concepts of structural and cultural violence, defined by Johan Galtung (1969), with reference to ingroup-outgroup dynamics (which shall be later illustrated) and their impact on socio-cultural interactions; and second, the *conflict between civilizations*, which of course refers here to the human/non-human relationship. It is not exaggerated to call

[1]Much should be said about the symbolic value of sacrifice in the main religions, and how much this value is implanted in our minds during everyday interactions. There is no point in glossing over the situation: still nowadays, meat stands for feast and affluence, and the fact that we eat turkey for Christmas and lamb for Easter, the fact that we practice ritual slaughtering, and so on, still remain a clear metaphor of the offer to the gods, a characteristic of both pagan and monotheistic cultures.

that relation a conflict: a brief mention of various instances should make the point clear (intensive breeding, hunting, bull-fighting, etc.).

5.2.5 *Standardization of Diversity and Binarization of Values*

A few more points can be added to Tarasti's list, with the suggestion, once again that they are implicit in his arguments:

1) *The standardization of diversity* – not to be confused with hybridization. The reference here is not to the (healthy) encounter/mixture of differences. The point here is the *replacement* of diversity with prototypes. A trivial yet significant example is the progressive disappearance of biodiversity in supermarkets: throughout a place like Italy there are (or used to be) hundreds of kinds of apples. Yet, what we find in a supermarket is never more than four or five types, one for each of the *official* colors (green, yellow, light red and dark red), and all with the same, very round, shape. A true heritage of goods – each carrying their own taste and flavor – is here replaced with four balls of the same size;
2) *The binarization of values*: the complexity/diversity/asymmetry of reality is invalidated through the claim that it requires too much effort to be completely grasped, and – rather than being simply softened or simplified – that complexity reduced by an extreme degree of generalization, i.e., binarization. Political discourse is increasingly adopting this strategy, as well. Most of the rhetoric employed during the recent Afghanistan and Iraq conflicts is based on a pretentious distinction between good and evil. Another example is the tendency, more and more fashionable, to stick aesthetic and artistic discourses into two huge, anti-critical, Duke-Ellingtonian cauldrons: "There are only two types of music, good music and bad music."[2] The illusion of a total mastering of reality (who, after all, is not able to tell the black from the white, the cold from the hot, etc.?) anaesthetizes minds and consciences, and makes it easier to convey messages of standardization and dominance.

   The crucial implications of these two points in animal ethics will be seen later.

5.3 The Human–Animal(ist) Relationship in Everyday Discourse

As it can be easily guessed semiotics is central to all the points in the above discussion. In the instances previously explored, Nature and its inhabitants are not seen as such, but as symbols, or, more precisely, as social conventions. The *living being*

[2]To be fair, that sentence was attributed not only to Duke Ellington, but to a lot of other musicians and composers, including Richard Strauss, Louis Armstrong and others.

becomes a projection of a general interpretation of reality. Reality, in turn, is subjected to a process of crystallization, which is typical of social conventions. The result is a stereotype, and stereotype is the cognitive basis of prejudice.

Sticking to the dynamics and the use of communication, but turning our attention to everyday discourse, it is even more interesting to discuss those instances where the non-human animal is directly involved, in those increasingly recurrent instances where somebody with a particular interest for animal ethics (e.g., an activist of Greenpeace) interacts with somebody who is either uninterested in this, or perhaps even a bit against (being against meaning that s/he either sees certain values as a threat to his/her lifestyle – as in the case of vegetarianism – or that s/he maintains that human-related ethical issues deserve an absolute priority in people's preoccupations, and s/he sees anti-whaling or anti-corridas campaigns as effort stolen from more important cases). In such interactions, one can witness the exchange of an almost fixed set of rhetorical arguments. A couple of examples may be mentioned.

1) "So many animals kill other animals, why should this devoted respect for life concern only *us* towards *them*?" Why do lions not care about the gazelles they kill on a daily basis (for some reason, the example tends always to be about lions and gazelles)? Here we are into the delicate realm of the difference between moral agents and moral patients. Delicate in the sense that no matter how clearly defined the problem is in ethics (agents have moral responsibility towards patients at the very moment in which the latter are proved to be sentient beings), there is a curious resistance in everyday discourse in (a) accepting that responsibility, when it comes to other animals (especially by arguing that they are not "sentient"); and (b) avoiding confusion between the agents' duties and the patients' rights (the whole enterprise is perceived as unfair, hence the lion-gazelle example, although hardly the same principle is applied to borderline human cases such as small infantss or mentally-disabled). Despite the astounding resemblance of this argument to the attitude taken by a young kid complaining to his mother about the greater number of toys owned by his schoolmate, the remarkable recurrence of arguments of this type (or analogous variants) makes it necessary to point out a couple of things.

In the first place, any ethical discourse (including the ecological or animalist ones) is – so far and until counterproof – a discourse that concerns the way the human being should behave towards the surrounding environment and the creatures that inhabit it (including the human being itself, of course). It is *not* a discourse on how mushrooms should behave towards the trees they get nutrition from, or on how koala bears should behave towards eucalyptus leaves. Here, we are talking about how *Homo sapiens* should interact with other animal species, particularly those with whom a specific (economic, semiotic, affective, etc.) relation exists. The rest does not count, and does not constitute a counter-argument. If it is laid down in volleyball rules that during a game the players must not catch hold of their opponents, it is silly (indeed: childish) to say that we should allow that because in rugby it is allowed, and rugby is also a sport played with a ball. What kind of an argument is that?

In the second place, even assuming that this rhetorical strategy is legitimate (and it hardly is), its supporters should follow the entire and coherent line of their own reasoning, that is, positioning the two variables (the human killing another animal, and the lion killing the gazelle) on the same moral ground. So, if it is true that we are biologically doomed to kill specimens of other species (and, why not specimens of our own, one may ask, since there are animal species that practice cannibalism and other forms of intraspecific killing?), just because that is the general rule, then we should follow the rule entirely. And that means catching our prey by ourselves (alone or in small groups), doing it only for survival purposes, etc.

In addition, does the animal (as individual, not in the evolutionary perspective) that kills another animal have any alternative? The physiology of carnivores, as lions are, unlike that of omnivores, as humans are, suggests that an alternative does not exist. Plus, anatomy makes it clear that the former are naturally equipped for hunting, eating and digesting their prey, while the latter are missing claws, sharp teeth, and a digesting apparatus that is short and fast enough to expel meat from the organism before it starts putrefying.

Most of all: if our long-term goal is to *distinguish* ourselves from animals (as it seems so important for human beings – see Chapter 1 of this companion), in terms of ethics and civilization, how can we possibly think of using only the "they do it too" strategy to justify our own behaviour, exactly when the point of the discussion is how to behave ethically and in a civilized way?

2) A second rhetorical strategy against the threat of animal rights can take two forms (if some irony is allowed, in order to make the points clear): one is simply masochistic, the second is directly suicidal. In the first case, it is claimed that those who are against the killing of other animals should remember to apply the rule also in the less convenient cases, like that – notorious and always-mentioned – of mosquitoes. The ideal animalist, this principle implies, is thus somebody completely devoted to the entomological cause, decorated by bee and mosquito bites, gradually disintegrated by fleas and ticks, and in general a happy habitat of any small creature which has the pleasure or need to live in his/her organism.

The second case is related to the most advanced and irreversible degree of personal safety. Here, the animalist is rhetorically put before the traditionally furious and starving lion, and is asked whether – in the name of animal rights – s/he would have any right or will to kill the beast, if s/he ever had a chance. In this case, the ideal animalist is a person with the soul of the explorer, a passion for Africa, and a very short life.

There are several possible replies for these two arguments. Firstly, a – so to speak – methodological bias can be noticed. In any ethically-oriented discussion, extreme and marginal cases should always be the last topic of debate, those that one deals with when all other problems have been faced and solved. If, while discussing musical education, we considered the opportunity of teaching notation systems to young kids, it certainly would not be a good idea to start with an individual contemporary notation system (e.g., one of the several fascinating ones proposed by the Finno-Swedish composer Erik Bergman), while ignoring the traditional one that lasted centuries and is still in use for 99.9% of the music produced in the Western world.

Moreover, as a general rule, the use of extreme cases as a counter-argument is not generally productive, as its main result is the establishment of (unrepresentative) rules out of isolated, rare and in any case unusual instances. When we talk about extreme cases we should be fully aware of their (very limited) specific weight within the economy of an entire issue. Conversations among friends, rather than panels of scientists, are based on this principle: there, arguments are very often constructed and reconstructed on the basis of isolated cases. *Do you dare maintain that Renoir was one of the most representative French impressionists (a notion that is nearly universally acknowledged)? You must be kidding! Just the other day, I have read* one *article written by* one *guy who placed Renoir among the forerunners of Surrealism!* It may even be (but it is not sure) that our friend has really read a similar article (evidently written by someone who has no knowledge in visual arts), but the point is that such an opinion, although to be respected, cannot really qualify as a counter-argument, if it alone clashes with an entire tradition of art research literature. Maybe an occasion for reflection: this is the maximum one can afford conceding to such a statement. Certainly not something that, in the present state of things, can call into question the stylistical location of Renoir (and, most of all, that could relocate him right *there*, between Dalì and Magritte, with Andrè Breton left to go through his own writings in search of an even vague justification for such a move). First, we need to find at least a couple of more (authoritative and competent) essays maintaining the same position: then we can re-open the question of Renoir and Surrealism.

The actual implication of these counter-arguments is, quite simply, summarized in sentences like "you are either 100% coherent, or do not bother me with this animal rights issue". Now: absolute coherence is an overrated utopia, but even if it was not, we should not blame somebody who manages to achieve a respectable 70% of coherence with his/her own ideals (i.e., somebody who, despite occasional entomological murders and the impression that, yes, if his/her life depended on it s/he would probably kill the lion, still manages to observe a rigorous vegetarianism, to avoid fur clothes, and to buy exclusively cruelty-free cosmetics), especially if the criticism comes from somebody who does not possess that coherence, not even to a small degree. A 70% (or less, for the matter) coherent individual has all rights to discuss the ideals s/he defends, even if s/he wears a leather belt and spends half of the summer trying to get rid of flies and mosquitoes. In other words, stealing candies from our parents'cupboard, as kids, does not imply that we do not have the right to feel indignated towards our country's Prime Minister, if he is caught using public money and making laws for his personal aims.[3] If anybody (maybe our still-resentful parents) came to tell us that we are not entitled to complain about our Prime Minister because once in the past we stole candies, we would be entitled to laugh, as the gap between the two issues is so wide and unbalanced.

We can safely say that in a mature conversation interlocutors normally avoid using cheap rethorical tricks of this kind. If A talks about half a billion of animals killed just for testing a mascara, B should counter with coins of the same value, not by reminding A that the other day s/he was using a mosquito-repellent before sleeping.

[3] As in the sad case of Italian Prime Minister Silvio Berlusconi.

5.4 A Semiotics of Anthropocentrism

After this long, hopefully not too boring, digression on the constituents of the *condition animal* discourse, it is time to proceed with more systematic analysis. The leading question is: was there something in common among all the listed discoursive elements? Do the "mosquito" or "lion-gazelle" sarcasm, the use of the term "animal", expressions like "mood synchronization" and all the rest, present common denominators? The answer is yes, and it has to do with those two processes of "standardization of diversity" and "binarization of values" that was on purpose left a bit vague:

1) In all the cases mentioned, we behave *anthropocentrically*, that is – to the letter – we place the human being in the middle of an entire system. Which is probably obvious, as we *are* human beings. But it remains to see how *fair*, not only ethically, but also methodologically (and this is a rather important point, because the goal here is not only to open a case of practical ethics, but also of research ethics);
2) In all cases we are (ideally and/or practically) classifying this entire picture according to a scheme that is strongly binary, with all the human beings ("us") on one side, and all the other animals ("them") on the other side. This might look less obvious, but somehow, we know that this is the way it happens.

According to the usual definitions (see for instance the interesting Bartolommei 1995: 40–83), anthropocentrism interprets Nature as (a) an entity existing *apart from* and *for the benefit of* humans, so that (b) nothing in Nature can be considered in itself, autonomously from humans; and (c) it is ethically acceptable for humans and non-humans to be treated in different ways. In other words, Nature is not of interest because of its hypothetically *intrinsic value*, but just because of its *instrumental value*, i.e., the values it has for and to humans.

In animal-related studies, typical forms of criticism tend to insist that a totally impartial interpretation of animal behaviour is not possible, for observations are external to the subject of study and cannot avoid frames of reference that are typical of the human interpretation of reality. In this sense, the approach is considered anthropocentric, i.e., concentrated on and mediated by the fact of being human.

However, such a statement deserves more specific reflection, as in some cases, it appears more like a comfortable, socially-shared (thus, to some extent, stereotypical) way out from facing a problem that is in fact rather more complex. It may be easy to speak of anthropocentrism as an apparently unavoidable interpretation of reality that influences scientific research; however, to pose the problem in a *a priori* form does not really help. Rather, it seems more adequate to dissect the question into all its components and ultimately re-interpret anthropocentrism in a hopefully more accurate way.

The very first question, therefore, is: Is there just one type of anthropocentrism, or are there more? In other words, How many ways exist to observe reality according to the criteria of interpretation and classification proper to the human being?

To start with, it may be reasonable to suggest the existence of a first elementary level, which consists in the banal consideration that the subject who observes Nature is evidently a human being, with all its resources, limits and modes of categorization. We may call this level *Default Anthropocentrism.* What we understand about a hedgehog is what we are *able* to understand, given the means that allow us to do so. Technology does not (yet?) allow us to understand the hedgehog the way, say, a pigeon would understand it. Such a consideration is not very different from a statement like "Alvar Aalto is a great architect". Quite evidently, in pronouncing such statements, we are reporting one of our forms of interpretation of reality, founded on personal experience, education, culture, perceptive sources and so on. Now, this looks obvious, inevitable, and not dangerous. In fact, it is the other way round that might in principle be dangerous, for it could mean the expressing of opinion without any point of reference or any code, resulting in a sort of perceptive anarchy. Now, the claim here is, as long as an anthropocentric attitude is reduced to this very basic, default, expression, no kind of scientific research risks not being taken seriously. Perhaps hardcore Derridians and Postmodernists will not really appreciate this last argument, but then again, one cannot make everybody happy. What is important to stress is that the existence of *a point of departure*, so to say, in the formation of *any* kind of (scientific or not) discourse (i.e., not only the animal-related one) simply cannot be prevented, and – as a matter of fact – the history of ideas *did not* bother to (try and) prevent it.

Things get trickier when we consider a next level, in anthropocentrism, what we may here call *binary anthropocentrism.* What happens here is that the fact of being a different entity from the object observed (human, rather than another animal) produces a dualistic (as opposed to pluralistic) interpretation of reality, based on criteria of *difference* (*qualitative anthropocentrism*) and/or a strongly *hierarchical identity* (*quantitative anthropocentrism*), which puts the observer, and the group s/he belongs to, in a superior position in relation to the group observed. In the case of qualitative anthropocentrism, the observer-human being tends to see him/herself as distinct from the non-human animal by means of "either/or" qualities, in a relation that is almost of causal fashion (i.e., "humans do, *therefore* animals do not": this shall be seen more in detail, later on). In the case of quantitative anthropocentrism (a post-Darwinian anthropocentrism, in a way), the difference between human beings and other animals is expressed by means of quantities: humans are more or less than other animals, to a given extent of comparison.

To make things clearer, and following the previous architectural example, a statement like "Unlike Gropius, Alvar Aalto *is* a great architect" is of qualitative type, while the statement "Walter Gropius is a good architect, but Aalto is definitely better" is quantitative. Analogously, the discourse about non-human animals, and the field of interactions between these and the human ones, is full of statements (and consequent reflections) of this type. To stick to the semiotic area, it shall be enough to recall what positions have been taken during the debate over whether language is a human species-specific feature or not. The position expressed by the Darwinian-oriented semioticians has always been of quantitative type: "Some animals possess a rudimentary form of language, but humans went much further with it". On the other hand, Sebeok and his followers have always been rather categorical in stating that no matter how complex and specialised a non-human communication system, it is always something else than language, the latter being an absolute human exclusivity.

Now, the major claim of this section is that the characteristics and the dynamics of anthropocentrism, both in scientific and non-scientific environments, are not only of an anthropological type, as it would be obvious and predictable, but also – perhaps mostly (especially in Western societies) – of social type. It is also, very possibly, in the formation of its *social* identity that the human being feels an urge of creating boundaries and distinctions with other animal species. In other words, and paradoxically enough, other animals end being treated like a *social group*, or at least treated with the mental frames and tools that are normally used for social groups.

Although the dynamics of intergroup relations have been studied very little, as applied to the human-other animal relation, they are anyway theorized at length, within the human sphere of interactions, in social psychology, with a distinct connection with the functioning of stereotypes and prejudices. If this hypothesis is acceptable, then the application of these theories to the anthrozoosemiotic area should proceed relatively smoothly.

5.4.1 Intergroup Relations

Following Gordon Allport (1954), categorizing people into groups by identifying some common attributes or characteristics is an adaptive necessity. It reduces the amount of information to be dealt with and thus reduces the complexity of the whole social world.

Whether we are talking about, say, an Arsenal or a Chelsea fan, persons interested in football fit into a category depending on whom they cheer for. Some consequences go along with this simplification strategy. When a football fan identifies him/herself with, e.g., Arsenal (or, better, with that group of people playing in or supporting Arsenal), then s/he perceives him/herself and the group members as different from other groups. That person may work with a Chelsea fan, have the same position, love the same western movies, and live in the same town. But the Arsenal fan will view the Chelsea fan as significantly different from him/herself. If another co-worker was also an Arsenal fan, but lived on the other side of town and hated western movies, s/he would still be seen as more similar to the Arsenal fan than to the Chelsea fan. This co-worker is in practice a member of the same *in-group* to which the Arsenal fan belongs.

To acceptable extents, such a categorization leads to the belief that one's own group is *different* from other groups in a kind of special way. People put themselves into the in-group and all other people in the out-group. Typical is the case of news distribution in newspapers and magazines. What is found there is not "Senegalese news", "Japanese news", "Venezuelan news" and so on. The distinction is rather between "domestic news" (in-group) and "foreign news" (out-group).

In the worst case, however, categorizations produce *intergroup biases*, i.e., a belief or beliefs that one's own group is *better* than all other groups. This leads to an increased attraction to in-group members and a devaluation of out-group members. This is called the *minimal intergroup paradigm*, a concept pioneered by Henri Tajfel (1981). In-group members tend to behave in more pro-social ways and be more cooperative towards other in-group members than they are to out-group members. Football fans are obviously more pro-social and cooperative towards the in-group members than those of the out-group. Rarely do fights occur between two Arsenal fans, but it is common to see fights between Arsenal and Chelsea fans. Of course, if the context shifts towards a conversation about western movies, the two Arsenal fans may easily be in competition with each other, as parts of a different in-group/out-group dynamic (loves-western versus hates-western).

Other important intergroup biases are the following:

1. *Out-group homogeneity effect*: out-group members are not only seen as being different/inferior from the in-group, but also as being more interchangeable with each other. They are perceived as being more similar in their characteristics, opinions, and behaviour than are in-group members. Arsenal fans look at Chelsea fans and think "they are all alike".
2. *Polarised appraisals of out-group members*: in-group members tend to evaluate out-group members as more extreme in a variety of positive and negative personal dimensions, thus creating greater social distance between in-group and out-group members.
3. *In-group favouritism effect:* the Arsenal fan instinctively tends to favour members of his/her own group over those of other groups.
4. *Differences in causal attributions across groups*: in-group members make different attributions of cause and blame depending on whether the target individual is a member of the in-group or not; in-group members make more favourable attributions about members of in-groups than those of out-groups; positive behaviour is attributed to internal causes rather than external ones;
5. *Differences in information processing and retention across groups*: people process and retain information differently with respect to in-group/out-group members; in-group members perform a more elaborate and careful processing about fellow in-group members; have a more accurate memory for personal information regarding fellow in-group members; remember more positive information about fellow in-group members; have favourable expectations of fellow in-group members; and unfavourable expectations of out-group members.
6. *Illusory correlation*: Infrequent, distinctive behaviour is associated with out-groups even though the behaviour is constant across all groups.

Let us now return to anthropocentrism, particularly what was here called Binary Anthropocentrism. It is hopefully clear that the whole theorization of intergroup relations operates clearly and easily within the human-animal relation. Anthropocentrism itself can be defined simply as a set of mental attitudes that consider human beings to be a distinct and independent part of the Animal Kingdom and of all Nature, to the extent of not considering humans as animals at all, but more a sort of unique entity, that escapes (because it transcends) biological classifications.

However simplistic it may seem, "homo sapiens" is the in-group humans choose to belong to, when involved in biological (and sometimes just cultural) contexts. Particularly in the human-other animal relationship, the distinction is *Homo sapiens* and all-the-other-animals (e.g., very rarely is it *primates-other animals*, almost never *mammals-other animals*). Such a categorization, which is clearly binary, involves the same intergroup biases emphasised by social psychologists.

In Fig. 5.1 there is an attempt to summarise these aspects: first, the tendency is to regard the human being as something different (and, by consequence, superior, as a result of intergroup biases) to the rest of the Animal Kingdom (minimal group paradigm). Secondly, as a result, any possible comparison with animals is literally displaced by one degree, so that the out-group is regarded in a far more generalised way (out-group homogeneity effect). In other words, *Homo sapiens* is related/opposed to the Animal Kingdom (rather than to another species, which would be the – at least taxonomic – equivalent), and an individual *Homo sapiens* is related/opposed to a whole species rather than to an individual specimen.

Occasionally, even a single part of one human – a character trait, typically – is related/opposed to a whole individual animal, as if its behaviour was characterised

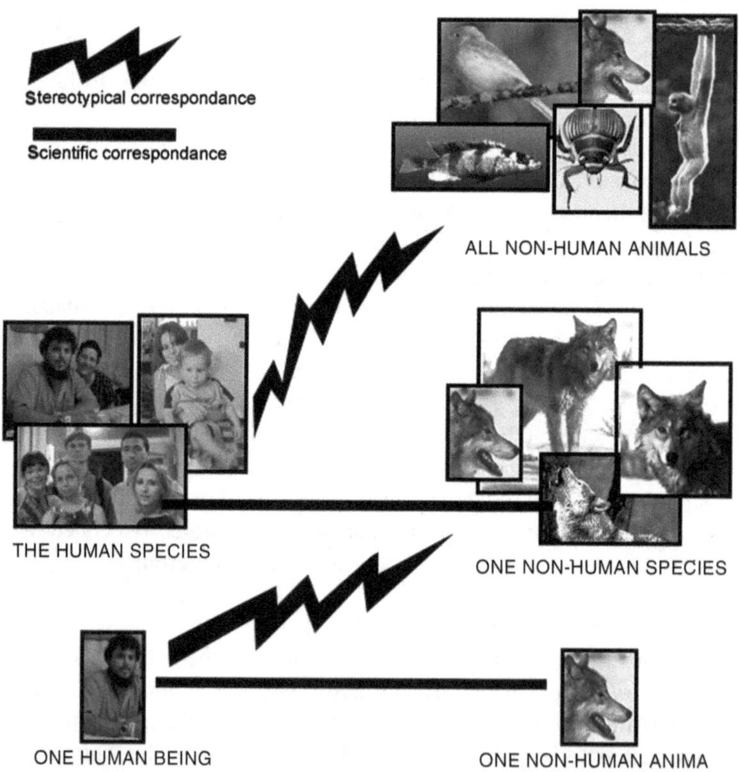

Fig. 5.1 Intergroup biases generated by binary anthropocentrism

by that trait only. For instance, it is a common belief that dogs, *all* of them, are devoted to their owners, without exception, and that *all* cats have an independent temper. And such attitude, it must be added, is not alien even to specialists of the field (veterinary doctors, for instance), who, on the contrary, should be the first to advocate the individuality of each animal's character.

5.4.2 Categorizing Anthropocentric Attitudes

After trying to define them, one shall now attempt to classify the different anthropocentric biases into a schematic model.

However, before proceeding, it is useful to give a short definition of *stereotype* and *prejudice*, as this will be a useful piece of information later. The construction of intergroup biases is strongly linked with these two notions, and moreover a certain popularization of the terms in everyday conversations has somehow reduced (by generalizing it) their semantic value.

Often used as synonyms, the two concepts are different but closely (and causally) related, and rarely does one exist independently from the other. Historically speaking, *prejudice* is a concept many philosophers have dealt with. The notion of *stereotype*, in the sense meant here, is much more recent, and was coined by the journalist Walter Lippmann in 1922, while the term itself existed already since the end of the eighteenth century, as used in typography for naming the technique of items' reproduction, through the use of rigid forms (stereòs = rigid, tùpos = print).

Starting from the 1980s, there has been a growing interest in the study of prejudice, especially in the fields of social psychology (see, for instance, Tajfel 1981, Brown 1989). However, in the history of human thought, discussions about the definition and functioning of prejudice as an evaluation of reality not supported by facts have formed a leitmotif in philosophers' speculations.

In 1620, Francis Bacon had elaborated an impressive analysis on prejudices, whose completeness and precision are still appreciated nowadays. Bacon provided an explanation of what he strictly called "inherent and profound errors and superstitions in the nature of the mind", called *idola mentis*. Idols, maintains Bacon, keep humans far from true knowledge, and should simply be eliminated in order for souls to be provided with a better interpretation of reality. More than two centuries later, the very same categories were called *cognitive biases* by modern psychologists, and nowadays they probably constitute the real core of psychological interpretations of prejudice.

Idola tribus are the first of these errors, and concern human beings as such, as a species. Examples of *idola tribus* are: (a) to believe that superior and transcendental entities rule our world and our existence as humans (what is today known as the need for coherence and uniformity); (b) to believe there are always final causes to each event; (c) to affect the interpretation of reality by the most evident or influential data; (d) to affect the interpretation of reality by data corresponding to our expectations and/or opinions; (e) to affect the interpretation of reality by emotions, and so on.

The second group, called *idola specus*, concerns the individual in terms of personal and subjective experience, education, personality traits, etc. Thus, to be superficial or profound, to prefer the old or the new, to look at things in their totality or to seek the particulars, to look for similarities or differences – all these are features that concern individual personality and do not necessarily refer to the characteristics of a whole society or community. A particular interpretation and a consequent distortion of reality derive from each of these (combinations of) characteristics.

Further to this, *idola fori* designate those biases coming from interactions and conventions among humans, language first of all. Words, as created to communicate (even in a very simple way), may become a serious obstacle to real knowledge, since they reinforce barriers and boundaries between things. Typically, this occurs when highly complex issues (such as the emigration of people from poor countries, to mention one) are reduced to simple and simplistic explanations ("they came to steal our jobs", etc.).

Finally, prejudices may derive from tradition and/or debatable – if not false – past theories (provided the historical perspective was such that those theories could be true at their time). Bacon classifies this case under the name *idola theatri*, as referring to myths and tales. Once they have been crystallised, traditions tend to impose their standpoint, which also has an impact on scientific research.

Centuries later, social sciences completed the picture, by pointing out that prejudice (a) mostly refers to social groups, rather than to facts and events, and that prejudice (b) definitely tends to consider the group observed in quite a negative light.

A definition of prejudice may therefore be placed on any point of a continuous axis, according to specific contexts. One side represents the level of denotation, which is the most general form (i.e., *to evaluate before or without empirical data or experience*), and the opposite side represents connotation, understood as extreme specificity (i.e., *tendency to a strongly negative evaluation of members belonging to a given social group*). All along the axis, each attitude is associated with the idea that actions are normally coherently consequent to evaluation, according to a typical sequence IF-THEN. Apparently, the more opinions lean towards the connotative extreme, the more social implications are evident. Since the social implications of human attitudes towards other animals are the topic of this section, an analysis of connotative aspects of such a relation play the leading role, although more general (neutral) aspects may warrant comment, as well.

As for "stereotype", in 1922, the journalist Walter Lippmann published a very interesting essay concerning the processes and dynamics of public opinion. He maintained that knowledge of the external reality is not achieved directly, but through representations that people, as they are strongly conditioned by the media, construct in their minds. These representations, a sort of symbolic environment with which people interact, are often approximate and fixed, basically because the human mind is not able to cope with the endless complexity and variety of reality. Although not a scientist, Lippmann anticipated some crucial points in further theories about stereotypes:

1. Simplification of reality is not casual, but rather motivated by precise cultural processes: stereotypes are part of a group's culture, and affect individuals in their interpretation of reality.
2. Stereotypes have a defensive function: they maintain the culture and the forms of social organization of a group, allowing individuals to keep their social position.
3. Stereotypes affect both experience and evaluation of experience – thus modifying interpretation – according to the dominant trends in a given culture.
4. Stereotypes are related to prejudices, being the *cognitive basis* of the latter. Thus, quoting from the Italian sociologist Bruno Mazzara, stereotype may be defined as "all information elements and beliefs concerning a given category of objects, constructed in a coherent and usually stable way, that support and reproduce a prejudice towards that category" (translated from Mazzara 1997: 19). That is the reason why, although basically in a cause-consequence relation, stereotype and prejudice are often considered as synonyms.

In addition to Lippmann's reflections, a few variables should be considered in order to provide a global understanding of stereotypes' functioning:

1. Degrees of *social sharing*: the way a certain mental representation is shared within more than one social group. In other words, stereotype X could be more common than stereotype Y, the latter being just the result of a limited cultural process.
2. Degrees of *generalization*: the way stereotypical characteristics that are attributed to a given group are considered to be spread homogeneously in that group. In other words, once a certain mental representation about a social group has crystallised, one may be convinced that every individual of that group possesses those characteristics.
3. Degrees of *rigidity*: This concerns the way a stereotype is or is not constant in time. Some mental representations are strongly related to limited events or temporal contexts, and tend to fade or disappear with time.

Now, as in the case of Francis Bacon's *idola mentis*, there are some common features in the way humans perceive and categorise their relationships with other animals. Individual variables are certainly significant, but these categorizations are mostly cultural, and thus it is possible to draw together constants and rules into a model.

The basic idea is to interlace causes and effects, i.e., where an attitude comes from, and where it tends to. That is, cultural presuppositions on the one hand (e.g., philosophical, psychological, or generally conventional), and practical consequences in actions/thoughts (according to the way they are *pro* or *anti* animals) on the other. However banal, this latter distinction (and its orientation towards one direction or the opposite) is in fact rather crucial: although one could say that the main tendency in anthropocentrism is to "favour" humans to the detriment of other animals (i.e., considering the former as an in-group in every respect, and opposing it to the out-group represented by the other animals), it seems that this is just partially true. In actual fact, a more or less fair share of what will be called *zoophobic* and *zoophilic* practical attitudes can be detected. As both these attitudes are opposite extremes of a continuum, they can be represented with a Cartesian axis. It must be pointed out that *Zoophilic* does not necessarily mean "good for animals", exactly as *Zoophobic* is not necessarily "bad" for them. We need to remember that these are stereotypical approaches, so any tendency is just *claimed* to be as such, while in fact its essence is totally unilateral, as it represents uniquely the perception/evaluation produced by one single group involved.

Let us return to the previously-mentioned value of causes, that is, what we may consider the cultural presuppositions of each attitude. Here, too, the suggestion is that two extremes on a Cartesian axis can be traced, although once again the connection is not between two discrete points. Although the present reflections are always expressed in terms of dualistic oppositions, it is rather important to point out that we are never dealing with black or white cases, but rather with complex grey areas, on which the most distinctive nuances are emphasized in form of oppositions. On one side, we may place a (self-styled) rational, or rationalistic, approach, here called *Scientific-Pragmatic*. On the other side, a so-called *Spiritual-Emotional* dimension. Once again, the attitudes are true in their intentions, but not necessarily in practice, so the qualifications noted on the previous paragraph apply here too.

From these dualities one can of course develop, one can develop a typical Cartesian plane, with one axis perpendicular to the other, and therefore reveal four different quadrants (Fig. 5.2). It is preferable to avoid, for these purposes, the use of the Greimasian square, for two reasons: there are not exactly contradictory and contrary terms here, which an orthodox application of the square would require; and in addition the main point is here conveying the image of two *continuous* axes, with analogical degrees of *more* and *less*, rather than a discreet, digital repartition.

Each quadrant is filled with the following categories of attitude:

1. Zoophobic/Scientific-Pragmatic: *Mechanism*, *Utilitarianism*, and *Zoomorphism*;
2. Zoophilic/Scientific-Pragmatic: *Conservationism/Environmentalism*, and *Zoophilism*;
3. Zoophobic/Spiritual-Emotional: *Hedonism/Indifferentism* and *Christianism* (not to be confused with Christianity, as it will be later explained);
4. Zoophilic/Spiritual-Emotional: *Anthropomorphism*, *New-Ageism*, *Pietism*.

At this point, it is important to qualify the use of these terms with some disclaimers. First, some of these words are neologisms invented for the occasion. All of them represent a mentality, an attitude that is nearly ideological, so we could regard them as a series of *-isms*. Secondly, as a consequence, individual relations with other animals normally involve more than one attitude: It is unlikely, for example, that you will find an example of a New-Ageist – life is not so simple as this. What seems possible is the appearance (and consequent analytical potential) of a New-Ageist attitude within the more complex relation that the person in question establishes with another animal (which may also include, for instance, a number of christianist and zoomorphist features). Thirdly, most of the terms that already

Fig. 5.2 Anthropocentric attitudes

exist are not meant literally. The term utilitarianism, for instance, does not exactly correspond with the ethical doctrine from the nineteenth-century, but rather refers to the kind of general attitude that identifies the good with the useful. Fourthly, these words should be considered as representing categories, which (again) can be divided into several sub-categories. Mechanism, for example, stands not only for Cartesianism, but also for other types of scientism or pseudo-scientism that share the basic assumptions (even in terms of simple frame of mind) of the Cartesian scheme.

What follows now is a brief and general (generic, more to the point) illustration of the ten categories just classified. It will include:

1. A brief presentation, with a synthetic introduction to the typology of attitude within a given category;
2. Historical and philosophical antecedents: in this case, too, the elements are simply mentioned and, most importantly, what will be pointed out is not necessarily the most ancient or the best known historical point of reference, but rather what could be considered the most influential one (provided that it is not always

possible to trace a precise cultural basis, since some attitudes originate mostly from social conventions and everyday interactions, and not from an identifiable intellectual speculation);

3. Functioning of stereotype: following the above-mentioned classification, the most influential variables for each category will be emphasised. In fact, it is evident that, in all the categories, all three variables – rigidity, generalization and social sharing – play a very precise role. Yet, in considering their dynamics, i.e., their increase or decrease, this task acquires more sense, allowing to sketch the evolution (or involution) of the given stereotype time after time;
4. Typology of prejudice: following Bacon's scheme, each category is put in one of the four possible *idola mentis* proposed by the British philosopher (this task, beyond its systematic advantages, also suggests a possible direction for eventual sociological analyses). Note that the choices made here are not to be considered exclusive, but simply dominant. If Mechanism is labelled as a form of *Idolum theatri*, that is because the latter is considered as the most relevant type of prejudice involved, not the *sole* one. Needless to say, an attitude like Mechanism includes at least some features from the categories of *Idola specus* and *Idola fori* (to start with, certain language habits should be mentioned). The same applies to the previously mentioned measuring of stereotypes: again, the goal was to underline relevance, not exclusivity;
5. Examples, selected mostly from everyday experience, which emphasise the type and degree of occurrence of a given category.
   Starting from the scientific/pragmatic and zoophobic quadrant, the first category to be analysed is *mechanism*. As already noted, such a term should not be understood only in its strictest sense. The Cartesian mechanism is certainly the theoretical point of reference, but in fact several post- pseudo- or para- mechanistic attitudes are quite often put into action (even in contexts totally foreign to the scientific model of the French philosopher).

The basic notion, as provided by Descartes himself, tends to be a devaluation or underestimation of non-human animals as sentient beings. According to Descartes, every living being, in its elementary functions, is comparable to a sort of automatic machine, humans included. What really marks the difference between humans and other animals is the possession of a soul. It is the soul that allows humans to have feelings and, more importantly, to be aware of them. Descartes finds no trace of hetero- or self-perception in non-human animals. Such a conception, in a mechanist view, implies that if a dog is beaten, it does not actually suffer any pain, and its cries do not represent pain, but rather "use", as with any other object beaten in a similar fashion, or, as the famous mechanistic example suggests, like a wheel that creaks when spinning.

This notion provides a mind-set that finds it acceptable to exploit non-human animals, for every "use" that is made of them is considered to be not perceived or felt. In addition, in the most extreme cases, it may be a mechanistic approach (in the sense meant here) that of labeling forms of philosophical research on animals that go beyond simple anatomical study as a merely a waste of time.

In the dynamics of stereotype functioning, one important aspect emerges: as time goes by, there is a significant decrease in the generalization variable of the mechanistic approach. Studies of language acquisition among primates made during the twentieth century, for instance, showed that, however problematic the idea itself may be – i.e., to attribute an extra-specific communication system to a given species – certain cognitive characteristics, starting from awareness, are evident in many non-human species. Hence, generalization decreases. Primatologists, for instance, try to protect apes from scientific experimentation, since they claim that apes are by all means sentient beings.

As for the theory of prejudice, it is not too difficult to identify mechanism as a perfect example of *idolum theatri*. Mechanism is a philosophical heritage with clear roots, and it certainly had the capacity to influence the following generations. Bacon himself highlighted a tendency in *idola theatri* to influence scientific research.

Examples of applied mechanism are animal experimentation, genetic engineering, pre-Darwinian zoology (or naturalism, as it used to be called), plus some behaviouristic models.

A second category to be examined in this analysis, included in the same quadrant as that of mechanism, is *Utilitarianism*. This way of thinking certainly has less pretensions from a scientific and philosophical point of view, and rather takes more practical and everyday positions. It regards the non-human animal as an economic good, a source of survival. As such, utilitarianism is not provided with specific historical bases, except for human evolution itself, starting from the first pre-historic animal breeding and hunting practices. The utilitarianistic process depersonalises the non-human animal, and perceives it as an object, a resource. Cows and sheep may thus be "cattle"; minks do not have any identity as animals, but just as furs; and so on (up to extreme points – like in the case of canned tuna fish – that results in the incapacity of recognizing certain species apart from their identity as economic products). Thus, a utilitarian study of animals is directed towards their exploitation as resources, not to the comprehension of their behaviour or constitution (and here, a typical case is pigs, the utilitarian approach to which designates their anatomy according to the parts useful for food production).

The evolution of this stereotype goes hand in hand with human technological progress. As a consequence, degrees of generalization and, above all, rigidity tend to increase. The idea is that animals are more and more subject to such forms of exploitation. Just a few decades ago, for instance, nobody could foretell such a massive development of intensive farms (increase of rigidity), while, within the same field, nobody could ever imagine finding ostrich meat in a supermarket (increase of generalization).

Like mechanism, utilitarianism is another example of *idolum theatri*, clearly related to tradition and its reinforcement throughout the centuries. Examples of utilitarian approaches are all the animal-related economic activities, from leather to furs, from cosmetics to fast-food. Anna Mannucci (1997: 13–56) provides a very interesting description of the typical day of a human being in relation to the use of products that, in one way or another, involve the exploitation of non-human animals (food, clothes, furniture, cosmetics, medicines, etc.).

Zoomorphism is the third and last category in the scientific/pragmatic and zoophobic quadrant. This bundle of attitudes is, properly and improperly, related to Darwinian and post-Darwinian theories, which are often re-interpreted according to syncretistic processes. Hence, zoomorphism is involved not only in its etymological form, but also as several forms of evolutionisms and positivisms common to scientific essays. Basically, zoomorphists accept the scientifically verified idea that *Homo sapiens* is in every respect a member of the animal kingdom, but the implications of this idea might end up having very little to do with science. In its most extreme forms, zoomorphism is also an easy justification for less-civilised behaviour, because this would be the result of natural "animal" instinct. All in all, zoomorphism is mostly a peculiar refashioning of the "law of the jungle", according to which life is an endless struggle for survival, in which the fittest live and the weakest die. As a consequence, it is a zoomorphistic claim, since we are the strongest species on earth – only one able to bring about the survival or the extinction of a species – it is only natural, so long as our strength allows us to dominate, to regard animals as being at our disposal.

Considering the increasing popularity of evolutionist theories – the view that humans are descended from apes is almost universally accepted, and not only in scientific environments – the zoomorphist stereotype has gained wider and wider currency, especially in the areas of rigidity and social sharing. Within the theory of prejudice, zoomorphism most naturally falls under the category of *idola fori*, i.e., those forms of prejudice whose consistency is mostly granted by interpersonal relation (such as language), which quite easily lead to simplified interpretations of reality. (Consider, for instance, how frequently the "instinct" black box is considered the cause of certain behavioural patterns).

Finally, examples of zoomorphism frequently appear in literary contexts; e.g., some Hemingway novels, Stevenson's *The Strange Case of Dr. Jekyll and Mr. Hyde* (and it is no counter-argument the fact that in these novels the zoomorphist allegory is indeed mostly and only an allegory). We also find zoomorphism in common myths; above all, the *homo homini lupus* principle, probably the best synthesis of the zoomorphist approach, not to mention hunting in all its manifestations, possibly the zoomorphist activity par excellence.

In the scientific/pragmatic and zoophilic quadrant, the first category is *Conservationism-Environmentalism*. Although undoubtedly similar in their interest towards the protection and improvement of the natural environment, conservationism and environmentalism definitely represent two different approaches to the problem, let alone different goals. The requirements of classification and categorization according to general rules are the only reason why they are dealt with as part of the same whole.

Having said that, the first thing worth remarking upon is the very young age of the category. It is hard to identify a precise historical and philosophical basis for it. More properly, it should be emphasized that the birth of conservationism and environmentalism is related to the very first ecological alarms, such as matters of endangered species, industrial revolution and similar. Of course, as ecological awareness developed, real conservationist and/or environmentalist theories and speculations took shape. Nevertheless, it is a widely-held opinion that a literary forerunner for environmentalism (as a movement, not particularly in the sense meant here) is Henry David Thoreau's *Walden*. This and similar works no doubt served as a "source of inspiration".

The main presupposition is that humans, as the most powerful beings on Earth, have the responsibility of saving bio-diversity from extinction, mostly in order to make Earth a better place for humans to inhabit. Such a position is thus anthropocentric, for it consists in of a sort of interspecific translation of the eighteenth century's figure of the enlightened despot: Nature is preserved mostly in order to preserve humans, which could easily disappear in as a consequence of a growing natural imbalance. In addition, such conceptions carry a distinction between species that "deserve" to be protected and those that do not. Since they mostly consider animal species as objects to be taken care of, it is no surprise that conservationists' and environmentalists' reaction to zoosemiotics is one of moderate skepticism.

In terms of stereotype functioning, conservationism-environmentalism found increased social approval especially since the second half of the last century (together, evidently, with the progressive worsening of environmental conditions); whereas in Baconian terms the category displays, in its most generic manifestations, the characteristics of the *idola fori*.

In the same quadrant with conservationism and environmentalism, we may also place *Zoophilism*. Here, animals are considered in terms of their aesthetic quality, like objects to preserve and improve until perfection is reached. Aesthetic purposes lead the zoophilists to cut their dogs' tails, to point their ears, to take them to exhibitions, to provide them with a pedigree, and so forth.

Similarly to utilitarianism, the process of animal depersonalization is rather evident. Once more, a living entity is treated more like an object – a good-looking object in this case, a useful object in the case of utilitarianism. Predictably enough, the zoophilist can identify him/herself only with "engineered", "pure-bred" animals, the sole category to be considered significant in his/her approach. A typical example is the accurate genetic selection of numerous species of songbirds, in order for them to perform catchier melodies from the human point of view.

Once more, it is not easy to identify the historical-philosophical basis of zoophilism, except for a general tendency, strengthened in the course of time, to modify nature and to genetically select plants and animals. It is nevertheless apparent that such a tendency became very strong during the times of the Scientific Revolution, Renaissance, and Baroque. Zoophilism bears witness to a new human being, one who no longer fears nature. Advances in genetic engineering were a natural consequence of these conceptions.

Examples of zoophilism, in addition to the above-mentioned pedigree dogs, are zoos and all other forms of captivity of animals (either dead or alive, or in between, as happens more often) for exhibition purposes; the collecting of butterflies and other insects; taxidermy; and all kinds of physical modifications of whatever species.

Starting from these examples, and their constantly wider currency, it is possible to deduce an important trend in the stereotype functioning of zoophilism: the increase of generalization. More and more species are involved in this form of human interest. To mention one example, just 20 years ago an animal shop selling iguanas and salamanders would have been regarded as fairly odd, while nowadays it seems quite trendy to have a reptile as a pet, in place of a banal cat or canary bird. As for the prejudice element, it seems appropriate to insert zoophilism in the category of *idola tribus*, since the human-animal relation is established according to totally human criteria (aesthetic criteria in this case). For example, songbirds are genetically bred to sing in ways more pleasing to humans, tail-cutting and ear-pointing in dogs are performed to make them more "aerodynamic", and so on.

Let us now turn to the intersection between the zoophobic and spiritual/emotional axes. The first category to be considered is *christianism*, so-named as to avoid confusing it with Christianity *tout court*. Christianism is comprised of attitudes related to and departing from (some, possibly many) Christian precepts, but it does not usually represent the whole complex Christian doctrine (although it may occasionally do so). Indeed, in its most embryonic form, the christianist attitude has much to do with Aristotelian ideas (particularly that of the *scala naturae*, which was already referred to in this book) and pre-Christian, Jewish traditions (such as ritual slaughtering). The most effective synthesis of the christianist conception of the human-animal relation can be traced to Genesis 9:2: "The fear and dread of you [Adam] will fall upon all the beasts of the earth and all the birds of the air, upon every creature that moves along the ground, and upon all the fish of the sea; they are given into your hands".

Christianism tends to consider the human being to be a divinely selected species, and the rest of Nature as a system explicitly created for human use and – within certain limits – abuse. Similarly to the Cartesian framework, non-human animals are regarded as having no soul, thus they are incapable of any form of sensibility beyond the basic levels of exploration and perception. Though Mechanism and Christianism share certain conceptions, there is a very crucial difference in their respective purposes and motivations. Mechanism claims a scientific basis for its ideas, whereas Christianist conceptions are framed as an exclusively spiritual philosophy.

Christianism, as a behavioural stereotype in the human-animal relationship, is becoming less and less consistent (decrease of rigidity), possibly because of a more and more scientific approach towards nature, thanks to significant revisions in certain interpretations of the Bible, and possibly thanks to a constant process of cultural secularization of our societies (although, judging from everyday news from certain parts of the world, it does not necessarily look this way). Obviously, christianism belongs to the category of *idola tribus*. Examples of applied christianism are certain superstitions and traditions, which vary according to place and time, and a generally allegorical interpretation of certain species (lamb, snake, etc.).

The next category, *hedonism/indifferentism* is totally alien to any philosophy, but at the same time it is widespread. As with conservationism/environmentalism, we may here include two different categories under the same label. The hedonist/indifferentist subject approaches animals simply the way s/he prefers to, according to his/her needs and wishes, without any ethical or other kinds of implication. Somewhat opposed to utilitarianism, this category identifies the good with the good, more than with the useful, where good stands for pleasant, tasteful, and so on.

It is thus the case of people who are not very much interested in the problem of the human-animal relation. This, among other things, means that Hedonist/Indifferentist behaviour may change according to contexts, people, and so forth. Of course, we are now dealing with a peculiar kind of category, because of its instability and variability, but Bacon typically considered total subjectivity in attitudes a fundamental character of *idola specus*. In terms of stereotype functioning, social sharing and generalization are on a very high level; although, possibly, the growing interest in environmental matters is causing a decrease in both of them.

There are numerous examples of hedonism/indifferentism: from the commercial, leisure-related use of animals (an aspect which is shared with utilitarianism), circuses, bullfights, the use of meat as a (often aesthetic) metaphor of luxury, and so on. It is hence appropriate to emphasise the consumer and luxury society as the basic foundation of Hedonism/Indifferentism.

The last quadrant zoophilic-spiritual/emotional, includes three categories. The first is *Anthropomorphism*. In this context, the term designates an ideal opposite of the above-illustrated zoomorphism. Whereas the latter defines a tendency to attribute every human characteristic to vague natural laws, in the case of anthropomorphism, the interpretation of non-human animal behaviour is based on exclusively human criteria, applied also in improper contexts. A typical example – which also constitutes a historical foundation of the whole category – is the long tradition of fables and fairy tales, in which animals are mostly used as allegories of humans: dressed in clothes, living in houses with doors and windows, speaking human languages, dealing with human problems, and so on.

This typology of attitudes is widely disseminated in the media, and also carries some peculiar consequences, starting with the red tartan pullovers worn by many dogs. On TV and in many books – especially those with didactic purposes – anthropomorphism is the most common descriptive technique. Yet, while on the one hand such a technique helps us understand certain topics (through processes of empathy), on the other hand it contributes to the implicit idea that animal-related studies are surely interesting, but not as serious as, say, physics, chemistry, or electronic engineering.

Examples of anthropomorphism are found in most cartoons, comics, and fiction in general that involves non-human animals. In particular, the commercialization of the main characters of this fiction – from Lassie to Donald Duck – is based on evidently anthropomorphic processes. The same also applies to any marketing form related to non-human animals, such as commercials advertizing dog food.

Given its deep-rooted relation with myths and fables, the anthropomorphist stereotype clearly belongs to the *idola theatri* category, while the growing level of generalization is definitely the most interesting aspect of its functioning as stereotype (e.g., a huge revisionist trend has introduced into popular imagery plenty of animals formerly not considered for such parts, such as the orca movie-star Willy, a member of a species that is also known as "killer whale").

Beside anthropomorphism we may place *New-ageism*, a very recent but popular category. The very rapid success of firstly New-Age philosophy then New-Age trends, followed by the uncontrolled New-Age commercialization has finally encouraged – at least in the Western world – a generally laudable tendency towards a more spiritual relationship between ourselves and nature, resulting in a huge business made up of astrology, exoticism, and books teaching us how to become happy. This tendency to exoticism has developed in the new-ageist culture a sort of new version of the seventeenth century *myth of the noble savage*, now applied to ancient civilizations, native populations, and non-human animals. In New-Age imagery, all these categories share a philosophical simplicity, a mystical and, at the same time, down-to-earth kind of wisdom, and that close contact with nature that Western civilization seems to have lost forever.

The origins of new-ageism, like New-Age itself, can be traced in many Eastern philosophies, Buddhism first of all. The attitude described here, though, is quite far from those foundations, so that it is more relevant to refer to recent authors, such as Richard Bach, Anthony De Mello, and others. Supermarket CDs incorporating natural sounds (including those of many animals) and ambient or classical music, other recent ideas like dolphin therapy, the already-mentioned books, and quite a few movies (e.g., the fourth episode of *Star Trek*) constitute the best examples of the new-ageist approach to non-human animals. Considering its essentially (originally) spiritual character, new-ageism probably belongs to the *idola theatri* category, and like all trends, is doomed to a decrease of rigidity.

The last category in this brief compendium is *Pietism*. To some extent, it can be considered the zoophilic side of christianism, mostly because they both share a clear religious basis. In this case, the main point of reference is the type of animal compassion inspired by figures like Saint Francis of Assisi. The concept, once more, is that non-human animals are weak and innocent victims of human power, and it is thus a human moral duty to be moved by a sense of charity and compassion, and to take care of these creatures. Results change, but evidently not the premises: the human being is still the master of nature, by divine choice, and animals are subordinate entities which, in this case, the "master" takes care of.

Although strongly based on very personal experiences and interpretations, thus belonging to the *idola specus*, the pietistic attitude is quite common, and is gaining wider social acceptance. Significant examples include the habit of feeding city animals in a park or, in particular, certain movies or novels featuring animals. The clearly anthropocentric nature of pietism is emphasised in those cases when people rescue dog or cat pups, which have been abandoned by their mothers, because of diseases or animal over-population. It is evidently a gratuitous application of human ethical values to situations which, for ages, have been based on the opposite principle.

5.5 The Concept of Biocentrism

The question to reflect upon now is whether a discipline like zoosemiotics (or others) should resign itself to using anthropocentric methodological models, and therefore exclusively interpreting nature and natural phenomena according to this paradigm.

Well, hopefully not. Hopefully, a *bio*centric approach is possible, if one is able to:

1) emphasize and motivate the necessity of adopting a biocentric theoretical paradigm within the zoosemiotic discussion;
2) define the concept of biocentrism, in the light of the different theoretical interfaces;
3) establish a relationship between biocentrism and other methodological approaches to Nature; and
4) build up a semiotically-based model of biocentrism.

However, one last step needs to be made before focusing the attention exclusively on the biocentric model. As animal activism was often mentioned in this companion, one may justly ask where to locate animalism in this analysis. It seems quite clear that such a complex ideology as animalism cannot be confined solely to the zoophilic categories (either pragmatic or emotional) which have been illustrated above. First, it is methodologically reductive to interpret every nuance of animal activism only within the anthropocentric dimension. Indeed, animalism – at least part of it – appears as a conscious attempt at emancipation from anthropocentric dynamics. Secondly, if we compare the zoophilic attitudes listed in the previous paragraphs with all the manifestations of animal activism that we experience every day, we notice that several pieces are missing from the puzzle. Where, for instance, should we locate all those associations that adopt very explicit and "creepy" images for their campaigns? And what about semi-terrorist activities such as those of the Animal Liberation Front? It is rather clear that the above-illustrated model is not fully satisfying.

At the same time, there is a certain difficulty in locating some forms of animalism straightforwardly within the biocentric approach, and the reasons for this will emerge quite clearly during the discussion. Here, it should be anticipated that the dynamics of some forms of animalism tend centrifugally towards the human point of reference (one cannot, for sure, list A.L.F. semi-terrorism as a pro-human attitude), and – in this sense – such forms cannot be classified as anthropocentric. However, the point is that, because of their very nature, such manifestations cannot be called biocentric either, since their "action" is conceptually an escape *from* something (i.e., the human being as a centre of universe), but not necessarily *towards* something else (i.e., the zoological, rather than simply anthropological, dimension of life). Still in the same example, one can easily define ALF semi-terrorist actions as actions against the human being, but animal activists themselves, the moderate and non-violent ones, often criticise those actions as being hardly ever *in favour* of other animals.[4]

[4] A typical case are ALF's raids in fur breeding farms, where hundreds of specimens (like minks, foxes, etc.) that were born and have spent their whole lives in cages, are liberated. As they are not used to a life in the wild, most of them starve to death very soon or die for other reasons. Now, these raids certainly represent enormous losses (not only economically speaking) for a fur farm,

It is thus necessary to create a theoretical "niche" in between anthropocentrism and biocentrism, something that illustrates more precisely the portion of animalist attitudes that are left out of the anthropocentric area. While waiting for a less complicated term, we could name this niche *anthropodecentrism*, in order to circumscribe both the point of departure of such an attitude (the human being), and the direction it takes (a centrifugal one in relation to the human being, but not necessarily a centripetal one in relation to other animals or nature, in general).

Anthropodecentric are thus the following categories:

1) Shockism – This is probably the type of animalism most people think about when hearing the word. Considered rather extreme, both ideologically and operatively, shockism is a typical attitude in those associations that adopt shocking and explicit communication and promotion strategies. These normally aim to describe accurately (thus condemn very strongly) the violent forms of exploitation towards non-human animals, with the implicit goal of provoking a certain feeling of guilt in people. Such is the case in much animalist literature, promotion campaigns (fliers, ads, etc.), and simple conversations. It shall be useful to remind that shockistic attitudes are also common within other forms of social and civil protest, like pacifism, feminism, anti-racism, and so on.
2) Terrorism – The already-mentioned Animal Liberation Front, just as some other activists that operate privately, belongs to a category that once and for all should be clearly distinguished from all other forms of animalism, including shockism. All legally-acknowledged animal activist associations have always condemned and dissociated themselves from those terrorist and fanatic actions carried out under the auspices of the ALF. It is important not to confuse radical but definitely pacifist associations with illegal organisations that base their actions on violence and the so-called strategy of tension. The basic idea of these latter organisations is that more respect towards non-human animals can be achieved only through threatening, damaging, and thus terrifying people and institutions with diverse acts of sabotage (towards pharmacological laboratories, fur farms, and so forth).
3) Misanthropism – It is not rare, in fact quite frequent, that animalist attitudes are animated first of all by a disillusionment (or a straight hostility) towards humankind. Misanthropism is a "the more I know men, the more I love animals" philosophy. The misanthropist sees in the non human animal an uncorrupted creature, innocent victim of human arrogance and stupidity. His/her support of animal rights is in this case also and mostly due to a wish for damaging structures and manifestations of the human domain.

---

and – because of this – we can definitely regard them as actions performed towards an opposite direction than the anthropocentric one. At the same time, however, it is difficult to conceive them as actions "in favour" of the liberated animals, at least when taking the idea of "life" as a point of reference, not to mention that – ecologically speaking – the sudden liberation of a huge amount of specimens that are basically alien to a given ecosystem, may impact on this latter in quite a traumatic way. Of course, one can always object that few hours of freedom are always better than a whole life in a small cage, but perhaps – for the purposes of this very argument – this is not the point.

4) Reactionism – Finally, there are also those who see in the violation of animal rights the consequence of an excessive, unwise human technological development. The reactionist maintains that some manifestations of human evolution are – apart from being useless and dangerous for humankind – unacceptable in terms of the exploitation of animals and natural resources. S/he generally hopes for the return to a past that is more respectful of the ecosystem, where – for instance – intensive breeding is abolished.

It soon seems clear that the last two categories are forms of – so to say – indirect animalism, since their manifestation belongs to a way of interpreting reality that is somewhat wider, where the promotion of animal rights is an important but not exclusive part. This applies particularly to the reactionist attitude.

To these categories, one clearly has to add the previously-described anthropocentric but zoophilic ones. For example, the WWF is an animalist association of the conservationist type; and "animalist" we should also call our pietistic neighbour who takes care of feeding the stray cats of the neighbourhood.

One last remark, before proceeding: as is true of all forms of classification, this entire model (anthropocentrsim and anthropodecentrism) is too simplistic. Very different attitudes may be grouped in a single category, without really being convicing. However, it is not productive to set apart the *raison d'être* of such schemes, i.e., to point out the most evident characteristics in the various attitudes, then group them by common denominators. Every human being has its own way of perceiving and interpreting the non-human animal: this is beyond discussion. All that was attempted here was to illustrate more specifically socio-cultural traits, applying to both scientific and everyday discourse.

Another evident limitation of the model is its western-centric tendency. The attitudes analysed and their categorization refer almost exclusively to western contexts (European, American, Christian, etc.). It is therefore important to take this aspect into account, in the perspective of applying this model to further studies.

Once these aspects have been pointed out, it is time to focus the attention on the theoretical goal of this part of the book, i.e., so-called *biocentrism*.

5.5.1 Why Biocentrism?

An issue such as biocentrism (this term being preferable to zoocentrism, as it is more important to stress the centrality of the concept of *life*) deserves more than the general scrutiny that will be given to it in this section, most of all because concepts that are complex and crucial in order to understand and regulate the human role in the ecosystem, could be made to seem a little generic and banal.

Having accepted that, one can move on to other considerations. As a first step, the concept of anthropocentrism must be reconsidered. Indeed, the real core of the biocentrism-anthropocentrism problem is not, as one might expect, the dichotomy between instrumental and intrinsic value. Anthropocentrism is certainly an interpretation of Nature on the basis of its instrumental value, but this conception is not necessarily opposed to the idea of intrinsic value.

> The concept of intrinsic value can be interpreted at least in three ways: a) as a quality or property that human conscience attributes to something, and that characterises this object as possessing a value in itself, not related solely to human interests, or solely to the value of the conscience that gives birth to it; b) as a property that emerges from the relation between an event and a conscience, in a relational-phenomenological fashion; c) in Platonic terms, as a totally objective value, inscribed in the objects themselves and independent of human evaluation (e.g., Nature's value existed before human conscience and will keep on existing when the latter disappears") (translated from Bartolommei 1995: 42).

In environmental philosophy, these three interpretations may be separated or combined. If kept separate, mostly excluding the third, ontological interpretation, there is little contradiction with a default anthropocentric attitude. It is not problematic for an anthropocentrist to accept that Nature, in whatever form or manifestation, may have value in itself, if a distinction is made between humans as *source* and as *centre* of values (Bartolommei 1995: 43). In a way, default anthropocentrism is nothing else than anthropogenesis: what is said or thought by humans starts from the human interpretation of reality, but – and this is the point – does not necessarily have to be confined to humans. What is said and thought by the Arsenal fan in a previous paragraph is not a priori "Arsenal-centric".

> In other words, the condition of the subject who speaks must not be confused with the contents of her statements: anthropocentrists should easily acknowledge that if it is true that there is no value without someone valuing, it is also true that the value of the object is not reduced to the sole value it has for the one who values, nor to the value of this latter herself. In short, it is one thing if anthropocentrism is meant as negation of values independent and separated from human acts of evaluation; anthropocentric prejudice is another thing (i.e., the idea that everything on Earth is only a function of human values) (translated from Bartolommei 1995: 43–44).

The main implication here is that, hermeneutically speaking, Nature has to be divided into levels, organised as follows (not before recalling once again the golden rule that these forms of categorization digitalize a reality which is in fact analogical: there will be no clear-cut transition from one level to another, but simply a gradual merging of each of them into one another). Beyond a general common basis, which we might call *ecological*, whose constraints – such as being subject to gravity – are shared by just everything on Earth, there is a second, *biological* level, in which every living being is included. Eating and reproducing, for instance, typify humans and birds, insects and flowers, and so on. The third level is called *zoological* or *transpecific* and concerns aspects that are shared by the whole animal kingdom. Fourthly, there is the level of characteristics that concern a single species (*species-specific*). An activity such as printing a street-map must be considered an exclusively human skill, just as giving the exact position of a flower by a figure-of-eight dance is a skill specific of bees. From the next level on, the course is quite clear for humans, but has yet to be defined more clearly for other species. This is because the categories are now species-specific, and each species has developed a unique process.

For most animals, including humans, this level is mainly social, but many species are not organized into societies at all. Thus, it seems fair to stop here.

There is no need to say that the study of animals in a biocentric perspective can be very useful for the comprehension of many species-specific human aspects, as well.

Coming back to the theoretical approaches previously illustrated, it becomes clear that both discontinuity and gradualism present a few basic problems. The discontinuist attitude is by definition opposed to the intent of any biocentric research, which in fact defends the thesis of the specificity but not the *special* uniqueness of the human being. At the same time, the gradualistic attitude promotes a monolithic interpretation of evolution, as if life developed in a unique and indivisible way. Both approaches are evidently problematic for the goals set out here. For these reasons, and for others to be considered later, it becomes clear that a biocentric approach to zoosemiotics must necessarily be pluralistic.

5.5.2 The Pillars of Biocentrism

In the light of these premises (and keeping in mind several reflections distributed across this companion), one can now offer a (almost manifesto-like) formulation of the main methodological and theoretical elements that characterize a biocentric form of research:

1. *An Umwelt-implemented Darwinian approach*, the way it was defined in Chapter 1 (Section 1.1.4). To conceive animals in the light of Umwelt theory means at the same time to acknowledge biological, zoological and species-specific traits. It means, in other words, to defend exactly that thesis of a pluralistic interpretation of Nature, which takes into account the biological foundations of certain behavioural patterns, and the autonomous and peculiar (but never "unique") developments of other ones. Common bases and specific development thus must both be taken into account.
2. Semiosis is the result of an interaction between a subject and an object, between a structure and a counter-structure, between a receptor and a carrier of meaning. These two parts are in constant and reciprocal informational exchange. In fact, the exchange itself is the real generator of any semiotic phenomenon, since the latter would simply not exist if the subject was not affected by it and did not affect it. Any zoosemiotic research, from bee-dance to whale songs, should take into account such a conception, otherwise it risks perverting the essence of the phenomenon itself of semiosis.
3. The acknowledgment and theoretical consideration of the *Zoosemiotic Universals*, the way they were discussed in Chapter 2 (see also Cimatti 1998: 179–190). These include: *Sociality*; *Arbitrariness of categorizations*; *Mind*; *Principle of complexity*; *Ritualization*; *Self- and Hetero-perception*; *Syntax of perception*. One must also add the eleven points of animal ontology as defined

in San Martin and Pintos (2001), and again discussed in Chapter 2. No research can be fully emancipated from anthropocentrism if it does not acknowledge and accept at least these departure points.

4. Adoption and application, as methodological standpoints, of *critical anthropomorphism and a purely semiotic* "canon", the way these formulations were analyzed in Chapter 3 of this book. In this sense, the suggestion is that we should find a wise balance between what was called an etic and an emic approach. If the scientific investigation on animal behavior certainly requires techniques and strategies in the emic direction, the etic part (as based on empathy and recognition) will be given a fair role of "advisor". A semiotic canon, moreover, will help avoiding rhetorical forms of aprioristic reductionism as the ones promoted by the Morgan's Canon.

These suggestions must not be interpreted as bold prophecy-like statements. The (very humble) opinion maintained here is that strategies like a pluralistic mind-frame, the acknowledgement of the intrinsic value of Nature, the practical and rhetorical acceptance of interpretive categories such as the universals or animal ontology are all cognitive factors that seriously call into question (and weaken) the anthropocentric tendencies to binarize the human-animal relationship. The thesis to be defended was thus that (1) there can and possibly must be an ethical agenda in zoosemiotics; and (2) this agenda can be *proactive* without being *militant*, maintaining the rigorous modes and objectives of a scientific program.

There is a curious inclination, within the scientific environment, to consider the explication of an opinion as a deterrent to the objectivity and reliability of research, or even as a shift towards pure propaganda. Such an assumption is possibly incorrect because approximate and, in fact, not fully aware of the epistemological nature of various sciences. It is approximate because propaganda is exactly an *introverted*, rather than extroverted, form of rhetoric: scientific and other forms of propaganda work precisely when the receiver is *not* aware of being exposed to it, and therefore his/her naturally critical attitude is kept quiet. An explicit (and, let us not forget, honest) set of opinions, like those delivered in Tarasti (2005) or Stefani (1985) (and in the present section), is on the contrary a form of rhetoric that consciously wants to *solicit* and *stimulate* a critical reaction, in that its main goal is to encourage critical inquiry with respect to a given topic. Furthermore, the biases of such an enterprise are much easier to spot than would be the case with a professionally propagandistic approach.

The criticism also appears to forget that several fields of inquiry, particularly the so-called human sciences, in fact originated as prescriptive disciplines. Philosophy and ethics do more than go hand in hand: they are often synonyms. And if semiotics is, as it is, a philosophical discipline (was it not Sebeok who defined semiotics as the name that the twentieth century gave to philosophy?), then it is probably true that as semioticians we have an ethical imperative. The dynamic tension that Existential Semiotics, Globality of Languages, Semioethics, hopefully Anthrozoosemiotics, and any other theories created in the attempt of *ethicalizing* the semiotic agenda, offers – or so it is claimed here – the right challenge for the future of this discipline.

It is not necessary that we all agree on each point, and in fact if we compare the aforementioned four models, differences are definitely more numerous than similarities. The important point is that these issues are addressed, humbly but firmly, and at the same time that sight is not lost of the scientific level of argumentation, which shall always be high enough for ethical considerations not to be dismissed as leading to *bad* science.

References

Akimushkin I (1988) Ethology. Mir, Moscow
Aldis O (1975) Play fighting. Academic, New York
Allport GW (1954) The nature of prejudice. Addison-Wesley, Reading, MA
Arluke A, Sanders C (eds) (2009) Between the species: a reader in human–animal relationships. Pearson Education, Boston
Armstrong EA (1963) A study of bird song. Oxford University Press, London
Arnold K, Zuberbühler, K (2006) Language evolution: semantic combinations in primate calls. Nature 441:303
Austern LP (1998) Nature, culture, myth and the musician in early modern england. J Am Musicol Soc 51(1):1–47
Baber C (2003) Cognition and tool use – forms of engagement in human and animal use of tools. Taylor & Francis, New York
Bacon F (1620) Instauratio magna. John Bill, London
von Baer KE (1828–1837) Entwickelungsgeschichte der Thiere: Beobachtung und Reflexion, 2 vols. Bornträger, Königsberg
von Baer KE (1876) Ueber Darwins Lehre, in his Studien auf dem Gebiete der Naturwissenschaften, 2 vols. Schmitzdorff, St. Petersberg
von Baer KE (1886/1986) Autobiography of Dr. Karl Ernst von Baer. Science History, New York
Bankov K (2004) Infinite semiosis and resistance. In: Tarasti E. (ed) From nature to psyche. Proceedings from the ISI summer congresses at Imatra in 2001–2002. International Semiotics Institute, Helsinki/Imatra, pp 175–181
Barbieri M (1985) The semantic theory of evolution. Harwood, New York
Barbieri M (2003) The organic codes: an introduction to semantic biology. Cambridge University Press, Cambridge
Barbieri M (ed) (2006) Introduction to biosemiotics. Springer, Berlin
Barbieri M (2008) Biosemiotics: a new understanding of life. Naturwissenschaften 95(7):577–599
Barbieri M (2009) A short history of biosemiotics. Biosemiotics 2(2):221–245
Barthes R (1957) Mythologies. Editions de Seuil, Paris
Barthes R (1967) Elements of semiology. Hill & Wang, New York
Bartolommei S (1995) Etica e natura. Laterza, Roma-Bari
Bateson G (1969) Metalogue: what is an instinct? In: Sebeok T, Ramsay, A (eds) Approaches to animal communication. Mouton, The Hague, pp 11–30
Bateson G (1972) Steps to an ecology of mind. University of Chicago Press, Chicago, IL
Bateson G (1976) A theory of play and fantasy. In: Bruner JS, Jolly, A, Sylva, K (eds) Play: its role in development and evolution. Basic Books, New York, pp 119–29
Battaglia L (1997) Etica e diritti degli animali. Laterza, Bari-Roma
Batteux C (1746) Les beaux arts: réduits à un mème principe. Durand, Paris
Bayle P (1826/1697) Historical and critical dictionary. Hunt and Clarke, London

D. Martinelli, *A Critical Companion to Zoosemiotics*, Biosemiotics 5,
DOI 10.1007/978-90-481-9249-6, © Springer Science+Business Media B.V. 2010

Beck B (1980) Animal tool behavior: the use and manufacture of tools by animals. Garland, New York

Bekoff M (1975) The communication of play intention: are play signals functional? Semiotica 15:231–239

Bekoff M (1995) Cognitive ethology and the explanation of nonhuman animal behaviour. In: Meyer JA, Roitblat, HL (eds) Comparative approaches to cognitive science, MIT Press, Cambridge, MA, pp 119–150

Bekoff M, Jamieson D (eds) (1990) Interpretation and explanation in the study of animal behaviour. Westview Press, Boulder, CO

Bentham J (1789/1996) Introduction to the principles of morals and legislation. In: Burns JH, Hart HLA (eds) The collected works of Jeremy Bentham. Oxford University Press, Oxford

Black M (1968) The labyrinth of language. Pall Mall, London

Bonner JT (1980) The evolution of culture in animals. Princeton University Press, Princeton, NJ

Bopry J (2002) Semiotics, epistemology, and inquiry. Teach Learn 17(1):5–18

Borror DJ (1960) The analysis of animal sounds. In: Lanyon WE, Tavolga WN (eds) Animal sounds and communication. American Institute of Biological Sciences, Washington

Bouissac P (ed) (1998) Encyclopedia of semiotics. Oxford University Press, New York

Bouvet J-F (2000) La stratégie du caméléon : de la simulation dans le monde vivant. Edition du Seuil, Paris

Box HO (1973) Organization in animal communities. Butterworth, London

Bradbury J, Verenkamp S (1998) Principles of animal communication. Sinauer Associates, Sunderland, MA

Brockmann HJ (1985) Tool use in digger wasps. Psyche 92:309–30

Brooks D, Wiley, EO (1986) Evolution as entropy. Toward a unified theory of biology. University of Chicago Press, Chicago

Brooks D et al (1989) Entropy and information in evolving biological systems. Biol Philos 4: 407–432

Broughton WG (1963) Method in bio-acoustic terminology. In: Busnel RG (ed) Acoustic behaviour of animals. Elsevier, London, pp 3–24

Brower LP (ed) (1988) Mimicry and the evolutionary process. University of Chicago Press, Chicago

Brown R (1989) Group processes: dynamics within and between groups. Basil Blackwell, Oxford

Bruner J (1986) Actual minds, possible worlds. Harvard University Press, Cambridge, MA

Burghardt GM (1984) On the origins of play. In: Smith P (ed) Play: in animals and humans. Basil Blackwell, Oxford, pp 5–41

Busnel RG (ed) (1963) Acoustic behaviour of animals. Elsevier, London

Byrne RW, Whiten, A (1991) Computation and mindreading in primate tactical deception. In: Whiten A (ed) Natural theories of mind. Basil Blackwell, Oxford, pp 127–141

Cavalieri P (1999) La questione animale. Bollati Boringhieri, Torino

Chandler D (2002) Semiotics: the basics. Routledge, London

Chauvin R, Bernardette, M (1977) Le monde animal et ses comportements complexes. Plon, Paris

Cheney DL, Seyfarth RM (1990) How monkeys see the world: inside the mind of another species. University of Chicago Press, Chicago

Cimatti F (1998) Mente e linguaggio negli animali. Carrocci, Roma

Cimatti F (2000) Nel segno del cerchio. L'ontologia semiotica di Giorgio Prodi. Manifestolibri, Roma

Clayton NS, Bussey TJ, Dickinson A (2003) Can animals recall the past and plan for the future? Nat Rev Neurosci 4(8):685–691

Cobley P (ed) (2009) The routledge companion to semiotics. Taylor & Francis, London

Coetzee JM (1999) The lives of animals. Princeton University Press, Princeton

Correia SP, Dickinson A, Clayton NS (2007) Western scrub-jays anticipate future need states independently of their current motivational state. Curr Biol 17:856–861

Cousteau J, Paccalet Y (1987) Il pianeta delle balene. Fabbri, Milano

Cowles JT (1937) Food-tokens as incentives for learning by chimpanzees. Comp Psychol Monogr 14:1–96

Cullen JM (1972) Some principles of animal communication. In: Hinde RA (ed) Non-verbal communication. Cambridge University Press, Cambridge, pp 101–122

Darwin C (1871) The descent of man and selection in relation to sex. Murray, London

Darwin C (1872) The expression of emotions in man and animals. Murray, London

Davila Ross M, Owren MJ, Zimmermann E (2009) Reconstructing the evolution of laughter in great apes and humans. Current Biol 19(13):1106–1111

Dawkins R, Krebs JR (1978) Animal signals. In: Krebs JR, Davies NB (eds) Behavioral ecology. Blackwell, Oxford, pp 282–309

Deely J (ed) (1986) Frontiers in semiotics. Indiana University Press, Bloomington, IN

Deely J (1990) Basics of semiotics, 1st edn. Indiana University Press, Bloomington, IN

Deely J (2000) The impact of semiotics on philosophy. Paper presented at Helsinki University at the 1st annual hommage to Oscar Parland, 1 Dec 2000

Deely J (2001) Umwelt. Semiotica 134(1/4):125–135

Deely J (2005) Defining the semiotic animal – a postmodern definition of human being. Tip-Top, Sofia

Deely J (2009) Basics of semiotics, 5th edn. Tartu University Press, Tartu

Delalande F (1991) Le condotte musicali. CLUEB, Bologna

Dennett DC (1996) Kinds of minds. Hypercollins, New York

De Waal FBM (1996) Good natured: the origins of right and wrong in humans and other animals. Harvard University Press, Cambridge, MA

Ditadi G (ed) (1994) I filosofi e gli animali. Isonomia, Este

Divers J (2002) Possible worlds. Routledge, London

Dobzhansky T (1962) Mankind evolving: the evolution of the human species. Yale University Press, New Haven, CT

Doyle AC (1887/1999) A study in scarlet. Demetra, Verona

Dreifus C (1998) She talks to apes and, according to her, they talk back. New York Times, 14 April: 15

Dyer FC (2002) Biology of the dance language. Ann Rev Entomol 47:917–949

Earle SA (1979) The gentle whales. Nat Geo 155(1):2–25

Eco U (1968) La struttura assente. Bompiani, Milano

Eco U (1976) A theory of semiotics. Indiana University Press, Bloomington, IN

Eco U (1997) Kant e l'ornitorinco. Bompiani, Milano

Eibl-Eibsfeldt I (1951) Nahrungserwerb und Beuteschema der Erdkröte. Behaviour 4:1–35

Eibl-Eibsfeldt I (1970) Ethology – the biology of behavior. Holt, Rinehart and Winston, New York

Eibl-Eibsfeldt I (1971) Love and hate: the natural history of behavior patterns. Methuen, London

Ellis MJ (1973) Why people play. Prentice-Hall, Englewood Cliffs, NJ

Emery NJ, Clayton NS (2008a) How to build a scrub-jay that reads minds. In: Itakura S, Fujita K (eds) Origins of the social mind: evolutionary and developmental views. Springer, Berlin

Emery NJ, Clayton NS (2008b) Imaginative scrub-jays, causal rooks, and a liberal application of Occam's aftershave. Behav Brain Sci 31:134–135

Engels F (1896) Anteil der Arbeit an der Menschwerdung des Affen. In Die Neue Zeit XIV/2: 545–554

Esch H (1967) The evolution of bee language. Sci Am 216(4):97–104

Evans EP (1906) The criminal prosecution and capital punishment of animals. Faber & Faber, London

Fagen R (1981) Animal play behavior. Oxford University Press, New York

Fagen R (1993) Primate juveniles and primate play. In: Pereira ME, Fairbanks LA (eds) Juvenile primates: life history, development and behavior. Oxford University Press, New York, pp 183–96

Fisher HE (1992) Anatomy of love: the mysteries of mating, marriage, and why we stray. Simon & Schuster, New York

Fisher JA (1990) The myth of anthropomorphism. In: Bekoff M, Jamieson D (eds) Interpretation and explanation in the study of animal behaviour. Westview, Boulder, CO, pp 96–116
Fodor JA (1986) Why paramecia don't have mental representationsMidwest StudPhilos 10:3–23
Fossey D (1972) Vocalizations of the mountain gorilla (Gorilla gorilla beringei). Anim Behav 20:36–53
Fossey D (1974) Observations on the home range of one group of mountain gorillas (Gorilla gorilla beringei). Anim Behav 22:568–581
Fossey D (1983) Gorillas in the mist. Houghton, Boston, MA
Fouts RS (1999) Next of kin. William Morrow, New York
Freud S (1899) Die Traumdeutung. Franz Deuticke, Leipzig and Vienna
Freud S (1959) Beyond the pleasure principle. (trans: Strachey J) Norton, New York
Frings H, Frings M (1964) Animal communication. Blaisdell, New York
von Frisch K (1967) The dance language and orientation of bees. The Belknap Press of Harvard University Press, Cambridge, MA
von Frisch K (1974) Animal architecture. Harcourt Brace Jovanovich, New York
von Frisch K (1992) Decoding the language of the bee In: Lindsten J (ed) Nobel lectures, physiology or medicine 1971–1980. World Scientific, Singapore, pp 76–88
Frith CD, Frith U (1999) Interacting minds: a biological basis. Science 286:1692–1695
Galdikas B (1995) Reflections of eden: my years with the orangutans of borneo. Little, Brown & Company, Boston, MA
Gallese V (1998) Mirror neurons, from grasping to language. Conscious Bull Autumn Issue:3–4
Gallese V, Goldman A (1998) Mirror neurons and the simulation theory of mind-reading. Trends cogn sci 2:493–501
Gallup GG, Jr (1970) Chimpanzees: self-recognition. Science 167:86–87
Galtung J 1969. Violence, peace and peace research. JPeace Res 6(3):167–191
Gardner AR et al (1994) The ethological roots of culture. Kluwer, Norwell, MA
Gardner AR, Gardner BT (1969) Teaching sign language to a chimpanzee. Science 165:664–672
Gardner AR, Gardner BT, Van Cantfort TE (eds) (1989) Teaching sign language to a chimpanzee. State University of New York Press, Albany, NY
Genette G (1972) Figures III. Seuil, Paris
Giannattasio F (1998) Il concetto di musica. Bulzoni, Roma
Von Glasersfeld E (1978) Les chimpanzés et le langage. La Recherche 9:725–732
Goodall J (1964) Tool-using and aimed throwing in a community of free-living chimpanzees. Nature 201:1264–1266
Goodall J (1968) The behavior of free-living chimpanzees in the Gombe stream reserve. Anim Behav Monogr 1(3):161–311
Goodall J (1971) In the shadow of man. Houghton Mifflin, Boston, MA
Goodall J (1986) The chimpanzees of Gombe: patterns of behavior. The Belknap, Cambridge, MA
Gould SJ (1986) The local map of honey bees: do insects have cognitive maps? Science 232: 861–863
Grafen A (1990) Biological signals as handicaps. J Theor Biol CX-LIV:517–546
Greimas A (1987) On meaning: selected writings in semiotic theory. Frances Pinter, London
Grier JW, Bunk T (1992) Biology of animal behavior. Mosby-Year, United States
Griffin D (1958) Listening in the dark: the acoustic orientation of bats and men. Yale University Press, New Haven, CT
Griffin D (1976) The question of animal awareness. Rockefeller University Press, New York
Groos K (1898) The play of animals. Chapman and Hall, London
Habermas J (1998) The inclusion of the other. studies in political theory. MIT Press, Cambridge MA
Hailman JP (1977) Optical signals. Indiana University Press, Bloomington, IN
Hamilton WJ, Marler P (1966) Mechanisms of animal behaviour. Wiley, New York
Harrington FH, Mech DL (1979) Wolf howling and its role in territroy maintenance. Behaviour 68(3–4):207–249
Harris C L (1996) Concepts in zoology, 2nd edn. HarperCollins College Press, New York

Hart S (1996) The language of animals. Henry Holt and Company, New York
Hartshorne C (1973) Born to sing: an interpretation and world survey of bird song. Indiana University Press, Bloomington, IN
Hawkins J (1776) A general history of the science and practice of music, London. Modern edn. Akademische Druck, Graz p 1969
Hausdorff F, As Paul Mongré (1897) Sant'Ilario. Gedanken aus der Lanschaft Zarathustras. C.G.Naumann, Leipzig
Head M (1997) Birdsong and the origins of music. J Roy Music Assn 122(1):1–23
Hebb DO (1949) The organization of behaviour. Wiley, New York
Hediger H (1965) Man as a social partner of animals and vice-versa. Symp Zool Soc Lon 14: 291–300
Hediger H (1968) The psychology and behavior of animals in zoos and circuses. Dover, New York
Hediger H (1974) Communication, between man and animal. Image Roche (Basel) 62:27–40
Hediger H (1981) The Clever Hans phenomenon from an animal psychologist's point of view. Ann NY Acad Sci 364:1–17
Herrick P (1999) The many worlds of logic. Oxford University Press, Oxford
Hicks SJ, Sales BD (2006) Criminal profiling. American Psychological Association, Washington, DC
Hill JH (1980) Apes and language. In: Sebeok TA, Umiker-Sebeok J (eds) Speaking of apes. Plenum, New York
Hockett CF (1960a) The origin of speech. Sci Am 203:89–97
Hockett CF (1960b) Logical considerations in the study of animal communication. In: Lanyon W, Tavolga WN (eds) Animal sounds and communication. American Institute of Biological Sciences, Washington, pp 392–430
Hoffmeyer J (1995) The semiotic body-mind. In: Tasca N (eds) Essays in honor of Thomas A. Sebeok. Almeida Foundation, Porto, pp 367–383
Hoffmeyer J (1996) Signs of meaning in the universe. Indiana University Press, Bloomington, IN
Hoffmeyer J (1997) Biosemiotics: towards a new synthesis in biology? Eur J Semiot Stud 9(2):355–376
Hoffmeyer J, Emmeche C (1991) Code-duality and the semiotics of nature. In: Anderson M, Merrell F (eds) On semiotic modeling. Mouton de Gruyter, Berlin, pp 117–166
Hölldober B, Wilson EO (1990) The ants. Harvard University Press, Cambridge, MA
Holley AJ (1993) Do brown hares signal to foxes?. Ethology 94:21–30
Honig WK (ed) (1971) Animal memory. Academic Press, New York
Horn E (ed) (1983) Multimodal convergences in sensory systems. Fischer, Stuttgart
Hsiao HH (1929) An experimental study of the rat's 'insight' within a spatial complex. Univ Calif Publ Psychol 4:57–70
Hume D (1740, mod edn 2003) A treatise of human nature. Dover, New York
Huxley JS (1923) Courtship activities in the red-throated diver Colymbus stellatus. J Linnean Soc 53:253–292
Ingold T (1988) What is an animal? Unwin Hyman, London
Jakobson R (1960) Concluding statement: linguistics and poetics. In: Sebeok TA (ed) Style in language. Cambridge University Press, Cambridge, pp 350–75
Jakobson R (1963) Essais de linguistique générale. Minuit, Paris
Janik VM et al (2006) Signature whistle contour shape conveys identity information to bottlenose dolphins. Proc Natl Acad Sci USA 103:8293–8297
Johnson WH, Delanney LE, Williams EC, Cole TA (1969) Principles of zoology. Holt, Rinehart and Winston, New York
Jowett B (ed) (1885/2008) Politics, by Aristotle. Kessinger, Whitefish, MT
Kant I (1790, mod. ed. 1978) Critique of judgement. Oxford University Press, Oxford
Kaplan G (2009) Animals and music: between cultural definitions and sensory evidence. Sign Sys Stud 37(3/4):75–101
Kawai M (1965) Newly-acquired pre-cultural behavior of the natural troop of Japanese monkeys on Koshima Islet. Primates 6(1):1–30

Kessel EL (1955) The mating activities of balloon flies. Syst Zool 4:96–104
Kircher A (1650) Musurgia universalis. [n.p.], Roma
Kleiber G (1990) La sémantique du prototype. Presses Universitaires de France, Paris
Köhler O (1943) Zähl: Versuche an einem Kolkraben und Vergleichsversuche an Menschen. Zeitschrift fur Tierpsychologie 5:575–712
Köhler O (1949) Zählende: Vögel und vorprachliches Denken. Zoologischer Anzeiger 13:129–238
Köhler O (1952) Vom unbenannten Denken. Zoologischer Anzeiger 16:202–211
Köhler O, Hockett CF (1960) Logical considerations in the study of animal communication. Zeitschrift fur Tierpsychologie 23:250–254
Kohts N (1935/2002) Infant ape and human child. Oxford University Press, Oxford
Kotov K (2002) Semiosphere: a chemistry of being. Sign Syst Stud 30(1):41–55
Kropotkin P (1902) Mutual aid: a factor of evolution. William Heinemann, London
Kull K (1998) On semiosis, Umwelt, and semiosphere. Semiotica 120(3/4):299–310
Kull K (1999) Biosemiotics in the twentieth century: a view from biology. Semiotica 127(1/4): 385–414
Kull K (2005) A brief history of biosemiotics. JBiosemiot 1:1–25
Kull K, Torop P (2003) Biotranslation: translation between Umwelten. In: Petrilli S (ed) Translation. Rodopi, Amsterdam, pp 313–28
Kurath GP (1977) Universals in dance. The World Music 19(1–2):43–62
La Bruyère Jd (1688) Les caractères, Later edn. Impr. Nationale Editions, 1998, Paris
Lee HDP (1965) Plato's Timaeus. Penguin Books, London
Lehr E (1967) Experimentelle Untersuchungen an Affen und Halbaffen über Generalisation von Insekten- und Blütenabbildungen Zeitschrift fur Tierpsychologie 24:208–244
Lestel D (1994) Symbols of discord: are Apes that talk trivia more interesting than apes that do not talk? Social Sci Infor 33(2):335–69
Lestel D (1995) Paroles de singe: I'impossible dialogue homme/primate. La Decouverte, Paris
Lestel D (1996) L'animalite: essai sur le statut de I'humain. Hatier, Paris
Lestel D (1998) How chimpanzees have domesticated humans: towards an anthropology of human/animal communication. Anthropol Today 14(3):12–15
Lestel D (1999) Portrait de 1'animal comme sujet. Revue de Synthese 1:139—64
Lestel D (2001) Les origines animales de la culture. Flammarion, Paris
Lestel D (2002) The biosemiotics and phylogenesis of culture. Social Sci Infor 41(1):35–68
Lewis D (1986) On the plurality of worlds. Basil Blackwell, Oxford
Liebermann P (1968) Primate vocalizations and human linguistic ability. J Acoust Soc Am 44:1574–1584
Liebermann P (1972) The speech of primates. Mouton, The Hague
Lindauer M (1961) Communication among social bees. Harvard University Press, Cambridge, MA
Lindauer M (1990) Botschaft ohne Worte – wie Tiere sich verständigen. Piper, München
Lippmann W (1922) Public opinion, 1934 edn. Macmillan, New York
List G (1971a) On the non-universality of musical perspectives. Ethnomusicology 15(3):399–402
List G (1971b) The boundaries of speech and song. In: McAllester D (ed) Readings in ethnomusicology. Johnson Reprint Corporation, London
Locke J (1690, mod. ed. 1825) Essay concerning human understanding. Samuel Marks, New York
Locke J (1692, mod. ed. 1823) Some thoughts concerning education. Thomas Tegg, London
Lomax A (1977) Universals in song. The World Music 19(1–2):117–129
Lopez B (1978) Of wolves and men. Scribner's (Italian edition: Lupi: dalla parte del migliore nemico dell'uomo. Casale Monferrato: Piemme, 1999), New York
Lorenz K (1949/1952) King Solomon's ring. Methuen, London
Lorenz K (1950/1954) Man meets dog. Methuen, London
Lorenz K (1963/1966) On aggression. Methuen, London
Lorenz K (1965) Evolution and modification of behavior. The University of Chicago Press, Chicago
Lorenz K (1973) Behind the mirror: a search for a natural history of human knowledge. Harcourt, New York

Lorenz K (1974) Civilized man's eight deadly sins. Harcourt, New York
Lorenz K (1979) The year of the greylag goose. Harcourt, New York
Lotman Y (1977) Primary and secondary communication-modeling systems. In: Lucid D (ed) Soviet Semiotica: an anthology. Hopkins University Press, Baltimore
Lotman Y (1990) Universe of the mind: a semiotic theory of culture. Indiana University Press, Bloomington, IN:
Louie K, Wilson MA (2001) temporally structured replay of awake hippocampal ensemble activity during rapid eye movement sleep. Neuron 29(1):145–156
Mâche F-B (1992) Music, myth and nature. Harwood, New York
MacCannell D (1976) The tourist: a new theory of the leisure class. Schocken Books, New York
MacLaren J (1926) My crowded solitude. R. M. McBride, New York
Mainardi D (1975) L'animale culturale. Rizzoli, Milano
Mainardi D (ed) (1992) Dizionario di etologia. Einaudi, Torino
Malacarne G (2005) Il linguaggio simbolico negli uomini e nelle scimmie. In Zucchelli M (ed) Il comportamento degli animali. De Agostini, Novara
Mannucci A (1997) Il nostro animale quotidiano. Il saggiatore, Milano
Maran T (2005) Mimikri kui kommunikatsioonisemiootiline fenomen. [Mimicry as a communication semiotic phenomenon]. Dissertationes Semioticae Universitatis Tartuensis 7. Tartu University Press, Tartu
Maran T (2007a) Semiotic interpretations of biological mimicry. Semiotica 167(1/4):223–248
Maran T (2007b) Structural and semiotic aspects of biological mimicry. In: Witzany G (ed) Biosemiotics in transdisciplinary contexts. Umweb, Helsinki, pp 237–243
Maran T (2008a) Deceiver's activity in biological mimicry systems. In: Tarasti E (ed) Global signs. ISI, Helsinki-Imatra, pp 231–238
Maran T (2008b) Towards an integrated methodology of ecosemiotics: the concept of nature-text. Sign Syst Stud 35(1/2):269–294
Maran T (2009) John Maynard Smith's typology of animal signals: a view from semiotics. Sign Syst Stud 37(3/4):126–144
Marchesini R (1996) Il concetto di soglia. Theoria, Roma
Marchesini R (1999) La fabbrica delle chimere. Bollati Boringhieri, Torino
Marchesini R (2000a) Bioetica e scienze veterinarie. ESI, Napoli
Marchesini R (2000b) Lineamenti di zooantropologia. Edagricole-Calderini, Bologna
Marchesini R (2004) Canone di Zooantropologia Applicata. Alberto Perdisa Editore, Bologna
Marchesini R (2005) Fondamenti di Zooantropologia. Alberto Perdisa Editore, Bologna
Marler P (1961) The logical analysis of animal communication. J Theor Biol 1:295–317
Marler P (1977) The structure of animal communication sounds. In: Bullock TH (ed) Recognition of complex acoustic signals. Springer, Berlin
Marler P (1978) Affective and symbolic meaning: some zoosemiotic speculations. In: Sebeok T (ed) Sight, sound and sense. Indiana University Press, Bloomington, IN pp 113–123
Marler P (1984) Animal communication: affect or cognition? In: Scherer KR, Ekman P (eds) Approaches to emotion. Lawrence Erlbaum Associates, Hillsdale, NJ, pp 345–365
Marler P, Slabbekoorn H (2004) Nature's music: the science of birdsong. Elsevier, London
Marler P et al (1992). Animal signals: motivational or both? In: Papousek H et al (eds) Nonverbal vocal communication. Cambridge University Press, Cambridge, pp 66–86
Marinoni A (ed) (1988) Leonardo da Vinci – Bestiario e favole. Tea, Milano
Martinelli D (1999) Il canto della megattera: Una musica bestiale. Musica/Realtà 60:37–55
Martinelli D (2000) Merenneidon laulut, laiskiaisen skaalat ja muita tarinoita. Synteesi 4: 14–32
Martinelli D (2001) Methodologies and problems in zoömusicology. Signs Syst Stud 29(1): 341–352
Martinelli D (2002) How musical is a whale? – towards a theory of zoömusicology. International Semiotics Institute, Helsinki/Imatra
Martinelli D (2004a) Liars, players and artists: a triadic zoosemiotic relation. Semiotica 150(1/4):77–118

Martinelli D (2004b) S for Sebeok – Introducing anthropological zoosemiotics. In: Bauters, M, Ipsen, G, Martinelli, D (eds) Semiotics from S to S. Umweb, Helsinki

Martinelli D (2005) Entry Zoomusicologia. In: Stefani G, Lisi SG (eds) Dizionario della Musica nella Globalità dei Linguaggi. LIM, Lucca

Martinelli D (2006a) Zoosemiotics: proposals for a handbook. ISI, Helsinki-Imatra

Martinelli D (2006b) Language and interspecific communication: a case to re-open. In: Barbieri M (ed) Introduction to biosemiotics. Springer, Berlin

Martinelli D (2006c) Animal experimentation and the (im)possible equation: an Italo-Finnish comparative analysis. In: Martinelli D, Navickaite L (eds) Finland-Italy: a few comparisons. Umweb, Helsinki

Martinelli D (2006d) State of the art report: zoosemiotics. Semiotix [online] 6. Available from: http://www.semioticon.com/semiotix/semiotix6/sem-6-05-1.html [Accessed 7 June 2010]

Martinelli D (2006e) The musical circle – the umwelt theory, as applied to zoomusicology. Sign Syst Stud 32(1):2

Martinelli D (2008) The ethic imperative in Eero Tarasti's semiotic path: reflecting on the relationships between resistance and biocentrism. In: Hatten RS, Kukkonen P, Littlefield R, Veivo H, Vierimaa I (eds) A sounding of signs – modalities and moments in music, culture and philosophy. International Semiotics Institute, Helsinki/Imatra

Introduction. Sign Syst Stud 37(3/4):9–22

Martinelli D (2009) Of birds, whales and other musicians. University of Scranton Press, Scranton

Martinelli D, Bankov K (2008) Bankov's razor versus martinelli's canon. A confrontation around biosemiotics. Biosemiotics 1(3):397–418

Martinelli D, Lehto O (eds) (2009) Sign Syst Stud 37(3/4):Special issue on zoosemiotics

Masson JM, McCarthy S (1995) When elephants weep: the emotional lives of animals. Delacorte, New York

Matney W, Duncan K, Trimpe MA (2005) Aging of shoes and its effect on shoeprint impressions. J Forensic Ident 55(2):181–188

Maynard Smith J (1982) Do animals convey information about their intentions? J Theor Biol 97(1):1–5

Maynard Smith J (1999) The idea of information in biology. The Quart Rev Biol 74(4):395–400

Maynard Smith J (2000) The concept of information in biology. Philos Sci 67(2):177–194

Maynard Smith J, Harper D (1995) Animal signals: models and terminology. J Theor Biol 177(3):305–311

Maynard Smith J, Harper D (2003) Animal signals. Oxford University Press, Oxford

Mayr E (1942) Systematics and the origin of species, from the viewpoint of a zoologist. Harvard University Press, Cambridge, MA

Mazzara B (1997) Stereotipi e pregiudizi. il Mulino, Bologna

Mead GH (1934) Mind, self and society. University of Chicago Press, Chicago

Meeker JW (1996) The comic mode – the biology of comedy. In: Cheryll G, Fromm H (eds) The ecocriticism reader. University of Georgia Press, Athens, Georgia, pp 155–69

Miles HL (1990) The cognitive foundations for reference in a signing orangutan. In: Parker ST, Gibson KR (eds) Language and intelligence in monkeys and apes: comparative developmental perspectives. Cambridge University Press, New York, pp 511–539

Miles HL (1994) Me Chantek: the development of self-awareness in a signing orangutan. Self-awareness in animals and humans: developmental perspectives. In: Parker ST, Mitchell RW, Boccia ML (eds) Self-awareness in animals and humans: developmental perspectives. Cambridge University Press, New York, pp 254–272

Mitani JC, Marler P (1989) A phonological analysis of male gibbon singing behavior. Behaviour 109(5):20–45

Mitchell RW (1990) A theory of play. In: Bekoff M, Jamieson D (eds) Interpretation and explanation in the study of animal behaviour. Westview, Boulder, CO, pp 197–225

Mitchell RW (ed) (2002) Pretending and imagination in animals and children. Cambridge University Press, Cambridge

Mitchell RW, Thompson N (eds.) (1986) Deception: perspectives on human and non-human deceit. State University of New York Press, New York
Modica M (1997) Che cos'è l'estetica. Editori Riuniti, Roma
Montaigne Md (1580/1992) Essais. Adelphi, Milano
Morgan CL (1903) An introduction to comparative psychology. W.Scott, London
Morris CW (1937) Logical positivism, pragmatism and scientific empiricism. Hermann et Cie, Paris
Morris CW (1938/1970) Foundations of the theory of signs. Chicago University Press, Chicago
Morris CW (1946) Signs, language, and behavior. Prentice Hall, New Jersey
Morris CW (1967) The naked ape. Jonathan Cape, London
Morris CW (1969) The human zoo. Jonathan Cape, London
Morris DJ (1963) The biology of art. Methuen, London
Mortenson J (1987) Whale songs and wasp maps. Duttons, New York
Mulcahy NJ, Call J (2006) Apes save tools for future use. Science 312:1038–1040
Myrberg AA Jr, Raggio RJ (1985) Acoustically mediated individual recognition by a coral reef fish (*Pomacentrus partitius*). Anim Behav 33:411–416
Nagel T (1974) What is it like to be a bat? Philos Rev 83:435–50
Nattiez JJ (1977) Under what conditions can one speak of the universality of music? In The World of Music 19(1/2): 92–105
Nollman J (1999) The charged border. Henry Holt, New York
Nöth W (1990) Handbook of semiotics. Indiana University Press, Bloomington, IN
Nöth W (1998) Ecosemiotics. Sign Syst Stud 26:332–343
Owen D (1980) Camouflage and mimicry. Oxford University Press, Oxford
Pain S (2002) Biorhetorics: an introduction to applied rhetoric. Sign Syst Stud 30(2)
Pain S (2008) Biosemiotics. Semiotix [online] 12. Available from: http://www.semioticon.com/semiotix/semiotix12/sem-12-02.html [Accessed 7 June 2010]
Pain S (2006) Inner representations and signs in animals. In: Barbieri M (ed) Introduction to biosemiotics. Springer, Berlin
Pain S (2009) From biorhetorics to zoorhetorics. Sign Syst Stud 37(3/4):146–156
Pallasmaa J (ed) (1995) Eläinten Arkkitehturi. Suomen Rakennustaiteen Museo, Helsinki
Parker ST, Gibson KR (1977) Object manipulation, tool use and sensorimotor intelligence as feeding adaptations in cebus monkeys and great apes. J Human Evol 6:623–641
Partan S, Marler P (1999) Communication goes multimodal. Science 283(5406):1272–1273
Patrick GTW (1916) The psychology of relaxation. Houghton Mifflin, New York
Payne RN (1982) New light on the singing whales. Natl Geo 161(4):463–477
Payne RN (1996) La vita segreta delle balene. Mondadori, Milano
Payne RN, McVay S (1971) Songs of humpback whales. Science 173:585–597
Peirce CS (1935–1966). Collected papers. In: Hartshorne C et al (eds) Harvard University Press, Cambridge, MA
Petersen AF (1988) Why children and young animals play. K.D.V.S., Copenaghen
Patterson FG (1977) Linguistic capabilities of a young Lowland Gorilla. Paper presented at a symposium of the American association for the advancement of science: An account of the visual mode: man vs. Ape. Printed in Peng FC (ed) Sign language and language acquisition in man and ape. Westview Press, Boulder CO, 1978, pp 161–202
Patterson FG (1978) The gestures of a gorilla: language acquisition in another pongid. Brain Lang 5:72–97
Patterson FG (1981) Ape language. Science 211(4477):87–88
Patterson F, Linden E (1981) The education of Koko. Holt, Rineheart and Winston, New York
von Pernau FA (1716) Unterricht: was reit dem lieblichen Geschöpff, denen Vögeln auch ausser dem Fang, , nur durch die Ergründung deren Eigenschaften und Zahmmachung oder anderer Abrichtung man sich vor Lust und Zeit-Vertreib machen könne. Johann Carl Bohn, Hamburg
Petrilli S (1998) Teoria dei segni e del linguaggio. Edizioni B.A.Graphis, Bari
Piaget J (1962) Play, dreams and imitation in childhood. Norton, New York

Pike K (1954) Language in relation to a unified theory of the structure of human behaviour, vol I. Summer Institute of Linguistics, Glendale, CA
Plotnik JM, de Waal FBM, Reiss D (2006) Self-recognition in an Asian elephant. Proc Natl Acad Sci 103(45):17053–17057
Podberscek AL, Paul ES, Serpell JA (2000) Companion animals and us. Cambridge University Press, Cambridge
Poe GR et al (2000) Experience-dependent phase-reversal of hippocampal neuron firing during REM sleep. Brain Res 855(1):176–180
Ponzio A, Petrilli S (2008) Bodies, signs and values in global communication. In: Petrilli S (ed) Approaches to communication. trends in global communication studies. Atwood Publishing, Madison, WI, pp 113–141
Portmann A (1952/1967) Animal forms and patterns: a study of the appearance of animals. Schocken Books, New York
Portmann A (1959) Animal camouflage. University of Michigan Press, Ann Arbor, MI
Portmann A (1961) Animals as social beings. Hutchinson, London
Prior H, Schwarz A, Güntürkün O (2008) Mirror-Induced behavior in the magpie (Pica pica): evidence of self-recognition. Public Lib Sci Biol 6(8):1642–1650
Prodi G (1983) Lingua e biologia. In: Segre C (ed) Intorno alla linguistica. Feltrinelli, Milano, pp 172–202
Raby CR, Alexis DM, Dickinson A, Clayton NS (2007) Planning for the future by western scrub-jays. Nature 445:919–921
Rackham H (ed) (1938) Pliny: natural history (volumes I–X). William Heinemann, London
Regan T (1983) The case for animal rights. Routledge, London
Reimarus HS (1762) Allgemeine Betrachtungen über die Triebe der Thiere. Johann Carl Bohn, Hamburg
Reiss D, Marino L (2001) Mirror self-recognition in the bottlenose dolphin: a case of cognitive convergence. Proc Natl Acad Sci 98:5937–5942
Rensch B (1965) Die höchsten Lernleistungen der Tiere. Naturwissenschaftliche Rundschau, 18:91–101
Rensch B, Dücker G (1959) Versuche über visuelle Generalisation bei einer Schleichkatze. Zeitschrift fur Tierpsychologie 16:671–692
Richard B (1967) Le determinisme de la construction des barrages chez le castor du Rhone. La Terre et la Vie 4:339–470
Robinson JG (1984) Syntactic structures in the vocalizations of wedge-capped capuchin monkeys, Cebus olivaceus. Behaviour 90:46–79
Rogers LJ, Kaplan G (1998) Not only roars and rituals. Allen & Unwin, St Leonards
Rogers LJ, Kaplan G (eds.) (2004) Comparative vertebrate cognition: are primates superior to non-primates. Springer, New York
Rorarius H (1548/1728) Quod animalia bruta saepe ratione utantur melius homine. Christian Frider, Helmstadt, p 1728
Rosch E et al (1978) Cognition and categorisation. Lawrence Erlbaum, Hillsdale, NJ
Rossi-Landi F (1972) Semiotica e ideologia. Bompiani, Milano
Rothenberg D (2005) Why birds sing. Basic Books, New York
Rothenberg D (2008) Thousand mile song. Basic Books, New York
Rothschild FS (1962) Laws of symbolic mediation in the dynamics of self and personality. Ann NY Acad Sci 96:774–784
Royce AP (1977) The anthropology of dance. Indiana University Press, Bloomington, IN
Ryder RD (1975) Victims of science: the use of animals in research. Davis-Poynter, London
Sachs C (1937) World history of the dance. Norton, New York
Sachs C (1943) The rise of music in the ancient world. Norton, New York
Salthe SN (1993) Development and evolution: complexity and change in biology. MIT Press, Cambridge MA

San Martin J, Pintos ML (2001) Animal life and phenomenology. In: Crowell S, Lester E, Samuel JJ (eds) The reach of reflection: the future of phenomenology. Electron Press West Hartford CT, pp 342–363

Sapir E (1921) Language: an introduction to the study of speech. Harcourt, New York

Sayers W (2009) Animal vocalization and human polyglossia in Walter of Bibbesworth's thirteenth-century domestic treatise in Anglo-Norman French and middle English. Sign Syst Stud 37(3/4):173–187

Schafer RM (1976) The tuning of the world. Knopf, New York

Schiller P (1951) Figural preferences in the drawing of a chimpanzee. J Comp Physiol Psychol 44:101–111

Schopenauer A (1840/2007) On the Basis of Morality. New York: Cosimo

Schopenauer A (1851/1974) Parerga and paralipomena. Oxford University Press, New York

Scott TA (1995) Concice encyclopaedia biology. Mouton de Gruyter, Berlin

Searle JR (1975) The logical status of fictional discourse, New Literary Hist 14:319–332

Sebeok TA (1963) Communication in animals and men, Language 39:448–466

Sebeok TA (1965) Animal communication. Science CXLVII:1006–1014

Sebeok TA (ed) (1968) Animal communication. Techniques of study and results of research. Indiana University Press, Bloomington, IN

Sebeok TA (1972) Perspectives in zoosemiotics. Mouton de Gruyter, The Hague

Sebeok TA (1979) The sign and its masters. University of Texas Press, Austin, TX

Sebeok TA (1981) The play of musement. Indiana University Press, Bloomington, IN

Sebeok TA (ed) (1986a) Encyclopaedic dictionary of semiotics. Mouton de Gruyter, Berlin

Sebeok TA (1990) Essays in zoosemiotics. Toronto Semiotic Circle, Toronto

Sebeok TA (1991) A sign is just a sign. Indiana University Press, Bloomington, IN

Sebeok TA (1994) Signs: an introduction to semiotics. University of Toronto Press, Toronto

Sebeok TA (1995) Semiotics and the biological sciences: initial conditions. Paper presented at the Collegium Budapest, November 1995. Printed in Perron P, Sbrocchi LG, Colilli P, Danesi M (eds) Semiotics and information sciences. Legas Press, Ottawa, 2000, pp 76–100

Sebeok TA (1996) Signs, bridges, origins. In: Trabant J (ed) Origins of language. Collegium Budapest, Budapest

Sebeok TA (1998) Come comunicano gli animali che non parlano. Edizioni dal Sud, Modugno

Sebeok TA (2001a) Global semiotics. Indiana University Press, Bloomington, IN

Sebeok TA (2001b) The Swiss pioneer in nonverbal communication Studies: Heini Hediger (1908–1992) Legas, New York

Sebeok TA, Danesi M (2000) The forms of meaning. Mouton de Gruyter, Berlin

Sebeok TA, Ramsay A (ed) (1969) Approaches to animal communication. Mouton, The Hague

Sebeok TA, Umiker-Sebeok J (eds) (1980) Speaking of apes. Plenum, New York

Seeley TD (1995) The wisdom of the hive. Harvard University Press, Cambridge, MA

Serpell JA (1996) In the company of animals: a study of human–animal relationships. Cambridge University Press, Cambridge

Shanahan MP (2006) A cognitive architecture that combines internal simulation with a global workspace. ConsciousCogn 15:433–449

Shanahan MP, Baars B (2005) Applying global workspace theory to the frame problem. Cognition 98:157–76

Sharov A (1998) What is biosemiotics? http://home.comcast.net/~sharov/biosem/geninfo.html. Accessed E-text June 17 2009

Sharov A (2001) Umwelt theory and pragmatism. Semiotica 134:211–228

Siapas AG, Wilson MA (1998) Coordinated interactions between hippocampal ripples and cortical spindles during slow-wave sleep. Neuron 21(5):1123–1128

Singer P (1989) All animals are equal. In:. Regan T, Singer P (eds) Animal rights and human obligations. Prentice Hall, New Jersey, pp 148–162

Slud P (1957) The song and dance of the long-tailed manakin. The Auk 74:333–339

Smith C (1995) Sleep states and memory processes. Behav Brain Res 69(1)–2:137–145
Smith WJ (1977) The behavior of communicating. Harvard University Press, Cambridge, MA
Smith WJ (1981) Referents of animal communication. Anim Behav 29:1273–1275
Smith WJ (1991) Animal communication and the study of cognition. – In: Ristau CR, Griffin DR (eds) Cognitive ethology: the minds of animals: essays in Honor of Donald R. Griffin. L. Erlbaum Associates, Hillsdale, NJ, pp 209–230
Soriau E (1965) Le sens artistique des animaux. Hachette, Paris
Sotavalta O (1956) Analysis of the song patterns of two sprosser nightingales, luscinia luscinia. Annales Zoologici Societatis Zoologicae Botanicae Fennicae 'Vanamo' 17(4): –31
Spalding DA (1873) Instinct. With original observations on young animals. Macmillan's Mag 27:282–293
Spencer H (1873) The principles of psychology. Appleton, NewYork
Stefani G (1985) Competenza musicale e cultura della pace. CLUEB, Bologna
Stefani G (1989) Musica con coscienza. Roma: Paoline
Stefani G, Guerra S (eds.) (2005) Dizionario della musica nella globalità dei lingaggi. LIM, Lucca
Styles EA (2005) Attention, perception and memory. Psychology, Hove, Sussex
Suits B (1977) Words on play. J PhilosoSport 4:117–131
Szöke P (1963) Ornitomuzikológia. Magyar Tudomány 9:592–607
Tagg P (1987) Universal music and the case of death. In: Pozzi R (ed) La musica come linguaggio universale. Olschki, Firenze
Tajfel H (1981) Human groups and social categories. Cambridge University Press, Cambridge
Tarasti E (1997) The emancipation of the sign: on the corporeal and gestural meanings in music. AS/SA 4:180–190
Tarasti E (2000) Existential semiotics. Indiana University Press, Indianapolis
Tarasti E (2005) Vastarinnan Semiotiikkaa: Oleminen, Muisti, Historia – Merkkien Vastavirta. Synteesi 1:2–29
Taylor H (2008) Towards a species songbook: illuminating the vocalisations of the Australian Pied Butcherbird (Cracticus nigrogularis). PhD thesis, University of Western Sidney, Australia
Taylor PW (1986) Respect for nature: a theory of environmental ethics. Princeton University Press, Princeton, NJ
Taylor T (1823) Select works of Porphyry. Thomas Rodd, London
Tebbich S et al (2002) The ecology of tool use in the woodpecker finch (*Cactospiza pallida*). Ecol Lett 5:656–664
Tembrock G (1963) Acoustic behaviour of mammals. In: Busnel RG (ed) Acoustic behaviour of animals. Elsevier, London, pp 751–786
Tembrock G (1971) Biokommunikation. Akademic, Berlin
Tembrock G (1993) Behavior programs, imperceptible messages and procedural learning. *Zeitschrift für Semiotik* 15(1–2):68–72
Tinbergen N (1951) The study of instinct. Oxford University Press, London
Tinbergen N (1953) Social behaviour in animals, with special reference to vertebrates. Methuen, London
Tinbergen N (1963) On aims and methods in ethology. Zeitschrift für Tierpsychologie 20:410–433
Tinbergen N (1974) The animal in its world. Harvard University Press, Cambridge, MA
Thompson KV (1998) Self assessment in juvenile play. In: Bekoff M, Byers JA (eds) Animal play: evolutionary, ecological, and comparative perspectives. Cambridge University Press, Cambridge, pp 183–204
Thorpe WH (1961) Bird song. Cambridge University Press, Cambridge
Thorpe WH (1963) Learning and instinct in animals. Methuen, London
Thorpe WH (1972) Vocal communication in birds. In: Hinde RA (ed) Non-verbal communication. Cambridge University Press, Cambridge, pp 153–176
Tomasello M, Call J (1997) Primate cognition. Oxford University Press, New York
Trask RL (1995) Language: the basics. Routledge, London

Turovski A (2000) The semiotics of animal freedom. A zoologist's attempt to perceive the semiotic aim of H. Hediger. Sign Syst Stud 28:380–387

Turovski A (2001) On the parasite's association as a vectorizing factor in biosemiotic development. Semiotica 134(1/4):409–414

Turovski A (2002) On the zoosemiotics of health and disease. Sign Syst Stud 30(1):213–219

von Uexküll J (1899–1940) Kompositionslehre der Natur. Biologie als undogmatische Naturwissenschaft. Ullstein, Frankfurt

von Uexküll J (1956) Streifzüge durch Umwelten von Tieren und Menschen. Rowohlt, Hamburg

von Uexküll J (1982) The theory of meaning. Published entirely as Special Issue in Semiotica 42(1)

von Uexküll T (1987) The sign theory of Jakob von Uexküll. In: Krampen M, Oehler K, Posner R, Sebeok TA, Uexküll Tv (eds) Classics of semiotics. New York: Plenum, pp 147–79

VanBlaricom GR (2001) Sea otters. Voyageur, Stillwater, MN:

Vauclair J (1996) La cognition animale. Presses Universitaires de France, Paris

Veitch J (ed) (2008) René Descartes: Discourse the method and mediation. Cosimo, New York

Vilkka L (1997) The intrinsic value of nature. Rodopi, Atlanta, GA

Visalberghi E (1993) Capuchin monkeys: a window into tool use activities by apes and humans. In: Gibson K, Ingold T (eds) Tool, language and cognition in human evolution. Cambridge University Press, Cambridge, pp 138–150

Vygotsky LS (1976) Play and its role in the mental development of the child. In: Bruner JS, Jolly A, Sylva K (eds) Play: its role in development and evolution. Basic Books, New York, pp 537–54

Wenner AM (1967) Honey bees: do they use the distance information contained in their dance maneuver. Science 155:847–849

Wenner AM (1969) The study of animal communication: an overview. In: Sebeok T, Ramsay A (eds) Approaches to animal communication. The Hague, Mouton

Wickler W (1968) Mimicry in plants and animals. McGraw-Hill, New York

Wierzbicka A (1989) Prototypes save: on the uses and abuses of the notion of 'prototype' in linguistics and related fields. In: Tsohatzidis SL (ed) Meanings and prototypes: studies in linguistic categorization. Routledge, London, pp 347–366

Willis T (1672) De anima brutorum, 2nd edn. Coloniae Allobrogum apud Samuelem de Tournes, 1680

Williston SW (1892) Note on habits of Ammophila. Entomol News 3:85–86

Wittgenstein L (1978) Philosophical investigations. Oxford University Press, Oxford

Witzany G (2007) The logos of the bios 2. Bio-communication. Umweb, Helsinki

Wallin N (1991) Biomusicology. Pendragon, Stuyvesant, NY

Wilson EO (1975) Sociobiology: the new synthesis. Belknap, Cambridge

Wolfe JB (1936) Effectiveness of token-rewards in chimpanzees. Compar Psychol Monogr 12:1–72

Yates FE (1985) Semiotics as bridge between information (biology) and dynamics (physics). In Recherches Sémiotique/Semiotic Inquiry 5:347–360

Zahavi A, Avishag Z (1997) The handicap principle. A missing piece in the darwinian puzzle. Oxford University Press, Oxford

Author Index

Glossary Index

Species Index

NOTE: The entries might also concern animals cited as mere examples and be classified by something else than species (family, order, genus, etc.). Moreover, the list excludes the human animal (*Homo sapiens*), whose recurrence in the text makes its indexing unnecessary.

FSC
www.fsc.org
MIX
Papier | Fördert
gute Waldnutzung
FSC® C083411

Kolibri 360 GmbH
Ferdinand-Jühlke-Straße 7
99095 Erfurt, Deutschland
produktsicherheit@kolibri360.de